Maintaining and Repairing
VCRs
Second Edition

Maintaining and Repairing
VCRs
Second Edition

Robert L. Goodman

TAB BOOKS
Blue Ridge Summit, PA

SECOND EDITION
FOURTH PRINTING

Copyright © 1989 by TAB BOOKS
FIrst edition copyright © 1983 by TAB BOOKS
Printed in the United States of America

Library of Congress Cataloging in Publication Data

Goodman, Robert L.
     Maintaining and repairing VCRs / by Robert L. Goodman.—2nd ed.
        p.   cm.
     Rev. ed. of: Maintiang & repairing videocassette recorders. 1st
  ed. c1983.
     Includes index.
     ISBN 0-8306-9003-4   ISBN 0-8306-9103-0 (pbk.)
     1. Video tape recorders and recording—Maintenance and repair.
  I. Goodman, Robert L. Maintaining & Repairing videocassette
  recorders.   II. Title.
TK6655.V5G66   1989   88-32316
621.388'332'0288—dc19   CIP

TAB BOOKS offers software for sale. For information and a catalog, please
contact TAB Software Department, Blue Ridge Summit, PA 17294-0850.

Questions regarding the content of this book should be addressed to:

     Reader Inquiry Branch
     TAB BOOKS
     Blue Ridge Summit, PA 17294-0214

Contents

Acknowledgments

Much of the information in this book was provided by VCR manufacturers, their technical personnel, and VCR service technicians. I am most grateful to the following companies and individuals:

General Electric Company (Mr. R.J. Collins); JVC (US JVC Corp.) (Mr. Paul E. Hurst); Matsushita Electric Corporation; Quasar (Mr. Charlie Howard); Magnavox Consumer Electronics Co. (Mr. Ray Guichard); RCA/Consumer Electronics Co. (Mr. J.W. Phipps); Sony Corporation of America (Ms. Paula V. Duffy); Zenith Video Tech Corporation (Mr. James F. White); and a big thanks to Greg Carey and Don Multerer of Sencore, Inc.

Maintaining and Repairing
VCRs

Second Edition

Introduction

This book begins with a brief history of video tape recording, early home VCR systems, and how the systems developed and were improved upon over the years. The first chapter contains a brief theory of operation of the Video Home System (VHS), the Sony (Betamax) video recording systems, and information on the helical VCR slant track tape recording principles.

Chapter 2 covers special Sencore test equipment for VCR service and how to use this equipment for fast troubleshooting.

In later chapters, eccentricity, torque gauges, dihedral head adjustments, and test and alignment tapes required for VCR servicing are discussed. Then some tips on making your own test tapes and the tools required for VCR machine adjustments are presented.

This book also contains service information for the Sony/Zenith Betamax format VCR machines, including circuit operation, mechanical operation and adjustments, block diagrams, and circuit schematics.

Another section offers a brief overview of the VHS recording system. The brands covered are General Electric, RCA, and JVC. Again, these chapters include theory of operation, circuit schematics, and mechanical and electronic adjustments. Some of these VCR machines are manufactured by the Matsushita Electronics Company in Japan, and the information contained in that section can be used for other models and brands of VHS machines. Similarly, much of the circuit operation and troubleshooting info found in the chapters that discuss Betamax machines can also be used for VHS machine service.

Other chapters contain information on the new camcorders, HQ and super VHS features, and stereo audio now used in TV and VCR equipment. Also, Sony's SuperBeta system circuits are covered.

The last chapter offers actual case histories of various VCR problems and solutions to aid in troubleshooting, plus a VCR glossary.

As you look through this VCR manual, there are some complex electronic and mechanical diagrams. Keep in mind that the video cassette recorder is a very sophisticated piece of equipment and should be handled and serviced with care. In this VCR book, I break the machine down into sections and then delve into each of these to explain basic operation and troubleshooting. My goal is to make a very complex device as easy to understand and service as possible.

1

Video Cassette Recording

Space age electronic technology has made a new era of the video tape machine for home use. These video cassette recorders (VCRs) provide an acceptable color picture at a fairly low cost for the machine and the video tape that is used.

This chapter looks at some basic video recording considerations, a comparison of Beta and VHS systems, a brief history of video tape recordings, and the types of test equipment to consider for servicing.

In a very broad sense, video tape recording uses the same principles as audio tape recording. It is a magnetic way to record a signal onto tape. However, there is a great difference between audio and video signals. Audio has a frequency range of approximately 20 Hz to 20 kHz. Video has a much greater and higher frequency range, from about 30 Hz to 4.5 MHz. These high frequencies are required for picture quality, and the lowest one is for the sync pulses. The sources of the video signals can be from a TV camera, another tape VCR machine, a pre-recorded tape, or an "off the air" TV station signal.

FREQUENCY LIMITATIONS

The gap effect of the heads is the most important limitation on the range of video frequencies that can be recorded and played back. Figure 1-1 shows an illustration of this effect. It is a curve of the playback frequency response. As the signal to be recorded rises in frequency, a steady increase in the playback signal occurs. This rise is at the rate of 6 dB per octave, an octave being a doubling of the frequency. This rise in output continues until a maximum is obtained. The frequency where this occurs is indicated in Fig. 1-1 as f_m, the maximum frequency. Beyond this frequency value, the output will rapidly drop, reaching zero at twice the frequency of maximum output. The points f_m and $2f_m$ are determined by the size of the head gap and tape speed. The more narrow the gap is, the higher the frequency will be for maximum output. But a very narrow gap restricts the output at the low frequencies.

The speed of the tape movement past the head also effects the frequency response of the

1

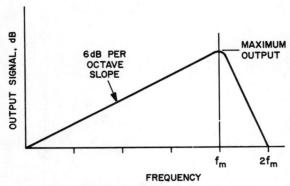

Fig. 1-1. *Playback frequency response.*

record-playback system. The greater this speed is, the higher the frequency of maximum output but the lower the signal output at the lower frequencies.

The audio frequency range of about 10 octaves (20 Hz to 20 kHz) applied to the output curve of Fig. 1-1 would show the 20 kHz limit equal to $2f_m$ and the low end, 20 Hz, at a very low level of output. The frequency of maximum output would be 10 kHz. Frequency compensation to raise the low and high ends of the curve and reduce the middle could produce a more constant output with frequency. Several different methods have been used in audio devices for this compensation. A frequency range of 10 octaves, with compensation, is about the maximum range that can be practically recorded and played back.

The approximate range of video frequencies to be recorded and played back for TV video is 30 Hz to 4.5 MHz, or 18 octaves. The signal output curve for a head designed to give maximum output at 4.5 MHz would show the 30 Hz output to be down about 110 dB. These extreme differences of signal level would make compensation and equalization of output at all necessary frequencies impossible. Thus, some other method must be used.

FM VIDEO RECORDING/PLAYBACK

If the range of frequencies for video recording and playback, an overall difference of approximately 4.5 MHz, is converted to a higher frequency spectrum, the ratio of the high and low video frequencies can be greatly reduced. For example, if the frequency spread were changed to a variation between 5 MHz and 10 MHz (the difference being about equal to the video frequency range used) the frequency spread becomes 1 octave rather than 18. If the head gap is designed for maximum system output at 7.5 MHz, the outputs at 5 MHz and 10 MHz would be down only a small amount, and equalization would be easy. The frequency variations in the video signal could occur within this altered frequency spectrum, that is, by *frequency modulation* (FM).

Frequency modulation of the video signal to be recorded and played back is then the answer to the video frequency response problem. For example, a high frequency sine wave strong enough to saturate the tape becomes the carrier for the FM process. This carrier is modulated by the video frequencies. The lowest frequency corresponds to the sync pulse tips, with the highest corresponding to peak white. See Fig. 1-2. The actual frequencies selected for this method of FM vary from recorder to recorder. FM systems also strictly control the signal amplitude levels, thus reducing noise problems. This is another advantage of the use of FM in the recording/playback process.

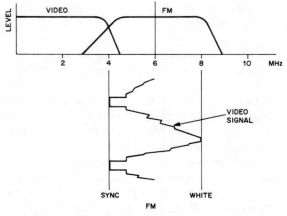

Fig. 1-2. *FM modulation of the video signal.*

The high carrier frequency used for the FM signal process requires a small head gap and high tape speed. The high tape speed, of course, means fast consumption of the tape and therefore larger tape reels. In most cases, this is an impractical situation. Also, the high tape speeds are difficult to control, complicating the drive mechanism. This method of recording by moving the tape past a stationary head is called *longitudinal recording*. This is the method still used successfully for audio recording, but not for video recording.

TRANSVERSE RECORDING

Longitudinal recording results in a recorded, magnetic track that is parallel to the length of the tape. Another method of recording that uses much less tape is *transverse recording*. This method produces tracks *across* the tape at right angles to the length of the tape, as illustrated in Fig. 1-3. The head is not stationary in this process; it moves across the tape at a relatively high speed, as the tape moves rather slowly past the head mount. This results in a high head-to-tape speed, also called *writing speed*. This idea of a moving head led to the development of a rotating head mechanism. Two heads are required for this operation. Head A is recording a track on the tape while head B is retracing or returning to the top edge of the tape for its next track. When that position is reached, head A is switched off and head B is switched on. Thus, properly timed

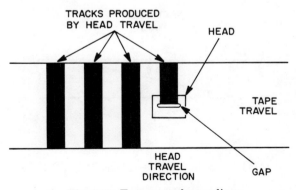

Fig. 1-3. *Transversed recording.*

switching is an important requirement of the rotating head and transverse recording process.

During the playback mode, proper switching and positioning are very important, to ensure that the heads retrace the exact paths made by the heads during recording. Servo systems have been developed to control the head positions — another requirement of video taping.

Small irregularities such as lumps or holes in the tape coating that are not much problem in audio recording become serious in video recording. They can cause loss of contact between the tape and the head resulting in incomplete tracks on the tape and temporary loss of the signal on playback. This signal loss is called *dropout*. It might appear as horizontal flashing on the screen — a portion of a line or up to several lines in duration. Surface bumps on the tape, dirt or residue on the heads, or non-regular tape motion can cause the dropout condition. Video recording requires the maintenance of precise contact between the head and the tape at all times. To insure this requirement, a slight penetration of the tape by the tip of the head is employed.

The best example of transverse recording is the quad-head recorder, a large and complex machine used much by the broadcast industry, but it is not suitable for the general video recording consumer market. The speed of the tape and the tension on the moving tape must be very steady to guarantee reliable signal recording and especially accurate, steady playback. Hence, tape tension control is another requirement of video recording/playback.

HELICAL VIDEO RECORDING

A video tape recorder that is simpler and smaller and less expensive than the quad-head machine is one that employs the helical scan method of recording and playback of the video signals. With helical scan, the tape wraps around a large drum that contains one of two heads that rotate in a plane parallel to the base of the machine. The tape leaves the drum at a different level than it approached it. The movement of the tape is along a spiral, or helical path.

The track that is recorded on the tape by the rotating head is "slanted" and longer than the path for transverse recording. The head drum diameter and the tape width can be designed to make the recorded track long enough to include a complete TV field for head control and switching is much simpler than for the quad-head recorder. Looking at Fig. 1-4, note the development of the slanted tracks on the tape with this helical scan method. These tracks contain the video information. The control and audio tracks are recorded in a longitudinal manner, along the two edges of the tape. As shown in Fig. 1-5, the recorded tracks slant to the right on the tape, as viewed from the side away from the drum. The same tracks, viewed from the other (oxide) side of the tape that contacts the drum and heads, appear slanted to the left as shown in Fig. 1-6. The spaces between the recorded tracks are the *guard bands*. Their purpose is to eliminate "crosstalk" between tracks.

Helical scan video tape recorders have become dominant in the non-broadcasting applications of video recording. Although many different models and formats have been developed and used, the basic principles are the same for all helical machines. The tape leaves a supply reel, passes over a tension arm, past an erase head, around the head drum, past a control head and audio head and through a capstan and pressure roller to the take up reel. Refer to Fig. 1-7 for this tape path. Figure 1-8 shows the relationship, timewise, between the tracks on the tape and the video information recorded on the tracks. Switching between the heads occurs during the few lines just before the vertical sync pulse.

COLOR RECORDING

We have looked at the general form and requirements and methods for recording and playing back the luminance video signals. The first video tape recorders were for black and white signals only. Later, methods were developed that permitted the recording of color signals. Figure 1-9 shows the National Television Systems Committee (NTSC) signal spectrum.

The basic requirements for a system that records and plays back color information would include the following requirements:

• Compatibility with black/white television
• Processing of the NTSC signal within the

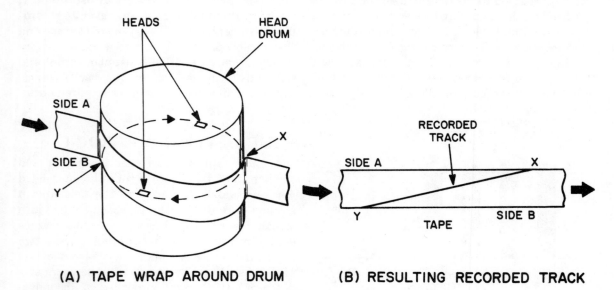

(A) TAPE WRAP AROUND DRUM   (B) RESULTING RECORDED TRACK

Fig. 1-4. *Helical scan development diagram.*

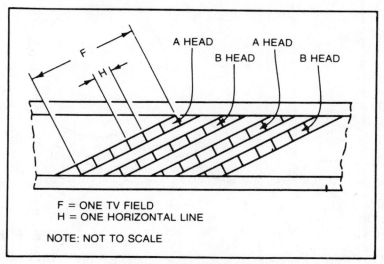

F = ONE TV FIELD
H = ONE HORIZONTAL LINE

NOTE: NOT TO SCALE

Fig. 1-5. *Helical recording tracks.*

same bandwidth as used in B&W recording, without changing the tint of the color signal

Two methods have been used in the recording of color signals onto tapes. One method is called the *direct method,* and the other is called the *color under method.* In the direct method, the NTSC signal is coupled to the FM modulator, just as is done with the black and white signal. The color signal consists of an ac, 3.58 MHz signal on a dc level. The dc portion of the signal determines, through the FM modulation action, the carrier frequency for the video and color signal modulation. The ac portion of the signal produces sidebands of this FM carrier. The color

signal can be easily demodulated as with the black and white signal. One problem with this direct method is that the sidebands on demodulation can cause interference beats in the picture.

The color under method separates the color and luminance signals from the incoming signal. Each portion of the total signal is processed individually. The color is hetrodyned down from 3.58 MHz to a lower carrier frequency and recorded directly onto tape. The FM carrier is used as a bias that is amplitude-modulated. During playback in this method, the color carrier is recovered and hetrodyned back up to 3.58 MHz. See Fig. 1-10.

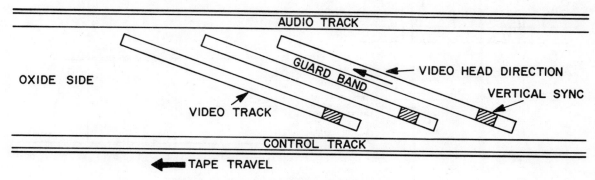

Fig. 1-6. *Oxide side of tape, helical scan.*

5

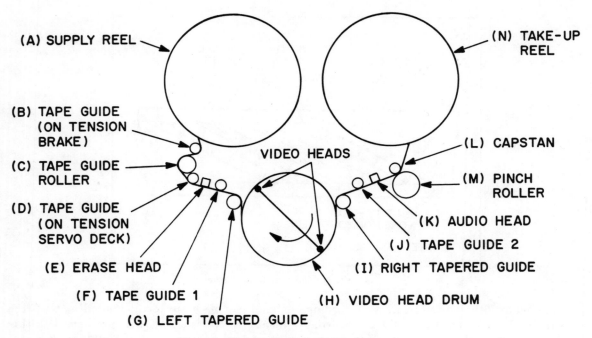

Fig. 1-7. *Tape path for reel-to-reel recorder.*

COLOR PLAYBACK

The successful playback of color signals recorded onto the tape is a more critical operation than that for B&W recording and playback. The major problems of color playback are as follows:

- tension changes in the tape movement
- tape stretch
- wow and flutter in the tape movement
- servo instabilities and corrections
- noise problems in the signal
- beats in the picture

If these effects occur, tint can change or color can be lost or interference bands appear in the

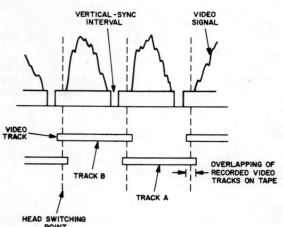

Fig. 1-8. *Video signal and recording tracks.*

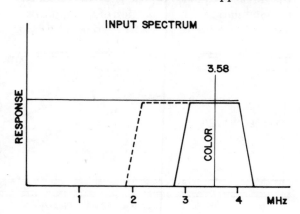

Fig. 1-9. *NTSC signal spectrum.*

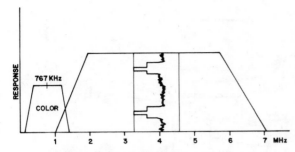

Fig. 1-10. *Color-under method.*

picture. Thus, some means of correcting for possible color problems is a definite requirement for color recording and playback machines.

VIDEOTAPE COLOR REQUIREMENTS

Among the basic requirements of a reliable, successful color video tape recorder machine are the following:

- properly timed switching of the heads
- properly positioned heads
- tape tension control
- continuous head-to-tape contact
- crosstalk reduction
- correction or avoidance of color recording or playback problems
- steady head-rotation speed
- steady tape movement speed

HISTORY OF VIDEOTAPE RECORDING

The recording of video information onto tape and the successful playback of this information has represented a great stride forward in communications. Development work that began in the late 1940s led to the first application of this technique in TV broadcasting. Amprex is credited with inventing, in 1956, a video tape recorder (VTR) that was used for this purpose. The machine was a model 1000, a quad-head recorder, using the transverse recording method and a 2-inch wide video tape. Improvements have been made consistently since then, in the original product and in other versions of video

recording. The VTR has definitely become an essential piece of equipment for all TV broadcasters. Tape recordings eliminate the live programs and can be used for delayed and repeated programs.

The development of "helical scan" video recorders in the early 1960s made possible the extension of the video recording into other fields, such as education, training, and industry. The first "helical scan" recorder was developed by Ampex, the model 660, in the early 1960s. It used a 2-inch-wide tape. One-inch machines then superseded the 2-inch version. Panasonic, as early as 1964, developed a VTR using a 1-inch wide tape that was used for medical purposes and also by industrial market.

The education field has been a major user of VTR playback units, connected into extensive closed-circuit TV (CCTV) systems. Pre-taped programs and information have become great aids for the teacher and trainer. Combined with TV cameras for on the spot, "live" input to the TV monitors, or recorded for later use, the VTR has given added dimension for many instructional activities. Among others, Sony has been particularly successful in this application of VTRs.

The early VTRs featured the "reel-to-reel" format, with manual tape "threading" from a supply reel, past the heads, to the take-up reel. This design is still used on some older video production systems. Automatic tape threading and smaller tape package are now possible with the VCR cassette tape units. VCRs were originally used by the teaching profession, but they are now enjoyed by the viewing public for home video recording. The compact design and easy operation of video cassettes opened the field for sales to the consumers market.

An example of early model VTRs using the helical scan technique was marketed by Sony in 1964. The unit was a reel-to-reel machine recorded black-and-white video on ½-inch-wide tape. Figure 1-11 shows the principles of this Sony recorder. The tape is wrapped around one-half or 180 degrees of the drum, which contains

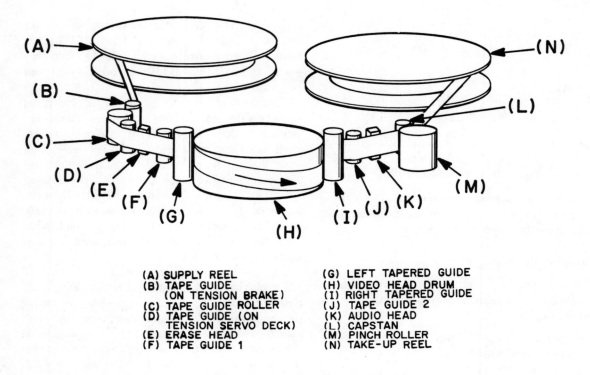

(A) SUPPLY REEL
(B) TAPE GUIDE
    (ON TENSION BRAKE)
(C) TAPE GUIDE ROLLER
(D) TAPE GUIDE (ON
    TENSION SERVO DECK)
(E) ERASE HEAD
(F) TAPE GUIDE 1

(G) LEFT TAPERED GUIDE
(H) VIDEO HEAD DRUM
(I) RIGHT TAPERED GUIDE
(J) TAPE GUIDE 2
(K) AUDIO HEAD
(L) CAPSTAN
(M) PINCH ROLLER
(N) TAKE-UP REEL

Fig. 1-11. *Early model Sony helical scan video recorder.*

two rotating video heads. The top portion of the picture details the angular path of the tape across the drum head. The tape enters the drum area from the supply reel on the left at a higher level than at its exit to the take-up reel. This tape movement past the rotating heads produces the magnetic patterns on the tape as shown in Fig. 1-12. The video tracks are slanted, hence the term "slant track recording." The audio track and the control track (on opposite edges of the tape) are produced from a stationary head, located in the tape path from the video head drum to the take-up reel.

The control track is part of an automatic servo system that ensures the correct positioning of the video heads relative to the recorded magnetic tracks on the tape during the playback mode of operation. Control track pulses are compared to the pulses developed by rotating heads.

Each slanted track on the recorded tape represents a TV picture field (two fields represent a complete frame). This early Sony machine included a skip field system, the result of recording by only one of the two heads at a time. During playback, both heads play the same track. This action provides the missing field (to complete the frame). However, the vertical resolution is only half the normal amount. It is necessary to switch between the heads so that the signal preamplifiers are always connected to the proper head—the one in contact with the tape. The switching is performed at a 60 Hz rate—a possible source of time-base errors in the playback signal. This is also the reason why it is desirable to have faster automatic phase control

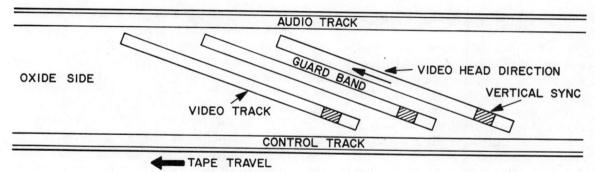

AUDIO TRACK

OXIDE SIDE

GUARD BAND

VIDEO HEAD DIRECTION

VIDEO TRACK

VERTICAL SYNC

CONTROL TRACK

TAPE TRAVEL

Fig. 1-12. *Slant track diagram.*

(APC) circuitry in the horizontal system of TV receivers that are used with VCR machines.

Shown in Fig. 1-13 are details of the signal frequency spectrum as processed by the early model Sony black-and-white recorder. For both recording and playback, the signal frequency band was moved to a higher range. The luminance or brightness signal frequency-modulated a carrier during the recording. Variations in brightness (white through shades of gray to black) became variations in frequency. This FM signal was present in the recorded slant tracks on the tape.

Guard bands are positioned between adjacent, recorded slant tracks. These guard bands contain no recorded information and represent wasted space on the tape. They are necessary to prevent crosstalk between the tracks produced by the VCR.

Some measure of standardization came into the picture of video tape recording about 1968 as the previously mentioned system evolved into the EIAJ type 1 format for ½-inch-wide reel-to-reel video tape recorders. EIAJ is an abbreviation for Electronic Industry Association-Japan.

Figure 1-14 shows the signal frequency spectrum for the color recorder system. The FM luminance signal was raised in frequency to the 3.2 to 4.6 MHz region. The color sub-carrier was shifted down during recording to 767 kHz, thus placing it below the luminance carrier. This conversion produces a "color under" system. The video track widths were reduced in width and the guard bands narrowed. Both video heads recorded, achieving full frame recording. These

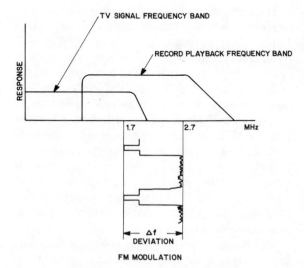

Fig. 1-13. *Early Sony black and white recorder.*

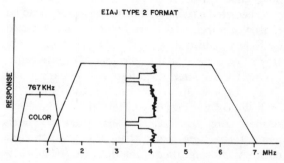

Fig. 1-14. *EIAJ color recorder.*

9

machines are used mostly in the industrial market.

The Philips VCR system, introduced in 1961, was one of the first ½-inch-wide tape cassette recorders. The cassette loading mechanism used by this model is widely used today. An example is that used in the Sony Betamax VCR machines. Also, the Philips VCR used a rotating transformer instead of slip rings to couple the rotating head signal to the preamplifier. This design eliminates the need for slip rings and the need for wear and pressure adjustments. This feature is also used on current VCR machines.

In 1971, Sony introduced the U-MATIC format-a ¾-inch cassette system, which has been very successful in the industrial market. It is characterized by high-resolution pictures and easy operation. This format represents a refined application of earlier technology, including the tape loading and rotating head transformer features of the Philips system.

In 1972, Cartrivision introduced a ½-inch-wide tape cassette system capable of two hours of playing time. This format used a skip field, three-head system, resulting in a reduction of tape consumption. Only every third TV video field was recorded. On playback, each of the three heads was played in order on the same track, producing a proper TV signal. Vertical resolution was one-half of a standard TV signal. Fast motion of the picture content could not be reproduced because of the loss of two out of three TV fields.

Cartrivision, in addition to recording live TV programs "off the air," promoted the playing of pre-recorded cassettes. These cassettes could be rented or purchased. Cartrivision was marketed by Sears, Admiral, and Wards, but sales were slow. The time probably was just not quite right.

One RCA format used four video heads, a 90-degree tape wrap of the head drum, and a ¾-inch-wide tape cartridge. The head wheel extended into the inserted cartridge to push against the tape, permitting the head-to-tape contact. One significant result of RCA's investigation of video tape recording at the time was a customer survey that determined a two hour uninterrupted playing time was desired by the TV viewers.

The next round of formats represent the present generation of home video tape recorders. All use ½-inch-wide tape cassettes and have built-in tuners and program turn-on systems. At this time, the VHS and Betamax are the standard format VCR systems.

The Sanyo V Cord Two format was introduced in 1976. It uses a ½-inch cartridge and is capable of one-hour full frame recording or two-hour skip field recording. The latter method results in one-half vertical resolution pictures. Tape loading is of the Philips type. The track widths and guard bands are reduced over earlier systems, thus reducing tape consumption. The video track width is 60 microns with a guard band of 37 microns.

The Matsushita VX2000 is a one-head machine using an "alpha wrap" of the tape around the head drum. Alpha wrap means that the tape almost completely encircles the head drum, permitting the use of only one head. This cassette type video recorder tapes up to a maximum of two hours per cassette. The first machine of this type in the United States had a Quasar brand name.

Through the years of development and improvement of video tape recording since the late 1940s, many variations have been tried in systems and component parts of the system. Several widths of tape have been used: 2 inch, 1 inch, ¾ inch, ½ inch, and ¼ inch. The 2-inch size has been a mainstay of the broadcast industry, although 1-inch tape width recorders have found a place in this market in recent years. Industrial and educational applications have employed the ¾-inch tape size. The ¼-inch-width tape is found in special machines such as portable color recording units. Of course, for the home video tape recorder market, the ½-inch-wide tape is now the standard.

The helical scan format is the most used system in video recording. Transverse recording is still used in the quad-head machines used by the

TV broadcasters. But even this field includes recorders with the helical scan format.

The reduced tape width, packaging it in small cassettes, and the helical scan format together make size reductions of the complete video tape recording machine possible, which in turn makes it very suitable for in-home use.

The success of the Sony Betamax-1 proves this point. An important element for consumer acceptance of the VCR is the reduction in the cost of tape usage. As Fig. 1-15 shows, the left curve, depicting the consumption of tape measured in square feet per hour, has been greatly reduced since 1968. The dots on the graph pin-

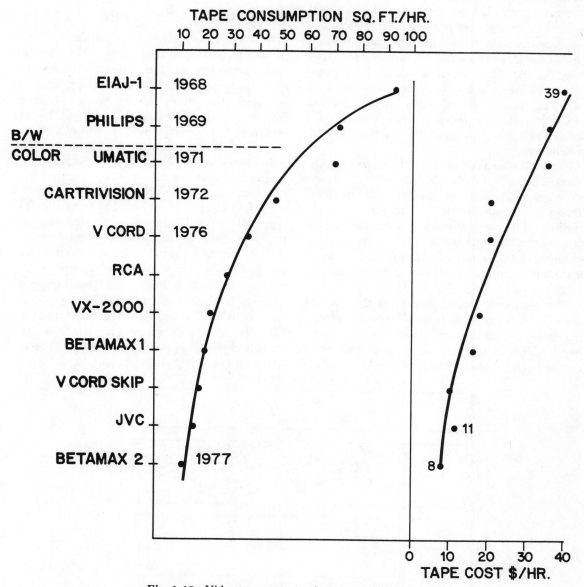

Fig. 1-15. *Video tape consumption comparison chart.*

point this usage for the various video tape recorders indicated. Thus, a machine using EIAJ-1 format in 1968 consumed more than 90 square feet of tape per hour.

The later-introduced Sony Betamax-2 consumes only about 10 square feet per hour. This reduced tape consumption, plus improvements in the tape, increased tape production, and the two-hour recording system have led to a sharp reduction in tape cost per hour. This curve appears on the right in Fig. 1-15. The graph shows a cost of $39.00 per hour in 1968 but only $8.00 per hour for the two-hour Betamax in 1977. Thus, the purchase price for blank tape cassettes that can hold two hours of video recording costs about $16.00. The Beta-3 speed machines consume 50 percent less tape than the Beta-2 VCR machines.

Sony's Betamax-1 was introduced in the U.S. market in 1975. It was a ½-inch-wide tape cassette system, capable of one hour playing time. Two video heads are used, with full frame recording. The tape threading (the method of pulling the tape out of the cassette and around the head drum) is of the Philips type. The Betamax-1 recorded tape pattern is shown in Fig. 1-16.

The recorded, slanted tracks of this helical scan machine are 60 microns wide with no spacing between tracks and therefore no guard bands. This is possible because of two well-devised arrangements that reduce crosstalk effects. Luminance signal crosstalk from track-to-track is avoided by slanting the two head gaps relative to the tracks. This is called the *azimuth technique*. Color crosstalk cancellation occurs because of a more involved signal phasing system involving a comb filter. The waveform drawing (Fig. 1-17) is the recorded signal spectrum for Betamax-1 tape speed.

TAPE FORMAT CONSIDERATIONS

Based on the fact that RCA had determined a three-hour VCR system was what the TV viewers wanted, JVC set about to modify the Sony Betamax-1 to a two-hour system. This was achieved by slightly reducing the capstan speed from 1.57"/s to 1.34"/s and increasing the tape length in the cartridge by 63 percent. This also increased the cartridge size by 31 percent. The video track width was left at 60 microns. To realize this, the track slant angle was increased by reducing the drum diameter. The azimuth recording and color processing are retained. The result was a two-hour version of video recording that JVC called VHS (Video Home System).

Sony in the meantime had produced its own two-hour Betamax. Sony chose to keep the same cassette and reduce the capstan speed by a factor of two. The slant track pitch was now 30 microns instead of 60. The head track was reduced from 60 to 40 microns. This results in a negative guard band and produces overlapping tracks. The azimuth and comb filter techniques, however, still

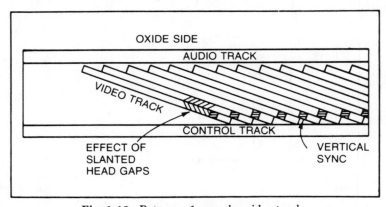

Fig. 1-16. *Betamax-1 recorder video tracks.*

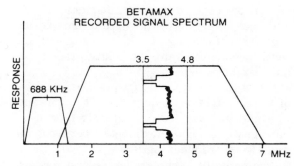

Fig. 1-17. *Betamax-1 recorder signal spectrum.*

provide adequate crosstalk rejection. The result is lower tape consumption.

The Sony threading produces very gentle tape handling. Note this tape threading path in Fig. 1-18. The tape guides are fixed and the guides are at 90 degrees to the tape motion and can be rotated to lower tape friction. The tape is simply wrapped around the drum. A longer piece of tape is removed from the cassette in the Sony machine, isolating the tape from cartridge feed irregularities and producing less time-base error. The Sony tape has a life of approximately 200 passes or plays. With the Sony VCR design, the tape does not leave the drum for fast forward or rewind, and precise location of a section of the tape is very easy to find.

VHS RECORDING OPERATION

Before delving into the VHS format, let's take a brief look at some video recording principles.

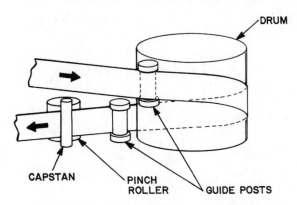

Fig. 1-18. *Sony Betamax recorder tape paths.*

VIDEO RECORDING BASICS

Like audio tape recording, video information is stored on magnetic tape by means of a small electromagnet, or head. The two poles of the head are brought very close together but they do not touch. This creates magnetic flux to extend across the separation (gap), as illustrated in Fig. 1-19.

If an ac signal is applied to the coil of the head, the field of flux expands and collapses according to the rise and fall of the ac signal. When the ac signal reverses polarity, the field of flux will be oriented in the opposite direction and continues to expand and collapse. This changing field of flux is what accomplishes the magnetic recording. If this flux is brought near a magnetic material, it will become magnetized according to the intensity and orientation of the field of flux. The magnetic material used is oxide-coated (magnetic) tape. Using audio tape recording as an example, if the tape is not continually moved across the head, just one spot on the tape will be magnetized and constantly re-magnetized. If the tape moves across the head, specific areas of the tape will be magnetized according to the field of flux at any specific moment. A length of recorded tape will therefore have on it areas of magnetization representing the direction and intensity of the field of flux.

As an example: the tape has differently magnetized regions that can be called north (N) and south (S), in proportion to the ac signal. When the polarity of the ac signal changes, so does the direction of the magnetization of the tape, as shown by one cycle on the ac signal in Fig. 1-20. If

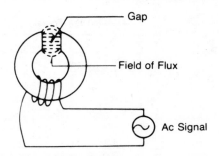

Fig. 1-19. *Head gap and field flux illustration.*

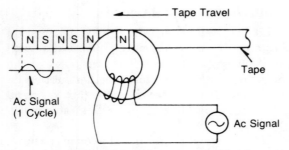

Fig. 1-20. *AC polarity change of head.*

the recorded tape is then moved past a head whose coil is connected to an amplifier, the regions of magnetization on the tape will set up flux across the head gap that in turn induces a voltage in the coil to be amplified. The output of the amplifier, then, is the same as the original ac signal. This is essentially what is done in audio recording, with other methods for improvement like bias and equalization.

There are some inherent limitations in the tape recording process that do affect video tape recording. As shown in Fig. 1-20, the tape has north and south magnetic fields that change according to the polarity of the ac signal.

If the speed of the tape past the head (head to tape speed) is kept the same, the changing polarity of the high-frequency ac signal would not be faithfully recorded on the tape, as shown in Fig. 1-21.

As the high-frequency ac signal starts to go positive, the tape starts to be magnetized in one direction. But the ac signal very quickly changes its polarity, and this will be recorded on much of the same portion of the tape, so north magnetic

regions are covered by south magnetic regions and vice versa. This results in zero signal on the tape, or self-erasing. To keep the north and south regions separate, the head-to-tape speed must be increased.

When recording video, frequencies in excess of 4 MHz might be encountered. Through experience, it is found that the head-to-tape speed must be in the region of 10 meters per second in order to record video signals.

The figure of 10 meters per second was also influenced by the size of the head gap. Clearly, the lower head-to-tape speed, the easier it is to control that speed. If changes in head gap size were not made, the necessary head-to-tape speed would have been considerably higher. How the gap size influences this can be explained as we look at Fig. 1-22.

Assume a signal is already recorded on the tape. The distance on the tape required to record one full ac signal cycle is called the *recorded wavelength*. Head A has gap width equal to one wavelength. Here, there is both north-and south-oriented magnetization across the gap. This produces a net output of zero, because north and south cancel. Heads B and C have a maximum output because there is just one magnetic orientation across their gaps.

Therefore, maximum output occurs in heads B and C, because their gap width is ½ wavelength. Heads B and C would also work if their gap width is less than ½ wavelength. The same is

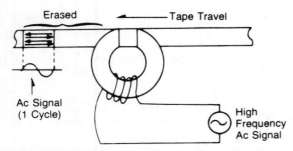

Fig. 1-21. *High-frequency ac signal considerations.*

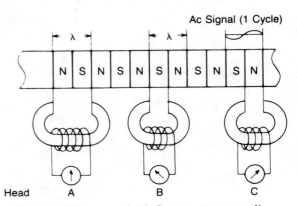

Fig. 1-22. *How gap size influences tape recording.*

also true for recording. A head-to-tape speed of 10 meters per second is a very high speed—too high in fact to be handled accurately by a reel-to-reel tape machine. Also, tape consumption on a high-speed machine is tremendous.

The method used in video recording is to move the video heads as well as the tape. If the heads are made to move fast across the tape, the linear tape speed can be kept very low. In two-head helical video recording (the only format discussed here) the video heads are mounted in a rotating drum or cylinder, and the tape is wrapped around the cylinder. This way, the heads can scan the tape as it moves. When a head scans the tape, it is said to have made a track. This is shown in Fig. 1-23.

In two-head helical format, each head records one TV field, or 262.5 horizontal lines, as it scans across the tape. Therefore, each head must scan the tape 30 times per second to give a field rate of 60 fields per second.

The tape is shown as a screen wrapped around the head cylinder to make it easy to see the video head. There is a second video head 180 degrees from the head shown in front. Because the tape wraps around the cylinder in the shape of a helix (helical), the video tracks are made as a series of slanted lines. Of course, the tracks are invisible, but it is easier to visualize them as lines. The two

heads A and B make alternate scans of the tape.

An enlarged view of the video tracks on the tape are shown in Fig. 1-24. The video tracks are the areas of the tape where video recording actually takes place.

There is one more point about video recording. Magnetic heads have characteristics of increased output level as the frequency increases, which is determined by the gap width. In practice, the lower frequency output of the heads is boosted in level to equal the level of the higher frequencies. This process is also used in audio recording and is called equalization.

Video frequencies span from 30 Hz to about 4 MHz. This represents a frequency range of about 18 octaves. Eighteen octaves is too far a spread to be handled in one system or machine. For instance, heads designed for operation at a maximum frequency of 4 MHz will have very low output at low frequencies. Because there is 6 dB/octave attenuation, $18 \times 6 = 108$ dB difference appears. In practice, this difference is too great to be adequately equalized. To get around this, the video signal is applied to an FM modulator during recording. The modulator changes its frequency according to the instantaneous level of the video signal.

The energy of the FM signal lies chiefly in the area from about 1 MHz to 8 MHz, just three octaves. Heads designed for use at 8 MHz can still be used at 1 MHz, because the output signal can be equalized. Actually, heads are designed for use up to about 5 MHz. Therefore, some FM energy is lacking but it does not affect the playback video signal, because it is resumed in the playback process. Upon playback, the recovered

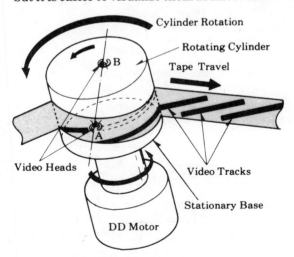

Fig. 1-23. *Head scan and tape tracks.*

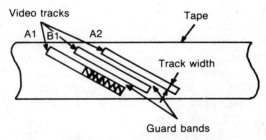

Fig. 1-24. *Video tracks and guard bands.*

FM signal must be equalized and then demodulated to obtain the video signal.

CONVERTED SUBCARRIER DIRECT RECORDING METHOD

In order to avoid visible beats in the picture caused by the interaction of the color (chrominance) and brightness (luminance) signals, the first step in the converted subcarrier method is to separate the chrominance and luminance portions of the video signal to be recorded. The luminance signal, containing frequencies from dc to about 4 MHz, is then FM recorded, as previously described.

The chrominance portion, containing frequencies in the area of 3.58 MHz is down-converted in frequency in the area of 629 kHz. Because there is not a large shift from the center frequency of 629 kHz, this converted chrominance signal can be recorded directly on the tape. Also note that the frequencies in the area of 629 kHz are still high enough to allow equalized playback. In practice, the *converted chrominance* signal and the FM signals are mixed and then simultaneously applied to the tape. Upon playback, the FM and converted chrominance signal are separated. The FM is demodulated into a luminance signal again. The converted chrominance signal is reconverted back up in frequency to the area of 3.58 MHz. The chrominance and luminance signals are combined, which reproduces the original video signal.

OTHER VHS RECORDER FUNCTIONS

Search-forward (Cue) and Search-reverse (Review)

In order to quickly find a particular segment on the tape during playback, the user can speed up the capstan and reel tables to nine times the normal speed, either forward or reverse, by pressing *cue* or *review* buttons. At this time, noise bars will appear in the picture because of head crossover. This is normal on some model VHS and Betamax machines. For example, some show four noise bars in cue and five noise

bars in the review mode. The bars for the cue and review modes are shown in Figs. 1-25 and 1-26.

Skew Correction

The various VHS machines have speeds that are referred to as SP, LP, and SLP. In the SLP mode and with the proper VHS cassette the playing time will be 6 hours.

Horizontal sync alignment on the tape occurs in the SP and SLP modes, but not in the LP mode as shown in Figs. 1-27 and 1-28. Thus, when using cue or review on LP recordings, severe skew or picture bending will occur at the top portion of the screen. Also, the color AFC will malfunction for this same reason. To correct this, the playback video is delayed by 0.5 H to compensate for skew, and the AFC frequency is shifted to maintain color lock.

Add-on Recording (Transition Editing)

Most VCRs allow pause during recording. But because of the arbitrary timing, there is most

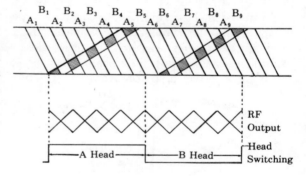

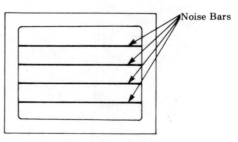

Fig. 1-25. *The cue mode.*

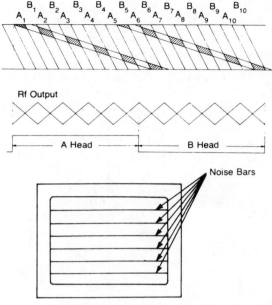

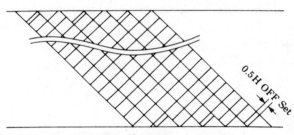

Fig. 1-26. *The review mode.*

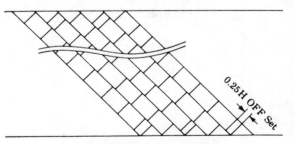

Fig. 1-27. *The SLP recorded mode.*

Fig. 1-28. *The LP recorded mode.*

likely a disturbance of the picture during playback at the place where the pause was used. To eliminate this disturbance, *transition editing recording* is used, which backs up the tape for 2.2

seconds during pause recording. When the pause is released, the deck will play back for about 1.2 seconds while aligning the control pulses already on the tape to the incoming vertical sync. After 1.2 seconds, the deck switches to the record mode, with the overall effect of no synchronization loss during playback (Fig. 1-29). Therefore, there will be no sync disturbance during playback regardless of how many times pause was used during recording. Refer to Fig. 1-30.

ABOUT THE VIDEO HEAD

Reduced track width requires the use of a smaller video head. But just making them smaller does not make them better. With less actual head material to work with, the magnetic properties of the head suffer. To offset this, a change in head material is in order. Because the VHS recorder is designed to be small, a reduction in the size of the head cylinder is called for.

A reduction in the size (diameter) of the head cylinder changes the head-to-tape speed. Remember, the head-to-tape speed affects the high-frequency recording capability of the head. To offset this problem, the head gap size was reduced.

The use of *hot pressed ferrite* as video head material in the VHS recorder helps improve the characteristics of the smaller heads. The hot pressed ferrite also has uniform domain orientation that further improves the head characteristics. It has been proven in many tests that the use of hot pressed ferrite material produces a superior video head.

From the preceding explanation, the need for smaller head gap size becomes apparent. In VHS recorders, the video head gap width is a mere 0.3 micrometers. This is quite a contrast from ordinary video heads used in other helical applications whose gap widths are typically in the area of micrometer.

HEAD AZIMUTH

Azimuth is the term used to define the left to right tilt of the gap if the head could be viewed

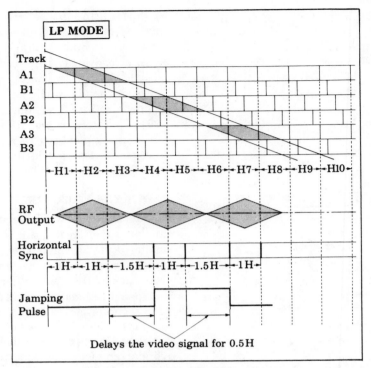

Fig. 1-29. *LP mode ½H correction.*

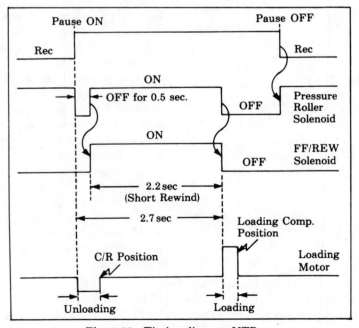

Fig. 1-30. *Timing diagram VTR.*

straight on. As azimuth recording is utilized in the VHS systems, the heart of the azimuth recording process is in the video heads themselves. This requires still another change in head design. In most VCR applications, the azimuth has always been set perpendicular to the direction the head travels across the tape, or more simply, the video track. Figure 1-31 helps to explain this. The gap is perpendicular (90 degrees) to the head's movement across the tape. We can think of this standard as a perfect azimuth of 0 degrees.

In VHS, the video heads have a gap azimuth other than 0 degrees. Also, one head has a different azimuth from the other. The 2 values used in VHS machines are azimuth of +6 degrees and −6 degrees. Refer to Figs. 1-32 and 1-33. These heads make the VHS format different from most other VCR formats. Exactly how the azimuths of ±6 degrees helps to keep out adjacent track interference is explained next.

AZIMUTH RECORDING

Azimuth recording is used in VHS to eliminate the interference or crosstalk picked up by a video head. Again, because adjacent video tracks touch, or "cross talk," a video head can pick up some information from the adjacent track when scanning. The azimuth of the head gaps assure

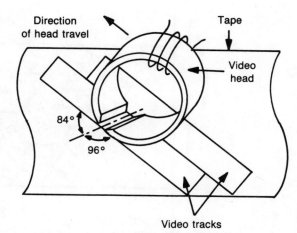

Fig. 1-32. *Head and tape alignment.*

that video head "A" only gives an output when scanning across a track made by head "A." Head "B," therefore, only gives an output when scanning across a track made by head "B." Because of the azimuth effect, a particular video head will not pick up any crosstalk from an adjacent track. Let's examine this more closely.

Figure 1-34 shows a VHS video recorder system in the SLP mode with the video tracks in a not-to-scale north and south magnetized regions on them. It can also be seen that these N and S regions are not perpendicular to the track; they have −6 degree azimuth in tracks A1 and A2, and +6 degrees azimuth in tracks B1 and

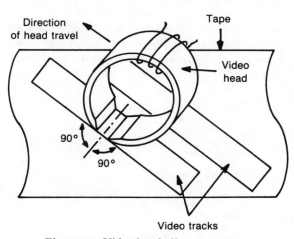

Fig. 1-31. *Video head alignment tape.*

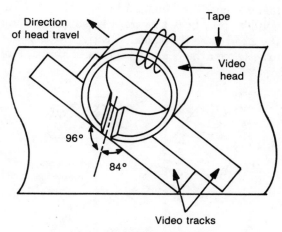

Fig. 1-33. *VHS head format.*

Azimuth in the LP mode

A1   B1   A2   B2

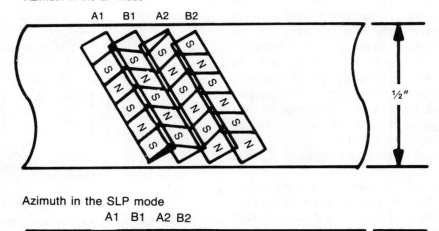

½"

Azimuth in the SLP mode
A1 B1 A2 B2

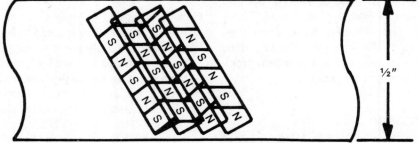

½"

Fig. 1-34. *Azimuth in the SLP mode.*

B2. If we take track A1 and darken the N region, it becomes easier to see. Refer to Fig. 1-35.

Figure 1-36 shows the information on track A made by head "A." Imagine now that head "A" is going to play back this track by superimposing the head over the track. Clearly, the gap fits exactly over the N and S regions, so that at any

moment there is either an N region or an S region or an N-to-S (or S-to-N) transition across the gap. This produces maximum output in head "A." Now, visually superimpose the "B" head over the track. Here there are N and S regions across the gap at the same time, at any given moment. Remember that simultaneous N and S

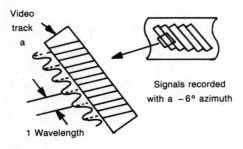

Video track a

Signals recorded with a −6° azimuth

1 Wavelength

Fig. 1-35. *Signals with a −6 = degree azimuth.*

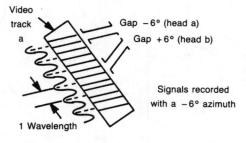

Video track a

Gap −6° (head a)
Gap +6° (head b)

Signals recorded with a −6° azimuth

1 Wavelength

Fig. 1-36. *Information made by "A" track.*

regions across the gap cause cancellation, and therefore, no output. Looking back at Fig. 1-33, see that the gap width is equal to ½ the recorded wavelength. Recall that this occurs at the highest frequency to be recorded. Therefore, the azimuth effect works at these high frequencies.

But what happens at lower frequencies? Figure 1-37 is a diagram similar to Fig. 1-36, except the recorded wavelength is longer, which represents a lower frequency.

Again, visually superimpose the heads over the track. Head "A" is the same as before. But look at head "B." There is much less cancellation across the gap, and its output is close to that of head "A." Therefore, you see where the azimuth effect is dependent on frequency. The higher the frequency, the better the azimuth effect. The lower the frequency, the lower the separation by azimuth effect.

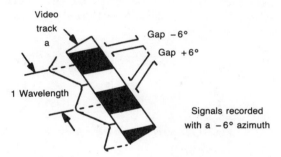

Fig. 1-37. *A longer recording wavelength.*

VHS COLOR RECORDING SYSTEM

Because there is insignificant azimuth effect at lower frequencies, a different type of color recording system was adopted. The fact that crosstalk occurs at lower frequencies cannot be changed, as it occurs right on the tape during playback. The method adopted processes the crosstalk component signals from the heads so that they are eliminated. It is important to realize that the crosstalk *does still occur*. It is the recording/playback circuitry that performs the crosstalk elimination.

In ordinary helical VCRs using converted subcarrier direct recording, the phase of the chrominance signal is untouched and is recorded directly onto the tape. The chrominance signal and its phase can be represented by vectors. Vectors graphically represent the amplitude and phase of one frequency. To keep it simple, assume the chrominance signal is only one frequency. For an example of vectors, see Fig. 1-38. The length of any vector represents its amplitude.

The azimuth effect does not work at the lower frequencies. And because the color information in VHS is recorded at low-converted frequencies, another technique for color recording was adopted.

Vector rotation in recording is actually a phase shift process that occurs at a horizontal rate of 15,734 Hz. The chrominance signal can be represented by a vector, showing amplitude and phase. In ordinary helical scan VCRs, the vector is of the same phase for every horizontal line on every track, as shown in Fig. 1-39.

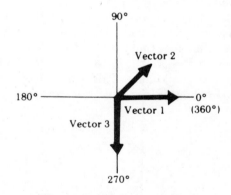

Vector 1 has a phase angle of 0°
Vector 2 has a phase angle of 45°
Vector 3 has a phase angle of 270°

Fig. 1-38. *Vector phase angles.*

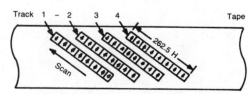

Fig. 1-39. *Normal helical scan VCR.*

In VHS, the 3.58 MHz is still converted down to a lower frequency, namely 629 kHz, but the color technique used in VHS format is a process of vector rotation. During recording, the chrominance phase of each horizontal line is shifted by 90 degrees. For head "A" (channel 1), the chrominance phase is advanced by 90 degrees per horizontal line (H). For head "B" (channel 2), the chrominance phase is delayed by 90 degrees per H.

- Channel 1 +90 degrees/H
- Channel 2 −90 degrees/H

VECTOR (PHASE) ROTATION

Now refer to Fig. 1-40 to see what this looks like on the tape. Now assume that as head "A" plays back over track A1, it will produce a vector output as such: head "A" when tracking over A1 will have an output consisting of the main signal (large vectors) and some crosstalk components (small vector).

Figure 1-41 is a vector representation of the playback chrominance signal from the head. One of the most important things done in the playback process is the restoration of the vectors to their original phase. This is done by the balanced modulator in the playback process. Note

the vector representation shown in Fig. 1-42. This restored signal is then split two ways. One path goes to one input of an adder. The other path goes to a delay line which delays the signal by 1 H. The output of the delay line goes to the other input of the adder. This can be more easily seen in Fig. 1-43.

As can be seen in Fig. 1-44, the crosstalk component has been eliminated after the first H line. The chrominance signal is now free of adjacent channel crosstalk.

The double output is not a problem because it can always be reduced. The process of adding a delayed line to an undelayed line is permissable because any two adjacent lines in a field contain nearly the same chrominance information. So, if two adjacent lines are added, the net result will produce no distortion in the playback picture.

In conjunction with the crosstalk elimination is the reconversion of the chrominance 629 kHz to its original 3.58 MHz. Now the color signal is totally restored.

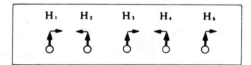

Fig. 1-42. *Vector for playback mode.*

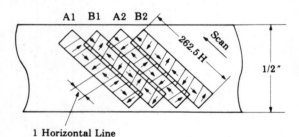

1 Horizontal Line

Fig. 1-40. *VHS helical scan.*

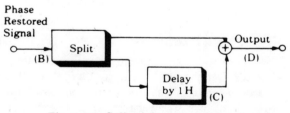

Fig. 1-43. *Split of the restored signal.*

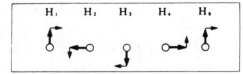

↑: Crosstalk Vector Component

Fig. 1-41. *Crosstalk vector component.*

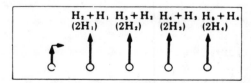

Fig. 1-44. *Double output signal.*

LP Tape Speed Mode

The recorded signal in the LP mode is considerably different from that used in the other VHS system tape speed modes. Like the SP mode, the chrominance and luminance signals are separated as covered earlier. However, from here on things are treated differently. Let's examine again the video tracks on the tape of an LP recording.

Notice in Fig. 1-45 that the tracks do overlap, and that any picture area of any track does not line up perfectly with the picture area of the adjacent tracks. (No horizontal sync alignment.)

Let's now pull several horizontal line segments off of the track for greater detail. As can be seen in Fig. 1-46, the horizontal sync portion of track B lies somewhere in the picture area of track A, for any given horizontal line segment. Assume that track A was recorded first. Then, as track B is laid down, the 3.4 MHz horizontal sync section of "A" will produce a beat with the portion of track B that covers it. Although the entire overlapping region will produce beats, the beat caused by the horizontal sync is most noticeable because the sync tip FM frequency

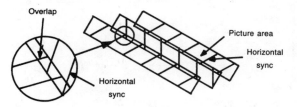

Fig. 1-45. *Overlapped tracks.*

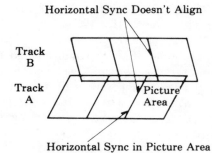

Fig. 1-46. *Horizontal sync does not align.*

never changes, whereas the FM frequency for the picture portion is constantly changing. This beat is visible on the screen, so measures must be taken to eliminate it. The method employed is called *FM interleaving recording*.

Note that beats are not the same as adjacent track crosstalk. Azimuth recording prohibits crosstalk pickup. But, the beat produced is a new frequency. It was not present in the video signal; it is the result of laying one track over another. The beat signal has no true azimuth, therefore, it will be detected by both video heads. The FM interleaving recording method does not actually eliminate the beat but rather places it at such a frequency so that no beat can be detected on the screen.

In the NTSC color TV system, video frequencies that are an odd multiple of half the horizontal line frequency have the property called "interleaving." Interleaving signals appear on the TV screen in a rather special way. Between any two adjacent lines, the signals are out of phase, as shown in Fig. 1-47 by the solid lines. Because the two lines are very close, the human eye tends to integrate them. The out-of-phase signals will virtually be cancelled (invisible to the viewer).

Now, when the frame is completed and the next frame begins, the signal on the top line will be out of phase with what was previously scanned. This is shown by the dotted line. This cancels any phosphor persistence from the previous scan. Thus, interleaved frequencies, for all purposes, do not create interference on the TV screen.

This interleaving is accomplished by raising the sync tip FM frequency in channel 1 by 15734/2 MHz or 7,867 Hz. For channel 1, then,

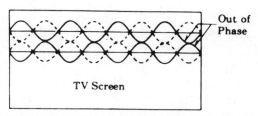

Fig. 1-47. *Adjacent lines are out of phase.*

23

sync tip frequency is 3.407867 MHz and peak white becomes 4.407867 MHz. Channel 2 remains the same as before — sync tip is 3.4 MHz, and peak white is 4.4 MHz. This displacement by 7.867 kHz causes the beat produced by the overlapped horizontal sync to become an interleaving frequency, which solves the problem.

Recovery of this shifted FM signal, although somewhat different, is essentially the same as before. The chrominance and FM signals that are mixed and then applied to the tape occupy a spectrum that is shown in Fig. 1-48.

SLP MODE FOR VHS SYSTEM

Like the other mode speeds, the video track on the tape of an SLP recording is shown in Fig. 1-49. Notice that the tracks do overlap, and any track picture area of any track will line up perfectly with the picture area of the adjacent tracks (horizontal sync alignment). Let's pull several horizontal line segments off of the track for greater detail.

As can be seen in Fig. 1-50, the horizontal sync portion of track B is in alignment with one of track A. Assume that the SLP recorded tape is played back. When the A head scans the A track, the A head picks up the B track signals on the overlapping region. Although the entire overlapping region will produce beats as in other

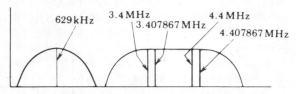

Fig. 1-48. *Chrominance and FM spectrum.*

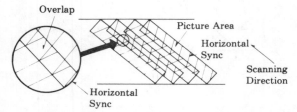

Fig. 1-49. *Video tracks in SLP mode.*

modes, the beat is eliminated by the FM interleaving recording.

TAPE M LOADING

In this type of VHS machine, the tape path is out of the cassette, across the stationary heads and around the video head cylinder forming a letter "M," thus the name "M" loading. Refer to this "M" type loading diagram in Fig. 1-51. The M loading has several advantages over previous, more complex loading formats.

- Less tape is pulled out from the cassette. This reduces the chances of tape spillage and tangles.
- Because the tape path in the M load pattern is short, loading time is only 3 seconds, including video muting.
- Fast forward and rewind are performed inside the cassette, further reducing the chances of tape damage.

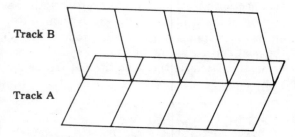

Fig. 1-50. *Horizontal sync is now in alignment.*

VCR TEST EQUIPMENT REQUIREMENTS

This section covers the types of test equipment needed for VCR servicing, test set-ups, and VCR test tapes.

The new home VCR market developed a new electronic service market. Video tape recorder service is no longer confined to a few service centers that specialize in just the industrial recorders, as was the case a few years ago. The VCR owner now expects to receive local service from the service shop that provides TV and stereo sales and service. Many dealers that sell

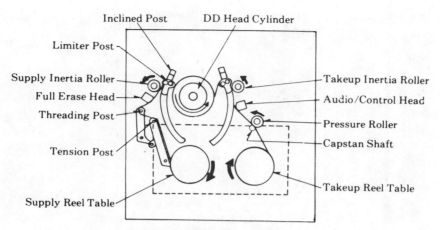

Fig. 1-51. *Drawing of "M" loading.*

VCRs believe that they must be able to service what they sell in order to have an advantage over their competition. This service is usually part of the same service network that services the TV receivers sold by the same dealer.

You might think you will need the same high-priced test equipment for servicing your VCR as that used for servicing broadcast-quality decks. Of course, the broadcast-quality test equipment can adequately service the new low-cost decks, but you need to look at what is the minimum equipment investment required to still ensure that the VCR is operating properly. In the case of a broadcast tape deck, the cost of the test equipment necessary to provide service is only a fraction of the cost for the machine, even if you consider only one machine. And, when you have more than one machine, this fraction becomes much smaller. However, in reference to the same equipment in relation to a household VCR, the cost of the test equipment can easily run from 15 to 20 times more than the VCR unit itself.

THE VA48 SENCORE ANALYZER

Before Sencore introduced the VA48 video analyzer, TV service technicians usually had to avoid VCR servicing. Or, most could only afford to equip the shop for VCR service and work with older test equipment for TV work. This video analyzer has eliminated this dilemma because it updates the shop for both TV and VCR servicing with one piece of test equipment.

Special considerations were made at the time the VA48 was designed to make sure the signals would be compatible with the new VCR systems. Extensive testing, both in Sencore's own lab and in working with the service managers of leading VCR manufacturers, has shown that these signals produce service results that are equal to, and in some cases superior to, those produced by an NTSC signal generator. This is especially true if the service shop is not equipped with an NTSC vectorscope for evaluating the output signals produced by this very high cost NTSC generator.

Using the VA48 Analyzer Patterns

The VA48 Sencore Analyzer is shown in **Fig. 1-52.** The two patterns used for VCR troubleshooting on the VA48 are the *bar sweep* and *chroma bar sweep.* There are a few special features built into each of these patterns that make them even more versatile in VCR service than you might find in standard TV service, because the TV receiver does not have all of the circuits that are part of a VCR machine.

The Bar Sweep Pattern. The bar sweep pat-

Fig. 1-52. *VA48 analyzer and dual-trace scope set up for a VCR bench check.*

tern should be used for all luminance (black and white) circuit testing. The first reason that this pattern is important is that it does not have a color burst. This gives a positive test of all automatic color-killer circuits found in both the record and playback VCR circuits. These circuits should automatically switch to the black and white mode when the color burst is not present.

The bar sweep pattern produces various amounts of B&W information. The grey scale at the left of the switchable frequency bars, for example, checks the black, grey, and white signal levels for proper amplitudes, linearity, and clipping. Each frequency bar after the grey scale alternates between pure black and pure white levels to test the circuits for proper frequency response. The bar sweep pattern looks much like the familiar "multi-burst" pattern used in many types of video testing. One difference between the bar sweep and the multi-burst is that the bar sweep is made up of square waves while the multi-burst is made up of sine waves. The key difference is that the bar sweep can show circuits that ring on fast signal transitions, where the

multi-burst may be passed by these stages without any noticeable ringing.

The various frequencies of the bar sweep pattern then check the video frequency response of the entire record/playback system. Certain applications of the bar sweep involve recording this pattern on a tape using a known good recorder. Many times you can use this tape in place of the expensive alignment tape and save wear and tear on your special alignment tape.

The Chroma Bar Sweep. The second pattern from the VA48 is the chroma sweep bar. This pattern contains a standard level color burst for operation of the color-killer circuits and the automatic color control (ACC) circuits. The color information is phase-locked back to the horizontal sync pulses so that this pattern produces a line-by-line phase inversion just like that of an off-the-air signal. This is very important because the comb filters used in almost all playback circuits require this phase-inversion to detect the proper color signal as opposed to the unwanted crosstalk information from an adjacent video stripe on the magnetic tape. The fact that the chroma bar sweep covers the entire frequency range of the color information (0.5 MHz either side of the color subcarrier) makes this pattern ideal for testing the operation of all the chroma conversion circuits to make sure that they are not restricting some of the color detail information.

Recording Your Own Personal Test Tape

If you have a VA48, you will want to record your own personal VCR test tape using the patterns from the analyzer. This test tape will not replace the need for the VCR test and alignment tape but rather supplements the alignment tape. The advantage of having a tape that you recorded yourself is that it will save much wear and prevent the possibility of damaging the expensive alignment tape. Now, should a defective machine damage your own test tape, you can replace it for the cost of a blank tape rather then having to purchase a new, costly alignment tape. However, the alignment tape allows you to confirm the various operations of the color circuits as well as the performance of the luminance circuits. A defective video playback head or preamplifier, for example, can be quickly determined by using your own test tape.

Any adjustments that affect compatibility with other machines should be made with the manufacturer's own test tape. This tape has been carefully prepared to provide a standard from one machine to another so that a tape recorded on one machine will play properly on another machine. This is especially important if your customer has more than one VCR and plans to play tapes on a different machine than was used for recording. The same goes for commercially prepared prerecorded programs on tapes that can be purchased and played back.

When you record your own VCR test tape, be sure to use a machine that you know is operating properly. New VCR machines are generally set up and aligned properly. Be sure to feed the signals directly into the CAMERA input jack of the VCR rather than feeding them through the antenna input terminals. This will assure you that an improperly adjusted IF stage or fine-tuning

adjustment will not cause the quality of your recording to be less than ideal for test purposes. Take advantage of the tape counter to allow you to locate specific portions of the tape so you can easily find the pattern you need to use. It is also desirable to include an audio signal for at least a portion of the tape to test the audio playback circuits. This audio signal is available from the "Audio 1000 Hz" position of the drive signal output from the Sencore VA48 video analyzer.

Perform the following outline of steps to record your VCR test tape.

1. Connect the VCR standard output from the analyzer to the CAMERA input of a known good VCR. Switch the input switch to the CAMERA position.
2. Set the tape counter to 000.
3. Select the faster tape recording speed on the machines that have more than one speed (i.e., Beta III, X1, or SP modes).
4. Select the BAR SWEEP position of the video pattern switch on the analyzer.
5. Place the VCR in the RECORD mode and record the pattern until the tape counter has progressed 50 counts.
6. Switch the video pattern switch on the analyzer to the CHROMA BAR SWEEP position and record this pattern for the next 50 counts.
7. Switch the video pattern switch on the analyzer to the COLOR BAR position and record this pattern for the next 50 digits on the counter.
8. Switch the video pattern switch on the analyzer to the CROSS HATCH position and record this pattern for the next 50 counts.
9. Switch the VCR to the slower tape recording speed (the Beta I, II, LP, or X2 mode).
10. Repeat steps 5 through 8 and record the same video patterns at each tape speed.

Your test tape is now recorded. If you want the audio tone recorded during one or more of these test sections, simply connect the drive signal output to the auxiliary audio input jack of the VCR while recording the pattern or use the

AUDIO DUB function to add the audio after the video is recorded.

The test tape now has the following patterns:

Tape Counter	Pattern
000–050	Bar Sweep
050–100	Chroma Bar Sweep
100–150	Color Bars
150–200	Cross Hatch
200–250	Bar Sweep (slow speed)
250–300	Chroma Bar Sweep (slow speed)
350–400	Color Bars (slow speed)
450–500	Cross Hatch (slow speed)

VCR Servo Alignment

The Sencore video analyzer signals are ideal for performing servo alignment adjustments on a VCR. The phase-locked, broadcast quality sync makes it perfect for this alignment application. Inject the signal from the VCR standard jack on the analyzer directly into the video or camera input jack on the VCR to eliminate the fine tuning errors that can occur when using an RF air signal.

There are several adjustments that call for counting the horizontal sync pulses just before the vertical sync pulse. When using an off-the-air signal, the equalizing pulses are present and can cause an error in the counting, resulting in setting a critical control setting wrong. The VA48 analyzer phase-locked sync signals do not have the equalizing pulses, and the counting of the horizontal sync can be done correctly without the equalizing pulses causing confusion, so the proper setting can be obtained easier. Equalizing pulses are not used in the VCR and are not important to servo alignment procedures.

VCR Audio Checks

The 1 kHz audio signal from the drive signal section of the analyzer can be used when making checks of the record and playback functions of the VCR machine. The manufacturer's align-ment tape should be used for making the play-back equalization adjustments. The following procedure allows injection of the audio test signal.

- Adjust the drive level control with the drive signals switch in the audio 1,000 Hz position to about 2 volts peak-to-peak reading on the VA48 meter.
- Inject the audio from the drive output jack on the analyzer into the audio input jack on the VCR. For VCRs without this jack, the signal can be injected into the audio circuits at a test point that is usually noted in the service data.
- The audio test signals can then be traced with an oscilloscope.

OTHER EQUIPMENT REQUIREMENTS

Besides a video analyzer or NTSC color bar generator, you will need a dual-trace oscilloscope, frequency counter, digital voltmeter, transistor checker, and perhaps a capacitance-inductance checker. Let's now look at a few of these test instruments that are now available for VCR troubleshooting.

SERVO ALIGNMENT PROCEDURES

In some VCR servo alignment procedures, a dual-trace scope with ADD channel capabilities might be called for. The ADD mode means that the two channels are added together in the scope. This feature is not found on some scopes, but it might not actually be needed in the alignment procedures. A Leader model LBO-515 dual-trace scope, shown in Fig. 1-53, is being used for servo alignment checks.

On scopes with no ADD mode, you can utilize the dual-chopped mode to provide proper waveform information for aligning the VCR servo systems as follows:

- Set both channels to ground and adjust the trace position control so that the two traces align in the center of the screen.

Fig. 1-53. *Leader LBO-515 scope used for VCR service.*

- Connect the A and B scope channels for the servo adjustment procedure and set the scope to the dual-chopped mode.
- Now make the adjustments for the servo alignment as stated in the service data.

Note: It is recommended that external sync be used in these procedures so that the oscilloscope will be locked to the channel that is not being adjusted. All you need do is determine which channel is stationary and which one is adjusted, and connect the external sync jack to the stationary channel test point.

DUAL-TRACE SCOPES

Many of the modern electronics service shops are now using wide-band dual-trace triggered sweep scopes for their everyday TV electronics troubleshooting. If you don't have a dual-trace scope, perhaps now would be a good time to review the need for one for VCR servicing. The greatest advantage of a dual-trace scope as compared to a single-trace scope is that it allows comparison of two waveforms at the same time.

This is very useful for tracing signals through inputs and outputs of various stages, but it is essential in troubleshooting critically timed circuits such as the servo stages. A dual-trace scope with a bandwidth of 20 to 50 MHz that has a good bright trace is your best bet for VCR and overall electronic circuit troubleshooting. The scope should also have a very stable trace with rock-solid lock-in of the waveforms. If you purchase a new scope, be sure the instrument has true TV sync separators built into the trigger circuits for both vertical and horizontal sweep rates. Some scopes that have a "TV" position do not use integrator filters to allow triggering on vertical sync pulses. One scope that has the features for VCR servicing and a 60 MHz bandwidth is the Sencore SC60 shown in Fig. 1-54. Another scope that can cover your electronic service needs is the Tektronix model 5403 (shown in Fig. 1-55), which has a bandwidth of 70 MHz. The one pictured is set up for dual-trace operation with one vertical plug-in unit (there is space on the front panel for another plug-in vertical amplifier).

Fig. 1-54. *Sencore SC60 oscilloscope.*

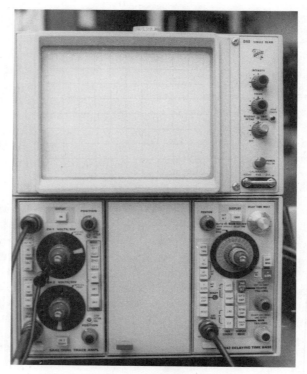

Fig. 1-55. *Textronix model 5403 scope.*

FREQUENCY COUNTERS

A stable frequency counter is required for setting the reference oscillators found in the color circuits of the VCR and for adjusting the servo circuits. The Sencore FC45 counter shown in Fig. 1-56 offers a high 25 mV sensitivity for measuring low-level signals with a full 8-digit readout for direct readings of the oscillator frequencies down to 1 Hz resolution.

The versatility of the FC45 can be expanded even more with the use of the PR50 audio prescaler, which lets you adjust the 30 Hz control track signal to an accuracy of .01 Hz. Some frequency counters use a "period measurement" mode for measuring these low frequencies. This can be very time consuming, because the frequency is shown as a time interval instead of a frequency. You must then use a calculator to figure the actual frequency by dividing the time measurement into 1 (frequency = 1/time). The PR50 provides a direct readout of frequency, which is updated every second, or 10 times a second with 0.1 Hz resolution, to eliminate the

Fig. 1-56. *Sencore FC45 frequency counter used for VCR troubleshooting.*

time-consuming calculations necessary with a period counter. The PR50 also has filtering built in to prevent false double-counting due to signal noise that is often present in these low-frequency signals.

DIGITAL VOLTMETER

The use of a digital voltmeter (DVM) is important for VCR servicing because most of the VCR circuitry is contained in ICs. The DVM is used for adjusting the regulated power supplies to the correct output voltages and for measuring the low-level signals found in some of the IC chips. The compact Hewlett-Packard DVM shown in Fig. 1-57 is ideal for troubleshooting those hard-to-get-at circuits found in some video recorders.

TRANSISTOR TESTER

Many of the circuits found in VCRs are controlled by discrete transistors. Some VHS video recorders, for example, have over 150 transistors that are used for various control functions. The use of an in-circuit transistor checker, such as the Sencore TF46 Super Cricket (shown in Fig. 1-58) greatly simplifies the troubleshooting of a transistor circuit, because the suspected transistors can be checked in-circuit to confirm whether they are defective or not. The TF46 has a high in-circuit testing accuracy for both transistors and FETs that helps to speed up circuit diagnosis. Also, the TF46 has an automatic gain test to allow grading or matching of transistors used in critical circuits.

OTHER VCR TEST EQUIPMENT

A capacitance and inductance checker can help you to locate leaky or off-value capacitors and check out any coils or transformers. The Sencore model LC53 shown in Fig. 1-59 will do a good job. Another frequency counter is the Re-

Fig. 1-57. *Hewlett-Packard DVM.*

Fig. 1-58. *Sencore TF46 transistor checker.*

gency model EC-175 shown in Fig. 1-60. This unit counts up to 175 MHz and down to 1/10 of a cycle. The Leader NTSC color bar generator shown in operation in Fig. 1-61 can also be used for VCR servicing. One corner of my VCR service bench is shown in Fig. 1-62 and illustrates the various pieces of test equipment I use to help solve VCR problems.

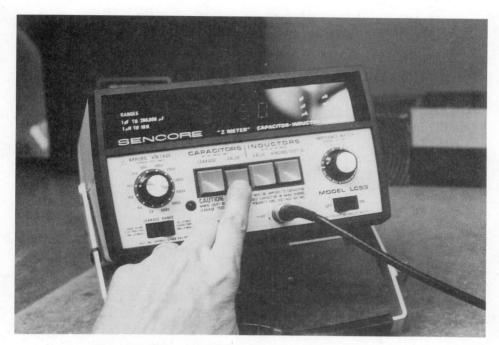

Fig. 1-59. *Sencore LC53 capacitance and inductance tester.*

Fig. 1-60. *Regency FC-175 frequency counter.*

Fig. 1-61. *Leader's NTSC color-bar generator.*

Fig. 1-62. *Partial view of my VCR service bench.*

2

Troubleshooting Tips and Techniques

This chapter contains some general VCR troubleshooting tips to help you locate problems in various types of machines. It is not practical to provide troubleshooting information for all VCR makes and models, so this is an attempt to give you an overview of troubleshooting techniques and how to use various test instruments. The same service techniques can usually be applied to both the VHS and Beta format VCRs.

This chapter features Sencore test instruments such as the SC61 waveform analyzer, the VA62 video analyzer, the VC63 VCR test accessory, and the NT64 NTSC color pattern generator. I explain how to use the test equipment for quick VCR alignment, trouble diagnosis, and VHS servo system analysis. A photo of the Sencore SC61 waveform analyzer is shown in Fig. 2-1.

GETTING STARTED

One of the important first steps is to have proper service information, such as schematics

and block diagrams. These should include scope waveforms, alignment data and the parts list. You should also look at the owners operating manual if you are not familiar with that particular brand of VCR.

The next step is to connect the VCR to a TV monitor and verify the complaint. Test the normal operation of the machine, using all of the controls. Make any required adjustments. Play back a good pre-recorded or test tape, and then make a recording on a blank tape.

Once the problem is identified, use logic to try and determine which section or block of the machine is at fault. The block diagrams for the unit, like those in Fig. 2-2 will be quite helpful in pinpointing the area of trouble.

When you have determined which block is at fault, you need to pinpoint the faults circuit and component(s). Remove the cover. A good way to begin is to use the old reliable look, listen, touch (carefully), and smell technique. Look for any obvious parts defect, listen for any unusual mechanical noise (or lack of such), carefully touch components that might be too hot or cold, and

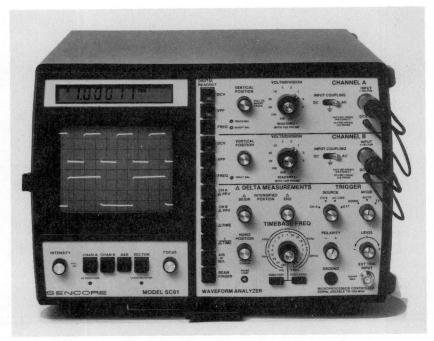

Fig. 2-1. *Sencore SC61 Waveform Analyzer. Courtesy Sencore.*

use your nose to ferret out any burned or overheated components. Always be on the alert for smoke.

After any obvious faults are cleared, use your test equipment to zero in on the defective component. Test equipment should include a video analyzer to inject signals, a triggered-sweep oscilloscope to look at waveforms and peak-to-peak voltages, a voltmeter, and a frequency counter.

OBSERVING SYMPTOMS

Use the following tips to determine the operating condition of the VCR. If one check does not produce the desired results, you should mentally (or physically) note some symptom of fault. Performing additional checks might add to a list of symptoms. This compiled information can then direct you to the area of the machine that is defective. What operating mode is not functioning properly? Is the playback mode or the record mode at fault?

Sometimes both record and playback processes are affected. The playback of a known, correctly recorded tape cassette in the suspected machine can aid in this determination. If the playback is normal, the record section should be investigated. However, if the attempted playback is not normal, check the playback system. A tape that has gone through a record process on the faulty machine will be able to be played back on a known good VCR. This test tells you whether the machine in question is recording properly. Some of the symptoms will be obvious, others will not.

PLUG-IN MODULES AND CHIPS

With some VCR brands, you can obtain new or rebuilt plug-in modules. With these types, you might want to isolate the problem by changing various modules. You then have the option

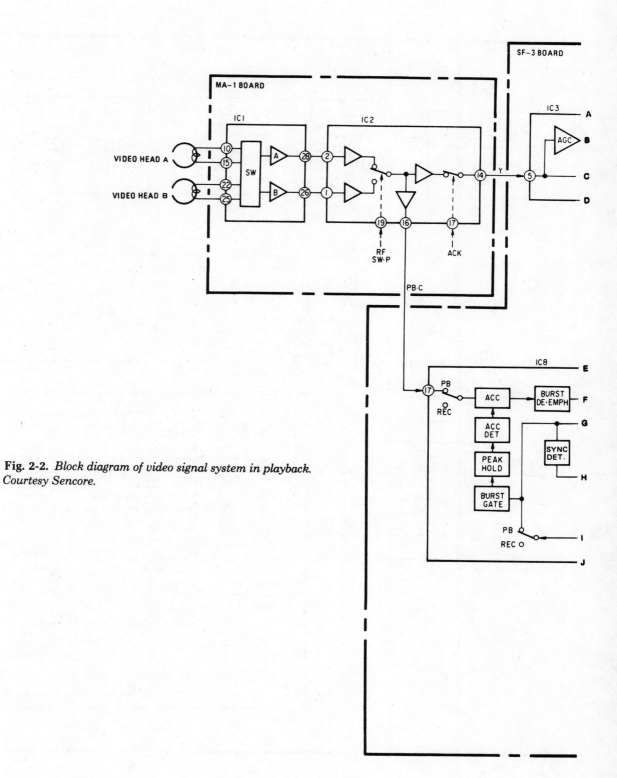

Fig. 2-2. *Block diagram of video signal system in playback.*
Courtesy Sencore.

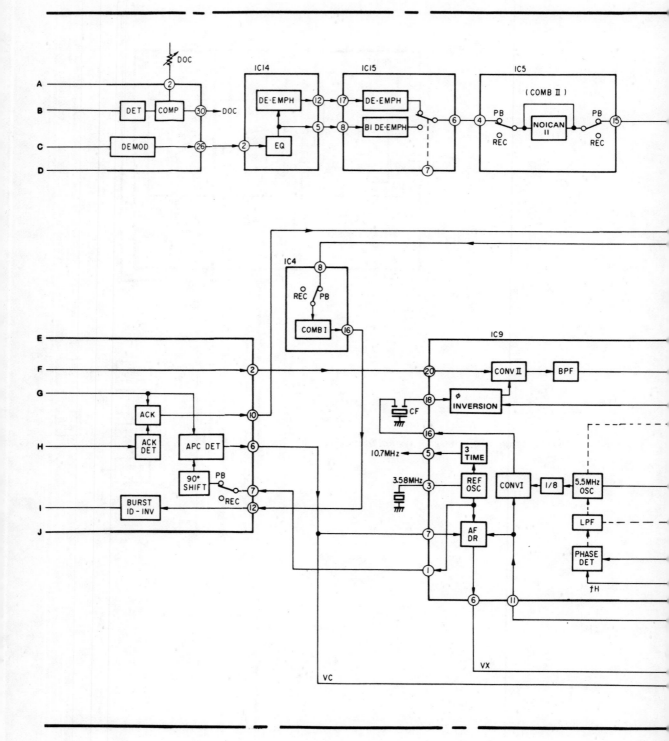

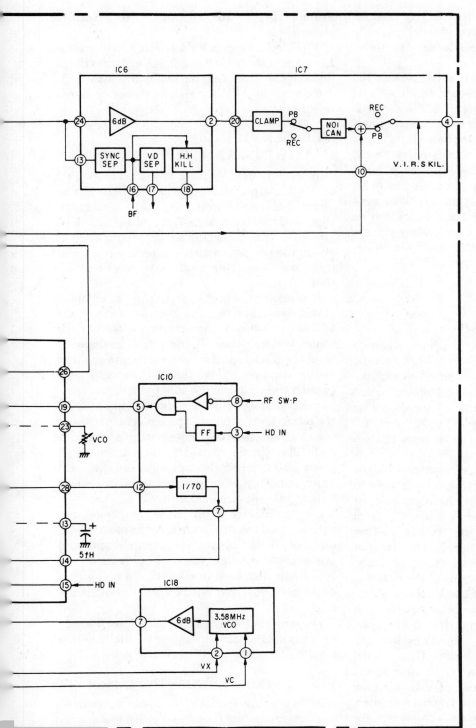

Fig. 2-2. *(continued)*

of repairing or replacing the defective module. This technique can allow you to quickly isolate the problem. Do not overlook loose or corroded module plug-in contacts.

Some VCRs have plug-in ICs that contain the complete playback video, record amplifier, and the chroma processing system. These are usually LSI (large scale integration) chips. Thus, should you suspect trouble in these chips, a new IC can easily be plugged in for a quick check. However, in most VCRs, these ICs are soldered into the board. In this case, you need to use a scope to check input and output signals or use a signal injection technique. Also, check the dc voltages at the chip pinouts. Only when you are certain a chip is defective do you want to desolder it and install a new one.

THE DIVIDE AND CONQUER TECHNIQUE

The "divide and conquer" technique can be used to quickly pinpoint defects in VCR circuits. This technique, sometimes referred to as *functional analyzing,* can save many steps in circuit diagnosis and bench time and will increase your troubleshooting efficiency. The "D and C" method can also be applied to all types of electronic troubleshooting.

To begin this technique, concentrate on the circuit operation to narrow the problem down to one stage or block, rather than individual circuit components. This method is different from other types of troubleshooting in that the first step is not finding which stage is defective, but finding out which ones are operating properly. Do this in a logical sequence until the defect is pinpointed to one block of the VCR block diagram. After the problem has been isolated with signal tracing or signal injection, use scopes, meters, transistor testers, etc. to isolate the faulty part(s) within the block. Thus, signal tracing and signal injection can be time savers for troubleshooting all types of VCR circuits. Also, keep in mind that signal tracing and injection techniques can be used simultaneously, for example in the RF, IF, and video amplifier stages.

The block diagram for a selected VCR amplifier stage is shown in Fig. 2-3. It illustrates the D and C technique. Note the test points 1 through 4 that can be used for signal tracing or signal injection.

VCR SERVICING CAUTIONS AND TIPS

When servicing VCRs or any other electronic device, be careful not to add to the original complaint. This can occur if a clip lead, for example, slips and shorts a signal or voltage level to ground or B+, damaging a component. Careful positioning of test instrument probes can avoid this time-consuming and costly service situation.

If measuring a signal or voltage at an integrated circuit pin, using an alternate test point is advised, because IC pins are very close to each other. Observe the foil trace that leads away from the pin to see if it is connected to another nearby component lead that is easier and safer to put a test probe on. Be sure there are no components in the circuit between the chip pin and the alternate test point. The best test probes to use in these cases are those with a spring-clip end. The plastic shroud insulates the small probe lead from adjacent pins or terminals. Finally, remove your wrist watch or other jewelry to help prevent circuit shorts.

Do not change an IC or other component without good reason. Determine the most probable fault in the circuit by measurements and other checks. It might not be the chip itself that is at fault. Also keep in mind that a faulty chip could have occurred due to another circuit defect.

Remember to make sure the TV set you are using as a monitor is operating normally before troubleshooting the VCR.

VCR CONTROL FUNCTION OPERATION

All of the early model VCRs used the mechanical control system that resembles a piano key-

Playback Circuits

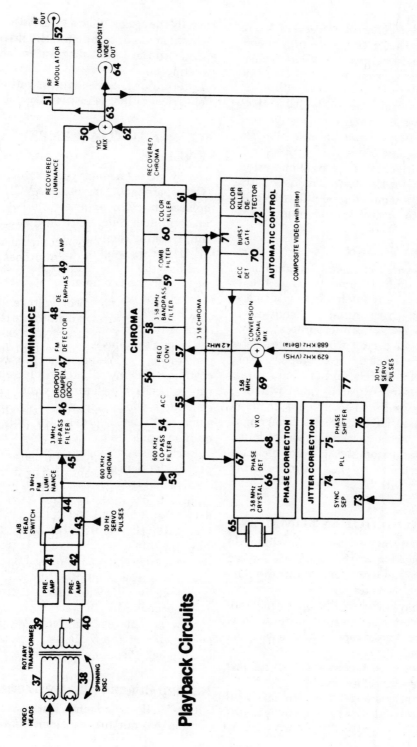

Fig. 2-3. *Diagram of the "Divide and Conquer" technique. Courtesy Sencore.*

board. The keyboard control moves leaf spring contacts together that activate the proper control function. These control functions include EJECT, PLAY, RECORD, FAST FORWARD (FF), REWIND, and PAUSE.

Many of the control function faults in these older machines were caused by defective contacts and bent springs. However, most all of the late model VCRs now use a microprocessor to control all of the machine's modes of operation. Just tap a button for the desired mode and the microprocessor transmits digital pulses to activate transistors to turn-on motors, relays, etc.

AUTOMATIC SHUT-DOWN

All VCRs have various sensors that automatically shut down the machine in case of a malfunction to prevent any further damage. In the newer machines, these automatic shut-down modes are controlled by the microprocessor. Automatic shut-down occurs when activated by sensors that detect slack or broken tape, end-of-tape, take-up reel problems, and dew (moisture).

These stop sensors are usually microswitches, photo transistors, and/or sensor lamps. In many of the older model VHS machines, all you might have to do is replace a burned-out sensor lamp. Check a suspected automatic shut-down circuit by bypassing the microswitches and/or covering the sensor lamps with black tape.

SERVICE PROCEDURES SUMMARY

- Collect as many symptoms as possible. Go through a checklist of the functions for the VCR to determine which operations are normal and which ones are not.
- If possible, have a normally operating machine available for comparison purposes and for the interchange of tape cassettes with the faulty machine.
- Have a known good recorded (in all modes) tape available as a reference.
- Use block diagrams to relate to the observed symptoms and to localize the area to be investigated.

- Study the circuit diagrams to determine which components or test points should be checked.
- Refer to the layout diagram to locate circuit test points.
- Determine if a mechanical rather than an electrical failure is involved.

YOUR FACTORY EXPERT: SENCORE

Say you are an employed technician in a TV/VCR repair shop. On your last VCR repair, you replaced the cylinder head, drive belts, idler, and capstan pinch roller. The mechanical tape tension, tracking, and head switching all checked OK. The work went off quickly with no hitches. Two days later the customer comes back with the unit and the complaint, "It just doesn't look as good as it used to."

Is it the VCR or the customer? If you challenge the VCR owner by saying, " . . . that's the best it's gonna do . . . " or " . . . it looks good to me," you stand a good chance of losing a customer and any of his/her referrals. Is there a way to positively eliminate doubt and prove to customers that their VCR does perform the same or better? Perhaps the only acceptable opinion would be that of a factory expert.

In the case of Sencore, your VCR factory expert is the Sencore Tech-Pak: the SC61 Video Analyzer (Fig. 2-1), the VA62 Universal Video Analyzer (Fig. 2-4), the VC63 VCR Test Accessory (Fig. 2-5), and the NT64 NTSC Pattern Generator (Fig. 2-6). All of these instruments have patents and exclusive tests that allow you to completely test VCRs to see if they meet manufacturers' specifications. In fact, the VA62 is accepted by all major manufacturers for factory-authorized VCR warranty work.

How can the Sencore Tech-Pak prove to the customer that the VCR really is as good as it used to be? There are many ways.

RECORD CURRENT BIAS ADJUSTMENT

One VCR performance test area that is often overlooked, but has considerable influence on

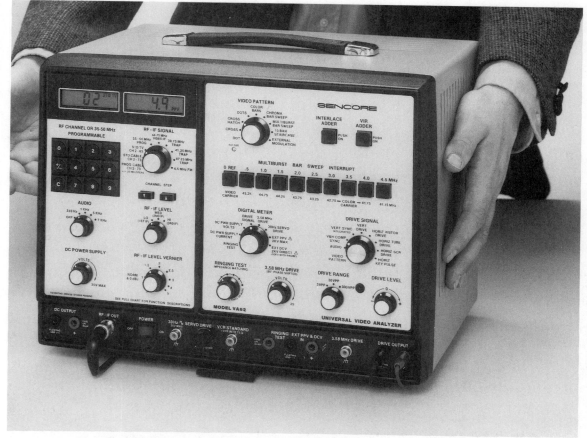

Fig. 2-4. *Sencore VA62 Universal Video Analyzer. Courtesy Sencore.*

whether a VCR looks as good as it used to, is the *record current bias adjustment.*

Do not assume that because the playback circuits are working fine that the record circuits are also working good. A poor luminance signal-to-noise ratio, degraded chroma, video overload, and "diamond beats" on the TV monitor screen can all be caused by inaccurately set record current bias adjustments.

The introduction of the new high-fidelity VCRs makes record current adjustments more important to accurately check and set than before. These new hi-fi VCRs deep bias the stereo audio (150 to 200 mV) right in the normal video tape path, but just before the video luminance information (100 mV average) is laid down (see

Fig. 2-7). The Azimuth recording principle (audio/video recording and playback heads placed opposite angles) enables the deep bias system to work.

A small change in the record bias level can alter the video, chroma, and hi-fi quality. The exclusive features of the VA62 Video Analyzer and the SC61 Waveform Analyzer enable you to quickly and accurately check VCR record bias.

Most manufacturers recommend that you set your record bias in the sequence of FM audio, chroma, and then luminance record level for the best performance. Because color is a common complaint, let's take a closer look at the chroma record level adjustment.

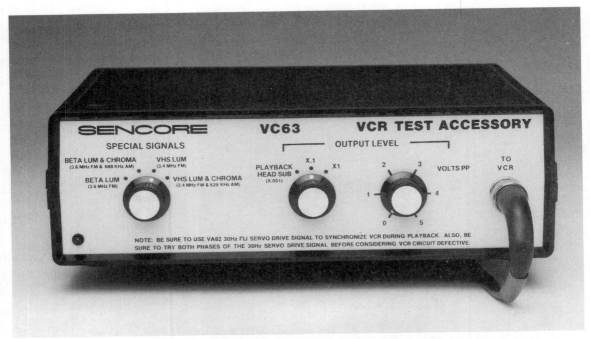

Fig. 2-5. *Sencore VC63 VCR Test Accessory. Courtesy Sencore.*

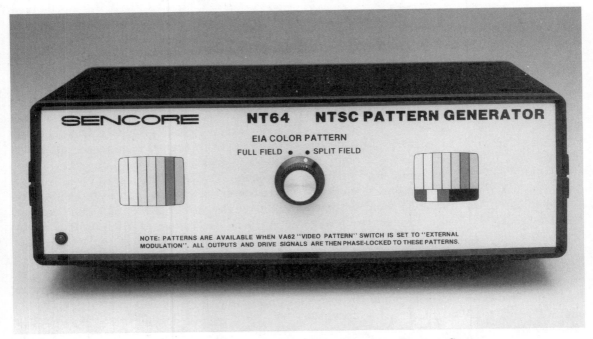

Fig. 2-6. *Sencore NT64 NTSC Pattern Generator. Courtesy Sencore.*

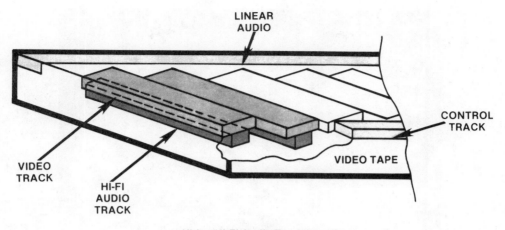

LINEAR
AUDIO

CONTROL
TRACK

VIDEO TAPE

VIDEO
TRACK

HI-FI
AUDIO
TRACK

Video/Hi-Fi Audio Track Locations
(Edge View of Tape)

Fig. 2-7. *Video hi-fi audio track locations shown from edge of tape. Courtesy Sencore.*

CHROMA LEVEL ADJUSTMENT

Chroma level bias adjustment accuracy is critical because too high a level can cause heterodyning or diamond beats in the picture, and too low a level causes poor color. You are working with a small visual area for the adjustment — less than one centimeter of scope signal (30 to 40 mV). But what makes chroma record level checking and adjustment even more difficult is the NTSC signal required.

The NTSC split field produces a maximum 100 percent white level positive pulse, which along with the standard negative sync pulse, drives most scopes nuts trying to decide which pulse to sync onto. At 35 mV, it's even tougher because the pulses are removed. Now your scope has to decide which low-level color bar or color burst signal to sync onto. Your adjustment accuracy is questionable at best. Plus, you have to adjust chroma bias current for the color bar having the highest modulation level, which is cyan at 75 percent modulation. Check your chromaticity chart first, because if you pick the wrong chroma bar, your chroma record bias level will be set wrong.

The Sencore VA62 NTSC chroma bar sweep saves the time you would spend guessing and looking. The chroma bar sweep pattern generates the 75 percent level cyan bar plus two 500 kHz sideband bars. You can switch out the two sideband bars and just use the one cyan bar for fast, easy lock-on for accurate chroma record percent adjustment.

First connect the VA62s VCR 1-volt peak-to-peak standard output to the VCR video input. Select the VA62s chroma bar pattern, and remove the two sideband bars. Then connect the SC61 scope with the DP226 direct probe to the chroma record test point.

The chroma record adjustment procedure is a simple three-step process:

- Load in a blank tape. Put VCR in SLP record mode.
- Turn the luminance record level adjustment down to view chroma record level signal.
- Push the SC61 peak-to-peak button and reset chroma record current level (30 to 40 mV average level).

That's all there is to it, and it's quick and accurate. Two hookups and three steps are all you need to do to know whether the VCR's critical chroma record current level is correct.

The patented chroma bar sweep pattern (shown in the Fig. 2-8) generates a 100 percent

Fig. 2-8. *The chroma bar sweep pattern generates a 100 percent white level, plus 75 percent cyan and two 500 kHz sideband bars.*

white level plus 75 percent cyan and two 500 kHz sideband bars.

SENCORE VC63 VCR TEST ACCESSORY

The VA62 brings the speed and effectiveness of signal substitution to VCR servicing. VCRs have two special signals that are quite different from signals in TV receivers, however. For this reason, the VC63 VCR accessory to the VA62 is the main signal source for VCR servicing. Let's now look at why VCRs have special signals, what these signals are, where to use the special VCR signals provided by the VC63, and how to use the VC63 signals effectively to isolate VCR problems. Figure 2-9 shows the engineering models of the VA62 and VC63 that the author of the troubleshooting manual used to evaluate these units in writing the manual.

SPECIAL VCR SIGNALS

The special VCR signals were developed to overcome the limitations of the magnetic tape recording process. A basic problem with using magnetic tape to store information lies in transferring information onto and off of the tape. When signals are read back off of the magnetic tape, the amplitude at the playback head increases as the frequency of the recorded signal increases, as Fig. 2-10 illustrates. This occurs because the amount of flux produced across the playback head gap varies with the wavelength of the recorded signal. For each octave or doubling in frequency of the recorded signal, the output voltage increases by 6 dB (quadruples). At some point, determined by the size of the playback head gap, the output sharply drops to zero.

If only a relatively narrow range of frequencies is to be recorded, this limitation can be tolerated. Audio engineers have worked around this problem for many years. Recorded audio frequencies range from about 40 Hz to 20 kHz, encompassing 10 octaves. The 6 dB/octave playback response produces a 60 dB change in output. Although recorders use elaborate equalization circuits to minimize the effects of tape playback response, 10 octaves is the limit for tape recording.

46

Fig. 2-9. *Photo of original engineer's model of the Sencore VA62 Video Analyzer and VC63 VCR test unit. Courtesy Sencore.*

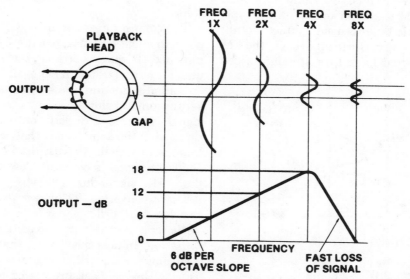

Fig. 2-10. *The voltage induced in the playback head increases at a 6 dB/octave rate as the frequency of the recorded signal increases. When the wavelength equals the head gap, the induced voltage is zero. Courtesy Sencore.*

FM Luminance Signal

The answer to this dilemma is to reduce the octave range of the recorded signal. This is accomplished, as Fig. 2-11 illustrates, by using the composite video signal to frequency modulate a carrier. Recording the luminance video information using FM is the first special VCR signal. In this FM conversion, the sync tips are the "at rest frequency" of the carrier and the peak white video signals are maximum deviation. How fast the carrier deviates corresponds to the frequency of the video signal. The range of video frequencies modulating the carrier is limited to about 2.5 MHz. This FM signal is recorded directly onto the magnetic tape.

Both VHS and Beta formats use FM to reduce the number of recorded octaves. With the FM modulation sidebands extending from about 1 to 6 MHz, less than 3 octaves are recorded. Though slightly different FM frequencies are used in VHS and Beta, frequencies around 4 MHz were selected as the best compromise between head gap size, writing speed, and bandwidth.

The FM signal offers a further advantage. An FM system is not affected by moderate signal amplitude changes. Thus, small amplitude variations in the FM signal picked up off the tape, caused by erratic tape-to-head contact, do not affect picture quality. During playback, the FM signal is converted back to a standard video signal.

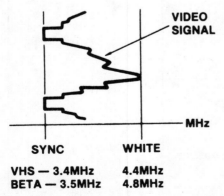

Fig. 2-11. *The video signal frequency modulates a 4 MHz carrier. Courtesy Sencore.*

Down-Converted Color

Using FM to record the luminance solves bandwidth problems, but it creates problems for recording color. The reason is conventional NTSC video signals use phase-referenced color information. The phases must be accurately maintained, or portions of the video picture will be the wrong color or continually change in color. All phase information, however, is totally lost in an FM process. The color information in a VCR, therefore, cannot be recorded as FM along with the luminance signal. Recording the color information directly as it appears in the 3.50 MHz NTSC signal would cause very noticeable and undesirable interference beats between the 4 MHz luminance and 3.58 MHz color signals and sidebands.

Because neither FM nor direct recording of color will work, the alternative is to convert the frequency of the recorded color, for example placing it either higher or lower than the FM luminance frequency. Placing the color frequency above the FM luminance causes problems with the head gap, so the color information is lowered in frequency. This becomes the second special VCR signal—down-converted color.

The color information in a VCR is converted down to about 600 kHz, placing it below the luminance frequency (thus the term "color under") as illustrated in Fig. 2-12. This signal is still amplitude- and phase-modulated for color saturation and tint. The down-converted color signal is recorded directly onto the magnetic tape using the same heads that record the FM luminance signal. As with the FM luminance, the color signal is converted back to the standard video signal during playback.

Down-converted color answers a second problem of recording color information on magnetic tape. As the tape moves through a VCR, variations in tape speed occur that cause severe problems for the critically phased color signal. During the playback process of converting the 600 KHz signal back to 3.58 MHz, these errors are easily corrected.

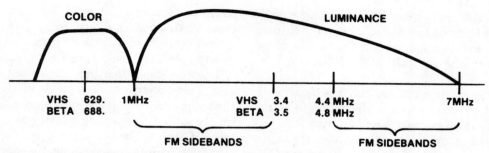

Fig. 2-12. *Color information is converted down to 600 kHz, placing it below the FM luminance frequency. Courtesy Sencore.*

Color recording requires one other change to the conventional color signal besides frequency conversion. To reduce color cross-talk from adjacent tape tracks, the phase of the recorded color information is shifted. During playback, color signals in the adjacent tracks have the "wrong" phase and are removed by a comb filter. Thus, only the correct color for each picture element is recovered off the tape.

SPECIAL SIGNALS CAUSE PROBLEMS

The FM luminance signal presents a major problem for VCR technicians who rely on signal tracing. To begin with, the FM signal picked off the tape by the spinning video heads is a minute 500 microvolts — well below the measuring capabilities of an oscilloscope. Yet, unless this signal is good, the VCR will not play back a good picture. Secondly, you cannot determine very much by looking at it as you can an FM signal with a scope.

Most of the symptoms caused by a problem in the circuits before the FM detector are the same — a raster filled with snow or a raster that is completely blank. How do you troubleshoot the luminance stages between the video heads and the FM detector? After measuring a few voltages and using a little head cleaner, you are faced with the possibility that the heads could be defective. So, to put an end to your doubt, you swap heads. If the VCR is now fixed, the heads were defective. But what if it's not okay? At least you know that the heads probably were not the

problem (you hope). Head swapping is a 50/50 proposition at best.

Using the recorded signal of the tape as the reference when troubleshooting VCRs leaves too many variables. You do not know at what point the signal goes bad, or if the signal is even being properly picked up off the tape. Is the problem electrical, or is the tape path alignment severely off, preventing the heads from picking up any signal? Perhaps the tracking is completely out of whack or the heads are not switching properly. The answer to these problems is to inject a known good reference signal into the circuits rather than trying to follow impossible-to-trace signals. This is the same efficient method of signal substitution that the VA62 makes possible for television servicing.

With signal substitution, you inject a good signal into the circuits. If you inject the good signal *after* the defect, the VCR operates properly, telling you that the circuits from that point forward are working. Then simply trace backward, step by step, into preceding circuits until the VCR stops functioning properly. At that point, you are injecting into the defective stage.

WHERE TO USE THE VC63

As you have learned, VCRs have special FM luminance and color-under signals. The VC63 VCR Accessory works with the VA62 Video Analyzer to provide these special signals for VCR troubleshooting. Not all the circuits in a VCR require the use of the VC63. The signals after the

FM detector are baseband video and are substituted for by using the VA62. Most of the chroma circuits involve important timing relationships and require the use of the SC61 Waveform Analyzer.

For review summation, the FM luminance and down-converted color signals begin at the spinning video heads during playback. The signal picked up by the heads is coupled through a rotary transformer to a head preamplifier. The output from the head amps is switched by the A/B head switcher so only the signal from the head that is in contact with the tape continues. This keeps the noise picked up by the other head out of the picture. After the switcher, the 4 MHz FM luminance signal is separated from the 600 kHz color-under signal by a combination of high-pass and low-pass filters. Both special VCR signals continue down separate paths until they are converted back to standard NTSC format signals and combined by the Y/C mixer. All of these circuits, from the video heads through to the FM detector and color-under frequency converter, require you to use the special signals provided by the VC63.

You troubleshoot a VCR by first connecting it to a television monitor and playing back a test tape. (Chapter 1 explains how to make a test tape.) Playing the tape can reveal one or more important symptoms on the monitor, depending on the defective stage in the VCR. Carefully analyzing the symptom tells you if the problem is in the servo, luminance, or chroma circuits.

Servo symptoms look like picture tearing or rolling. Sound that is "too fast" or "too slow" also indicates a servo problem. Symptoms other than these are called *video playback* and are caused by any of the non-servo circuits. Video playback symptoms include no video, poor video, no color, and poor color. The "trouble tree" in Fig. 2-13 summarizes these symptoms and possible causes. Use it as a guide for isolating playback problems.

To better understand how to use the VC63, let's use a Panasonic PV1225 VCR and the common VCR playback symptom of a blank raster as an example. A portion of the actual luminance block diagram supplied by the manufacturer for this deck is shown in Fig. 2-14. Always troubleshoot from a block diagram and trouble tree to help keep you on the right track.

The first step is to bypass the video mute circuit. This circuit (not used on all VCRs) mutes the video and audio if the servos are not locked in. Bypassing the muting allows you to determine if the problem is servo related or video circuit related. In this example, the monitor remains blank. Does this mean the problem is in the direction of the heads, or the video stages?

To determine which way to go, inject a video signal at the output of the FM detector, point (48). The signal here is a baseband video, rather than one of the special VCR signals, so use the VA62 for the signal source. Select the Cross Hatch Video Pattern on the VA62, because this pattern is different from any recorded on the test tape. Next, connect the VA62 DRIVE OUTPUT to the FM detector output. Increasing the drive level just above 0.3 volts brings a picture onto the monitor. This is a very important clue. It means the remaining video circuits function correctly and the problem lies in the direction of the heads.

The circuits in the path toward the heads require you to use the special VCR signals from the VC63. Because this is a VHS VCR, set the SPECIAL SIGNALS switch on the VC63 to VHS LUM, as shown in Fig. 2-14. This setting provides an FM luminance signal without the down-converted color signal (the presence of the color-under signal could cause misleading luminance symptoms). The LUM & CHROMA position provides the special color-under signal for troubleshooting color circuits. The setting of the VA62 VIDEO PATTERN switch determines the modulation of the VC63 signal that is set to provide a contrasting signal with those recorded on the work tape. Why? So you can quickly tell if the picture on the monitor is the tape signal or the VC63 test signal.

Now you want to see which of the circuits between the detector and heads are working, so inject an FM luminance signal into the output of the A/B Head Switcher, point (45). Injecting

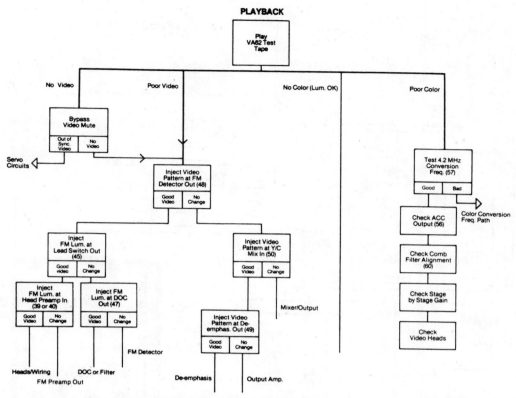

Fig. 2-13. *Troubleshooting trouble tree for VCR playback problems. Courtesy Sencore.*

here cuts the remaining stages in half, allowing you to quickly narrow down the problem. Point (45) is the combined output of both video heads after the preamplifiers. The substitute signal is supplied by the VC63 with an OUTPUT LEVEL control setting of X.1. Increasing the output level vernier between 1 and 2 in this example brings a pattern onto the monitor. Now you know that all the stages between test points (45) and (48) are also working.

As the trouble tree shows, the next logical step is to inject a signal into points (39) and (40). These points are the outputs of the video heads. The signal level here is unamplified and very small. The PLAYBACK HEAD SUB (X.001) setting of the VC63 OUTPUT LEVEL switch provides this low-level signal. With the X.001 setting, some setting of the OUTPUT LEVEL vernier should pro-

duce a picture. But in this example, increasing the vernier all the way to 5 mV does not produce a picture on the TV monitor. Can you see that this means the problem lies between points (39), (40), and (45)?

You could further isolate the problem to either the preamps or the A/B head switch by injecting a signal into test points (41) and (42), but as the block diagram for this particular VCR shows, the preamps and the head switch in this VCR are part of IC3002. Before you replace the chip, double-check that it is receiving B+ and the 30 Hz head-switching signal.

What if injecting at points (39) and (41) had returned the picture? This leaves the heads, rotary transformer, or associated connections in question. You can use the VC63 to isolate problems here as well. By using the VC63 and signal

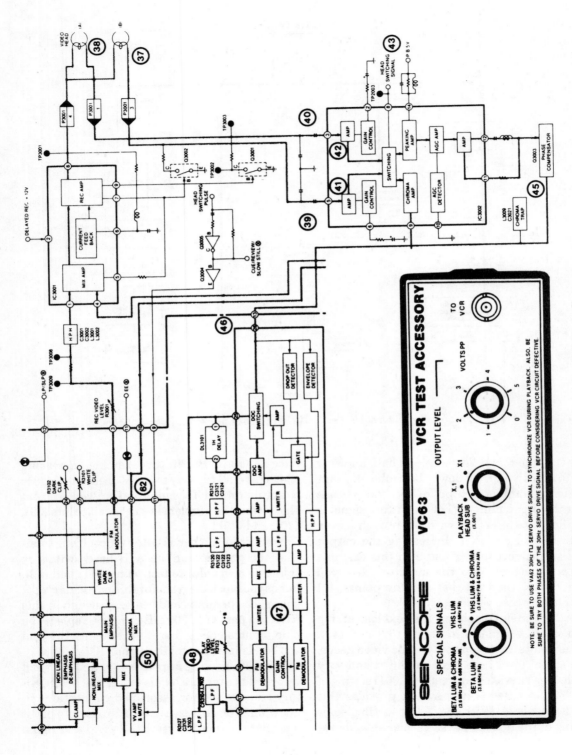

Fig. 2-14. *Use the manufacturer's service information for a more detailed block diagrams. Luminance playback section is shown here. Courtesy Sencore.*

Compliments of Panasonic

substitution to prove what stages work, VCR troubleshooting is reduced to a few, quick signal injections.

SENCORE VA48 VIDEO ANALYZER

Let's now perform some actual VCR electronic checks with the Sencore VA48 video analyzer. VCR electronic circuits require correct reference signals to perform the proper troubleshooting techniques. Various output signals and video patterns from the VA48 can simplify VCR servicing techniques.

THE VCR RECORD CIRCUIT

The first VCR recording stage is the AGC amplifier. This stage must be first tested with a reference input level of 1 volt p-p to make sure that the output is of the proper amplitude. To do this, inject the output of the VCR standard signal directly to the camera input of the VCR and adjust the AGC control for the proper output level using the bar sweep video pattern. The circuit is further tested by using the adjustable output supplied from the drive signals output jack instead of the VCR standard 1-volt jack. This adjustable output should be varied from 0.5 to 2.0 volts (negative polarity) while the output of the AGC is viewed with an oscilloscope. The top waveform in Fig. 2-15 should result if the AGC circuit is operating properly. The bottom trace is of a waveform produced by a faulty AGC circuit.

If the AGC stage is not working properly, the first step is to determine if the AGC stage, the dc amplifier, or the AGC detector is the cause of the defect. To do this, connect the bias and B+ sub supply voltage from the analyzer to the AGC line and vary the voltage. At the same time, feed the standard VCR signal into the camera input jack. The test set-up is shown in Fig. 2-16. The signal level at the output of the AGC stage should change as you vary this voltage. If it does not, you know that the trouble is in the gain-controlled stage. If the voltage does produce a change, you know to go back to the AGC detector and inject the dc voltage at its output.

Finally, you can substitute for both of the signals feeding the AGC detector itself by injecting the composite video signal at the VCR CAMERA input (using the VCR standard jack) and then feeding the composite sync pulses supplied by the drive signal output in place of the VCRs own sync separator output. This example lets you pinpoint an AGC defect to a single stage, which is especially important when you are troubleshooting defects caused by defective ICs or poor solder connections in the signal path.

By using the bias and B+ voltage subbing supply, you can tell if the color-killer signal is properly switching the color/B&W filters. Set the bias supply to 4 volts and inject it into the IC that controls these filters. The injection points and scope waveforms in Fig. 2-17 show the difference in the bar sweep pattern with the color

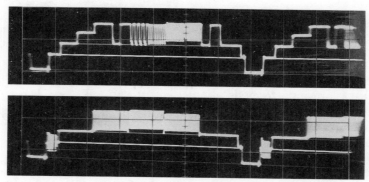

Fig. 2-15. *The top trace shows the normal AGC output while the bottom trace is that of an improper output.*

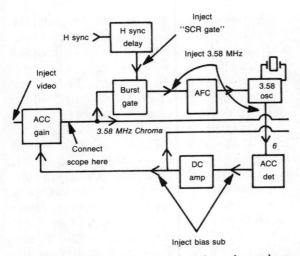

Fig. 2-16. *The driver signals output from the analyzer allows each stage of this AGC circuit to be checked.*

killer activated. Normal operation of this circuit should allow more high-frequency response during a black and white program as compared to one in full color. Thus, you now have proof that the circuits are working properly.

PREEMPHASIS CIRCUITS

The bar sweep allows the recording preemphasis circuits to be checked for proper operation. The newer two-speed VCRs require different amounts of preemphasis for each speed. The scope waveforms shown in Fig. 2-18 show what the bar sweep looks like in properly adjusted preemphasis circuits.

WHITE AND DARK CLIPPING

The clipping circuits located between the preemphasis network and the FM modulator must be properly set to prevent overmodulation, which causes the picture to tear out during playback. The adjustment of these circuits requires both a reference white level and a reference black level to make sure the limiters are not favoring one portion of the signal over another. Use the bar sweep pattern for these adjust-

ments. This pattern has a 3-step grey scale to check for proper video linearity and different frequency bars for a dynamic check of the clipping circuits at different video frequencies. This is needed when the signal is preemphasized, because the higher frequency content is boosted in amplitude. It is possible for the clipping circuits to be operating at the low-frequency range of the signal (like that produced by a 10-step grey scale) but they provide too much limiting to the compensated high frequency information. Thus, the bar sweep pattern allows all frequencies of the video signal to be checked at the same time (Fig. 2-19).

The VCR circuits from the FM modulator to the video heads are best analyzed by tracing the signals with the VA48 signal tracing meter, or with a scope. For general signal tracing, the high frequency response of the meter is usually faster, because the shape of the waveform is not as important as the peak-to-peak amplitude.

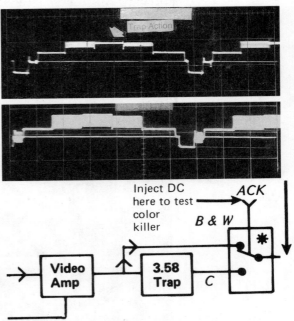

Fig. 2-17. *The operation of the color killer is quickly confirmed by using the bar sweep pattern and the bias and B+ subber voltages.*

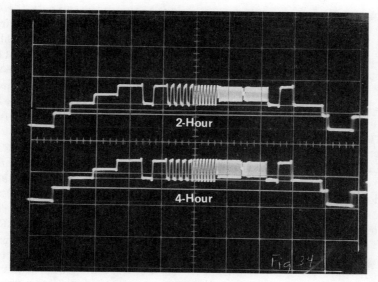

Fig. 2-18. *The recording pre-emphasis circuits boost the high-frequency content to reduce video noise.*

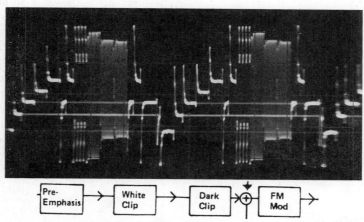

Fig. 2-19. *The bar sweep pattern provides both a grey scale and a "multi-burst" type signal that allows you to check the clipping circuits at all operating frequencies.*

COLOR CIRCUIT ANALYZING

Phase-locked signals produced by the VA48 let you quickly troubleshoot the chroma processing circuits. Direct signal substitution is used for checks of any signal up to the stage that converts the frequency of the chroma signal down from 3.58 MHz. Let's look at a few examples of how to find a defect in the VCR color processing stages.

The *automatic chroma control* (ACC) requires two input signals for proper operation. The first is the composite chroma signal. The important part of this signal that is required for ACC operation is the color burst. The amplitude of this burst signal is used to control the gain of the chroma circuits to maintain a constant color level with changing input signals. The second signal required is the "burst flag." This flag signal is the horizontal sync pulse that is delayed a small amount to place its timing exactly in line with the burst signal riding on the back porch of the horizontal blanking interval. The timing of this signal is very important because it determines what portion of the color signal is used to control the gain of the color circuits. If the burst flag arrives too late, for example, the burst gate will separate the first part of the picture (just after the blanking interval) instead of the color burst. The result is that the color levels will be constantly changing because the amount of chroma information will be different in each color scene.

Now use the drive signals from the analyzer to check out some faulty VCR color circuits. Start with a symptom of changing color levels when a tape is being played. First find out if the color levels are changing during the recording or playback of the color program. This can easily be confirmed by playing back a tape that has been recorded on the suspected faulty machine on a machine that is operating properly. If the color levels remain the same, the defect is in the playback circuits. In this case, however, the levels are changing when the signal has been recorded on the suspected machine and then played back on the good VCR. Thus, the recording circuits of the machine in question are faulty.

Changing color levels could be caused by a defect in any of the seven circuits shown in Fig. 2-20. These seven circuits make up the ACC circuit. A defect (such as a faulty IC or poor solder connection) anywhere in the stages would produce almost the same symptom: changing color levels with different input signals. A scope can be used to trace down the missing signals, but a substitute signal from the analyzer will give you a more positive check as it will duplicate the signals that should be produced at the output of each stage. The first step is to provide a reference signal at the input to the ACC stages. For this check, just use the VCR standard output of the analyzer and feed it into the VCR camera input jack. Now select the chroma bar sweep pattern to provide a reference color pattern these chromas check.

Let's first check the circuits that produce the burst flag signal. The input to this stage is the composite sync pulses that have been separated from the luminance signal. The V and H composite sync test signal from the analyzer is ideal for this injection signal check. Use the solid-

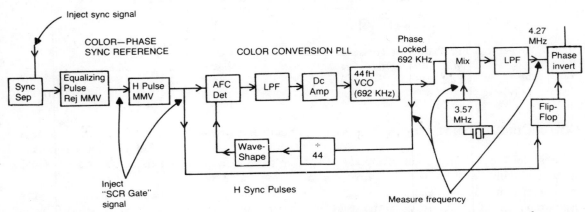

Fig. 2-20. *Each stage of the AGC circuit can be substituted with the drive signals from the analyzer.*

state mode of the drive signal switch to prevent the possibility of feeding too much signal in that could damage a solid-state device. The impedance of this mode is also matched to drive the low-impedance solid-state circuits found in these stages. Use the drive signal meter to monitor the amount of the injected test signal.

These checks are started by injecting the composite sync signal into the horizontal sync delay circuit to see if proper VCR operation returns. The best place to monitor the operation of this circuit is at the output of the ACC controlled stage.

Should the injected test signal *not* return the proper amplitude at the output of the ACC circuit, you can move one stage and substitute for the burst flag. For this test, change the drive signal switch from the composite sync position to the SCR gate drive signal. This signal provides a proper substitute for the burst flag because the pulse produced by the SCR gate signal is "stretched" the same amount as the burst flag. The pulse is present during the color burst and will therefore operate the burst flag just the same as the signal produced by the circuits inside the VCR's chroma processing stages. If the operation of the ACC circuits returns to normal, you know that the trouble is in the horizontal sync delay stage.

If the signal does not return proper operation, just continue the stage-by-stage injection at the output of the burst gate. This time, use the 3.58 MHz (phase-locked) signal. You can also check the dynamic operation of the ACC circuit with this substitute signal by just varying the amplitude of the injected 3.58 MHz test signal. As an example, if you increase the amplitude slightly, the ACC output signal should reduce in amplitude. If you see this dynamic change, you know that the ACC circuit is working and the trouble is in the burst gate stage.

The same 3.58 MHz signal can then be used at the output of the 3.58 MHz oscillator. Varying the amplitude of the substitute signal should again produce a change in the ACC output level. If not, go forward to the next stage.

The output of the ACC detector is a dc voltage whose amplitude is related to the amplitude of the burst signal. For this check, use a variable B+ or bias supply as a test of both the dc amplifier and the ACC controlled stage. Begin this check by setting the output of the bias and B+ sub to the level indicated on the schematic. Then raise and lower this voltage about 10 percent. The results should be a change in the ACC output signal level. Should you still not obtain proper operation, inject the dc voltage at the output of the dc amplifier (input to the ACC controlled stage) and again vary the voltage. If the output level still does not change, you know that the trouble is in the ACC controlled stage itself. With these checks, you can tie down all seven of the circuits in this rather complicated feedback system to locate a defect in one stage.

This might appear to be a long way to go, but in actual troubleshooting, you would not have to substitute for each and every signal. As an example, you could start at the output of the ACC detector and feed in the dc test voltage. This divides the circuit in half. If proper operation is obtained, you know that the defect is somewhere in front of this stage. If correct operation is not obtained, you know that the trouble is in either the dc amplifier or the ACC controlled stage. Thus, you now know which direction to go to further analyze the stages. The key point to keep in mind is you have a signal to sub-in for every input and output stage so you won't have to guess the cause of the circuit fault.

Signal substitution is very handy when combined with oscilloscope signal tracing. You just inject the substitute signal at the input to a stage and monitor the resulting signal at the output of the same stage or one that is supposed to be controlled by the substitute signal. Then, use the scope for circuits that require both the amplitude and the waveshape of the signals to be correct. The combination of scope signal tracing and signal substitution is the best team to use for VCR circuit analyzing.

Before leaving the VCR chroma processing stages, let's look at one more example where signal substitution can be used to locate a defective stage. In this case, use a frequency counter (in addition to scope and analyzer meter) to confirm that the stages are working properly. Recall that the frequency conversion stages mix the incoming 3.58 MHz chroma signal with a second signal that is referenced back to the horizontal sync pulses via a phase-locked loop arrangement. These steps follow the signals through a VCR Beta format machine, although the operations of the VHS conversion is similar. You might want to follow along with the block diagrams shown in Fig. 2-20.

The first step of our frequency conversion is to separate the horizontal sync pulses from the incoming composite video signal. These pulses are then formed into a series of pulses with a fixed amplitude (and pulse width) in two multivibrator stages. These clean pulses are then fed to the phase-locked loop to maintain the proper conversion frequency at the output.

The composite sync signals provided by the V and H comp sync output can be fed directly to the input or output of the sync separator stage. These pulses (being phase-locked to the composite video) will then replace the signals that should be at these two points. You could take the composite signals past the equalization pulse rejection multivibrator, but doing so would result in the wrong frequency at the output of the PLL. The reason is that the PLL would try to lock up to the vertical sync pulse (as well as the horizontal sync pulses) and change frequency of the equalizing pulse rejection multivibrator, which is to provide a constant pulse rate during the vertical blanking and vertical sync pulse intervals.

To eliminate this error, use the horizontal output (SCR gate) signal for injection after the multivibrator stages. This signal works well because it does not contain the vertical sync pulses. It is just a series of pulses that are phase-locked to the horizontal sync pulses. Therefore, it is an exact duplicate of the output of the "horizontal pulse MMV" and can be injected directly into the AFC detector, which is used to keep the

PLL output frequency an exact multiple of the horizontal frequency. Just remember when using the drive signals from the analyzer to feed in signals of the same polarity and amplitude as the signals normally found in the circuit.

The total operation of the PLL is determined by checking the output frequency with a frequency counter. The PLL output frequency should be exactly 44 times the horizontal sync pulse frequency of 15,734 Hz or 692,307 Hz. If this frequency is not correct, the VCR will record and reproduce color but that a color tape that has been recorded on another machine will not play back in color.

If you do not find the correct frequency at the output of the PLL, check the frequency at output of the divide-by-44 stage. At this point, you should find the horizontal sync frequency of 15,734 Hz. If this stage is dividing properly, the trouble could be in the low-pass filter or the dc amplifier. The dc subber supply can be connected to the output of these stages to see if the adjustment of the dc voltages changes the frequency of the PLL output. If there is no change in output frequency when the bias voltage is changed, you know the defect is in the *voltage controlled oscillator* (VCO) and that the IC is defective. As soon as you get the proper output frequency (with an injected signal) you know that the injection point is after the defective stage. Now, just check inputs and outputs until you find the stage that provides no improvement, and that is the one that is defective.

The operation of the remainder of the frequency conversion stages is analyzed with the frequency counter or scope. The second conversion frequency oscillator (in the Beta format, that is the 3.57 MHz crystal controlled oscillator) is just adjusted until you have the proper conversion frequency. The output of the mixer stage is measured with the frequency counter, and should provide 4.267918 MHz in the Beta format. If you have a scope with vector measuring capabilities, you can check for proper phase shifts in the frequency conversion stages. To make this check, connect the "A" channel of the scope to the 4.27 MHz signal before it is phase inverted and the "B" channel to the output of the phase inversion stage. The waveform shown in Fig. 2-21 shows what the patterns should look like if the stages are processing the phase properly. If they are not, the VCR will have "noisy" color, or no color at all on playback.

Proper operation phase

Improper phase shift

Fig. 2-21. *Scope vector mode can be used to test for correct phase shifting of the chroma conversion frequency.*

CONVERTED COLOR

The fact that each of the two VCR formats (Beta and VHS) uses a different converted chroma frequency means that direct substitution is not practical. The use of the chroma bar sweep pattern, however, allows you to check the resulting response at the output of the color conversion stages to be sure that you are not going to lose color detail in the converting processes.

An important point is that the pattern produced by an NTSC generator does not provide a check of the total color bandwidth of the color sub-carrier information. The actual frequencies occupied by the color subcarrier sidebands are determined by the size of the color information being represented. Several small colored objects

in the picture, for example, will represent a higher color sideband frequency than a large object. Remember that the chroma bandpass amplifier of a TV receiver is designed to accept all color information from 3.08 to 4.08 MHz or 500 kHz either side of the color subcarrier. This frequency range determines the amount of color detail that can be reproduced properly on the color picture. The output of the video tape recorder should be able to record and play back the same amount of color detail for good color reproduction.

The signals from the analyzer's chroma bar sweep produce a dynamic check of the entire color frequency response necessary for a good color picture with color detail in even the smallest objects on the TV screen. The three bars of the chroma bar sweep represent the color subcarrier and the points 500 kHz above and below the subcarrier frequency. Each of the bars is generated at the same amplitude, so they can be used as a reference of the total system's frequency response. Use a scope to trace the converted chroma bar sweep through the amplifier stages to make sure that the machine is not losing some of its color detail during the recording process. The scope waveform shown in Fig. 2-22 illustrates how these patterns are processed in a properly operating VCR. Note that there is a loss of the high-frequency color signal detail.

The following sections look at some more uses of the chroma bar sweep in troubleshooting the playback portions of the VCR. The key requirement is to have these signals phase-locked to the horizontal sync pulses for troubleshooting or alignment of the playback comb filters used to eliminate color crosstalk.

THE VCR PLAYBACK CIRCUITS

The test signals produced by the VA48 analyzer provide important checks of the playback circuits of the VCR. Some common VCR defect areas and some methods for troubleshooting them follow. A reference tape from the VCR manufacturer should be used for most of these playback checks. You'll see how to use this refer-

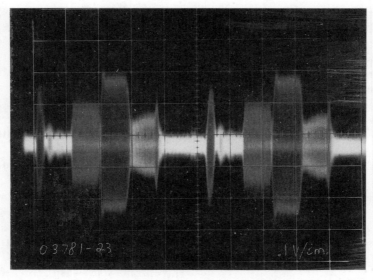

Fig. 2-22. *The chroma bar sweep tests full chroma bandwidth of the color circuits.*

ence test tape with a tape recorded with the VA48 test signals.

The big advantage of using your own test tape is the lower cost as compared to the pre-recorded alignment tape. Thus, you can easily record another tape if it is accidentally damaged during repair of the VCR. The chroma bar sweep pattern provides an additional check of the color processing circuits that is not found on the pre-recorded tapes. Finally, as you use the same patterns for testing the playback circuits as those used for the record circuits, you learn to interpret the different patterns. Let's look at some tests you can make with your own reference tape.

VIDEO FREQUENCY RESPONSE

The most important test of the playback system is to make sure that the entire system is providing the best possible frequency response. The use of the bar sweep pattern will produce a dynamic test of the entire system's video response. All you need do is connect a scope to the output of the VCR. Remember to terminate this output with a 75-ohm resistor to make sure that

the signal levels at the output are at the proper amplitude.

The top scope trace in Fig. 2-23 shows the output of a properly operating VCR. Notice that the output is flat to the 3 MHz bar, then drops off at frequencies above this level. If the bars dropped off more quickly as shown in the bottom trace, this would indicate a loss of frequency response

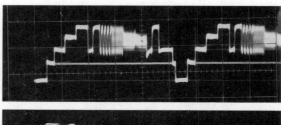

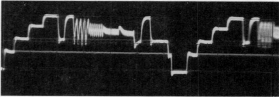

Fig. 2-23. *Top trace of bar pattern is of a properly operating VCR. Bottom trace shows defective head alignment that causes loss of high-frequency detail.*

somewhere along the line. The first place to suspect is the adjustment of the head equalization circuits that are used to compensate for the non-linear output of the video heads. The best way to check for proper equalization is to use the manufacturer's alignment tape and follow the recommended procedures for that VCR.

The chroma bar sweep provides a check of the chroma frequency response. This test is one of the advantages of using the analyzer patterns to record a test tape because there is no other test that is as complete for testing all of the circuits that are used to process the color signals. The chroma bar sweep checks the color circuits at both the upper and lower frequency limits that are necessary for good color details. The center (3.56 MHz) bar provides a reference level for a comparison of the frequency detail 500 kHz above and below the subcarrier. A key point about all three of the bars produced by the chroma bar sweep is that they are phase-locked back to the horizontal sync pulse and have a 180-degree phase shift every horizontal sweep line. This means that they will be properly separated by the comb filters used to cancel color crosstalk during the playback mode of operation. This test is not found in the NTSC color generator, but it is very important for good color detail.

Setting the comb filter used in the playback portion of the color circuits is easy to do if you use a scope with good vertical amplifier sensitivity. Start by using a direct scope probe to connect the scope to the output of the comb filter bridge that *is not* connected to the chroma amplifiers. This output point will show the crosstalk rather than the chroma output. The signal at this point does not have sync pulses, so the external trigger input should be connected to the video output jack for triggering. Set the scope to trigger at the horizontal rate. Scopes with built-in sync separators will give you a stable trace with the composite video signal used as a reference.

Now, play back the portion of the alignment test tape that has the chroma bar sweep pattern.

Adjust the comb filter's mixer control until the amplitude of the signal has the least amount of the second (3.56 MHz) bar as shown in Fig. 2-24. Be sure to use a direct scope probe and have the scope set for maximum sensitivity, as the signal level is very low at this point. Do not attempt to align the two phasing coils used in the comb filter of the VHS tape systems, because a broadcast vectorscope is required for alignment.

LOCATING VIDEO HEAD PROBLEMS

Most VCR service technicians say that one of the most difficult stages to analyze is the low-level input circuits associated with the video playback heads. The reason for this difficulty is that the signal levels produced by the spinning playback heads are so low that an oscilloscope is not effective in tracing a signal. The symptom for a defective head is the same as a dirty switch contact, a bad rotary transformer, or a defective

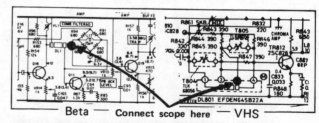

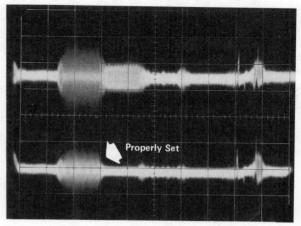

Fig. 2-24. *The chroma bar sweep lets you quickly check the comb filter in VCR playback circuits.*

head preamp. Thus, a technique is needed to determine which of these components in the low-level head signal circuit is actually causing the problem.

The symptom for a defective head circuit is easy to recognize: the picture (on playback) has a severe flicker and is very noisy. The cause of this symptom is that only every other video field is being viewed on the screen. Recall that the two-head system uses one of the heads to pick up every odd field and the other head to pick up every even field. When one of the signal paths is defective, you have one complete field followed by a period of noise information. The result, in an interlaced picture, is the symptoms just described.

Each of the two video heads has its own rotary transformer that transfers the signal from the head output to the preamplifier. This rotary transformer is actually made up of two coils — one that is part of the moving head disk, and the other that is part of the stationary portion of the video head assembly. As the video head picks up the signal from the tape, it is inductively coupled to the stationary coil by the moving coil. This eliminates the need for slip rings or brushes that could cause intermittent operation as they wear down.

The signal picked up by the stationary portion of the rotary transformer is passed on to the head preamplifiers. There are two of these preamplifiers, one for each head. Each preamplifier has a set of adjustments that allows any differences in the frequency response of the two heads to be compensated. The signal level at the input to these preamplifiers is approximately 1 mV. This low signal level means that there are several points in the video head system that can cause trouble.

The first possible cause of a defect is the video head itself. If the head wears too much, its output level drops down. Another possible defect is in the rotary transformer. A broken wire leading to the moving coil of the rotary transformer, for example, would mean that one of the two head signals would never reach the preamp. The next circuit item is a switch that is used to switch the heads between the record and playback circuits. A dirty contact here will result in the loss of the signal from one of the heads. There is also the possibility of a defect in the preamplifiers.

For these checks, you need a test signal that can be injected at any point in this low-level signal path. This signal should be the proper frequency to pass through the tuned head preamp circuits just as though it was being picked up by the spinning head. The amplitude of the subber signal must be low enough to duplicate the signals normally found in these stages. If too much signal is fed in, there could be enough to cause cross-coupling from one head channel to the other, and you cannot tell if the defective stage has been correctly located.

The VA48 analyzer has a test signal suited for this type of signal substitution. This test signal is normally used for troubleshooting the audio IF stages found in TV or VCR receivers. This signal is adjustable in amplitude and uses a 4.5 MHz carrier frequency. The use of a 100:1 (40 dB) attenuator drops the level of this signal to the 1 mV level required to troubleshoot the first stages of the playback circuits. The best place to look for the signal output is at the output of the resistive matrix that is used to mix the output of the two video heads.

Note that an electronic switch is used to switch between the "A" head output and the "B" head output during playback. This switch is normally switched by a pulse that comes from the servo circuits. When you are using the substitute signal to analyze the video head input circuits, connect the bias and B+ subber supply in place of the head switching pulse. When you supply the bias signal, the electronic switch will switch over to one of the head amps, and when the bias is removed it will switch to the other preamplifier.

The reason that the ability to analyze the head input circuits is so important is that the video head disk (which contains both of the video heads) is one of the single most costly parts to replace on the VCR. I have been fooled

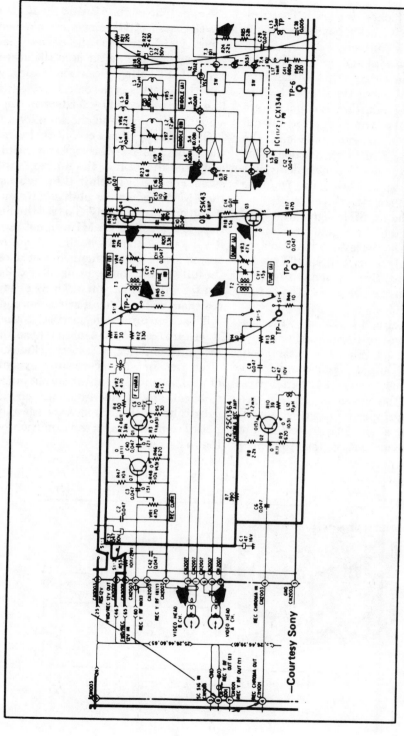

Fig. 2-25. The 4.5 MHz signal can be injected at any of these points for analyzing the video input circuits and dropout compensator.

—Courtesy Sony

by a severe picture flicker in a VCR that was diagnosed as a defective head, but it turned out to be faulty switching contacts.

The same output test signal used to troubleshoot the head preamps is also used (without the attenuator) to troubleshoot the drop-out compensator circuit. The injection test points are shown in Fig. 2-25. All you need is to feed a signal to the input of the DOC detector and look at the output of the DOC circuit with a scope. This circuit is designed to switch to the delay line output any time the signal level coming from the video heads drops to a certain level.

Then you inject the 4.5 MHz signal, the DOC detector should switch the signal around the delay line. When the signal level drops below the detector trigger level, the circuit should switch back to the delay line. Since you are no longer feeding a signal into the delay line, the output quickly drops to zero.

The DOC trigger circuit is tested by increasing the RF IF control to full output and then reducing the signal level. When the level is about 0.1 V (10 mV), the output signal should suddenly disappear. Increasing the signal level should then return the output. If the DOC circuit is not operating properly, use a dc voltmeter to check the output of the DOC detector. The detector should provide a dc voltage to control the switching circuits inside the IC. If this voltage changes as you change the signal level at the input, you know that the detector is working properly and the defect is in the switching circuits. If the voltage does not change, you know that the defect is in the detector itself. The delay line is also checked by feeding in the 4.5 MHz signal to its input and checking for an output.

One other playback circuit that is tested with the adjustable 4.5 MHz signal is the limited circuit. The function of the limiter is to compensate for changing levels in the playback signal so the FM demodulators always have enough signal to operate properly. To test the limiters action, just feed the 4.5 MHz signal into the limiter input and look at the output level. The output level should remain almost the same over the full range of the input signal (Fig. 2-26). If the limiter is defective, you will have a playback signal that varies in detail and noise content.

The color-processing circuits in the playback stages are treated the same as those found in the recording circuits. In fact, the same circuits are often used for both record and playback. The use of the bias and B+ subber output is the best way to check any of the automatic circuits as you substitute for the feedback voltage and see if you notice some change in the condition you are trying to correct.

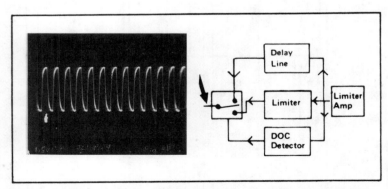

Fig. 2-26. *The limiter output should remain about the same with different input signal levels.*

3

Betamax Videocassette Recorders

The text and illustrations in this chapter are compliments of Zenith Electronics Corporation. This chapter covers the operation of the Betamax format videocassette recorders. The Zenith model KR9000 has been selected as a representative model. This is the same machine as the Sony SL-8600 (the Zenith KR9000 is built for Zenith by Sony). The following four chapters contain references to this model, but the circuits in those chapters (4 through 7) should also be used as a general guide when servicing VHS machines. However, later chapters describing specific makes might cover these circuits in further detail as to pertaining to that make.

INTRODUCTION AND SYSTEM OVERVIEW

Some features found in the KR9000 VCR machine are as follows:

- Built-in timer
- Tuner/IF block and UHF splitter
- Remote pause
- Pause solenoid
- New control system
- One record/playback speed (Beta 2)
- Smaller size and lighter weight
- New functional circuit boards
- Only one (ac) motor
- More shielding to reduce interference
- Reduced mechanical jitter
- No brake solenoid

The clock readout and timer is built into this unit. The buttons for setting the clock and timer are easy to reach above the timer, on the top-front of the machine. The built-in UHF splitter provides the ability to watch one UHF channel while recording another UHF channel. Note the block diagram in Fig. 3-1. The remote pause function is accomplished by connecting the special cable and the remote pause switch to the pause jack in the rear of the unit.

The control systems that control the video head drum and the tape speed have been simplified, since only one record/playback speed is available. The single speed of the tape (Beta 2)

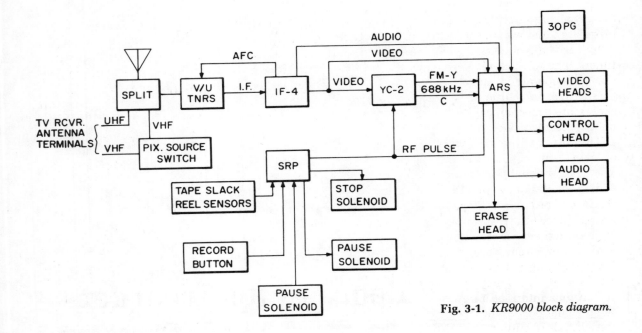

Fig. 3-1. *KR9000 block diagram.*

produces two hours of record/play time with L500 tapes and three hours when using the L750 tape cassettes.

The KR9000 features functional circuit boards. Each board incorporates functions and circuits formerly present on several boards of the older Sony VCR units. The functions of the three major circuit boards are as follows:

YC-2 Board. Y (luminance) signal and C (chroma) signal record/playback processing circuits.

ARS Board. RF record/playback amplifier, servo circuit, and audio signal record/playback circuit.

SRP Board. System control, dc power supply, and pause control circuits.

The older model machines used three motors (one ac and two dc motors) to drive the video head drum and the cassette reels, to move the video tape, and to actuate the threading ring. Only one ac motor is used in the KR9000 machine. A system of belts and pulleys connects this ac motor to the threading and capstan systems.

Extensive shielding is found inside this unit around the video head drum, the YC-2, and ARS board areas. Also added is improved filtering in the ac input circuitry. The purpose of this shielding and filtering is to reduce the interference effects from CB, ham, and police two-way radio transmissions.

SYSTEM OPERATION, BLOCK DIAGRAM, AND POWER SUPPLY

In normal usage, a program received from a TV station (VHF or UHF) is recorded by the machine and later played back via a TV receiver. Thus, the input to the machine is at a VHF or UHF frequency with luminance, chroma, and audio modulation.

RECORD MODE

Looking at the block diagram (Fig. 3-1), the TV input signal appears at top left. Each input signal is coupled through a splitter, producing two outputs for each of the VHF and UHF signals. One output is always connected to the appropriate tuner within the VCR unit. The other output is available directly to the TV receivers tuners. The two tuners (VHF and UHF) within the VCR operate in much the same fashion as those in a conventional TV receiver. These tuners are "mechanical" channel-selector/signal-processing devices.

The output of the VHF tuner (for either VHF or UHF input signals) is at the IF of the normal TV. The IF signal, with luminance chroma and audio modulation, is processed by circuitry on the IF-4 board in a fashion similar to that used with TV receiver IF systems. The signal is amplified and the three resultant signals developed by detection circuits are the outputs of this IF functional block. One signal is the recovered video signal, which includes both the luminance and chroma information. Another output is the audio signal portion of the program being processed. A third output is an AFC (automatic frequency control signal), developed by a circuit that senses a change in IF video carrier frequency caused by tuner oscillator drift. This dc voltage is coupled back to the tuner oscillator circuit, which then makes corrections of the oscillator drift, locking the oscillator to the correct frequency for the channel signal received.

The audio output is coupled to the ARS board for amplification and sent to the audio record/playback and erase heads, thus placing one part of the program on the video tape.

The YC-2 functional board processes the input video signal by splitting the signal into the luminance and chroma portions. As the name "YC" indicates, both Y (or luminance) and C (chroma) signals are handled on this board. In the record mode, there are two outputs from the YC-2 board:

• An FM luminance signal.
• A down-converted 688 kHz chroma signal.

These two signals become inputs to the ARS functional board. They are amplified and mixed together on the ARS board and coupled to the two revolving video heads through a rotating transformer. These two signals become inputs to the ARS functional board.

As the video head disc revolves during the recording process, two magnets on the bottom side pass over fixed coils to create the PG pulses that are directly related to the speed of the head rotation. These pulses are coupled into the ARS board, where they are compared with a signal developed from the video information, to control and maintain the correct speed of the video head disc.

PLAYBACK MODE

When the recorded video tape is played back, the several heads pick up the information as the tape moves past them. The audio head couples the audio information to the ARS board for amplification and connection through the CP-3 board to the audio output socket and the RF modulator (Fig. 3-2).

The control head sends the control track signal to the ARS board where it is compared with the 30 PG pulses from the revolving video head disc for maintaining the correct speed of the video head disc. The video heads track the recorded video signal on the tape and the rotating transformer couples this information to the ARS board. The video signal is composed of the FM luminance and 688 kHz chroma signals, which are amplified and separately coupled to the YC-2 board for further processing. This processing recovers the original luminance frequencies as well as the 3.58 MHz chroma, which are mixed together to form the normal NTSC composite video output that connects to the CP-3 board and the video output socket as well as the RF modulator. The audio and video signals are used to modulate either a channel 3 or 4 RF carrier is coupled to the TV receiver for viewing.

SYSTEM CONTROL

During both the record and playback modes of operation, the system control functions are performed to permit the normal operation of the

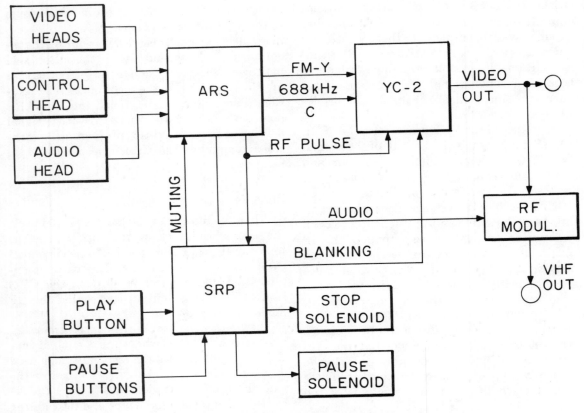

Fig. 3-2. *KR9000 block diagram in playback mode.*

machine as well as to protect electronic and mechanical elements should a malfunction occur. Most of these system control circuits are located on the SRP board of the KR9000.

Among the operation functions is the pause mode, engaged by either the pause function button on front of the machine or the remote pause cable button. Activating the pause mode causes the tape movement to stop; thus no information (such as unwanted portions of a program) will be recorded, and during playback no signal reaches the television receiver.

The protective function of the system control is concerned with automatically stopping the machine operation when continued operation could damage the machine (auto-stop).

TIMER SYSTEM

Another vital section of the KR9000 machine is the built-in electronic timer control. The timer can be programmed to operate the machine at some future time to record a desired program. The timer mode selection is accomplished by operation of the mode select (or power) switch with its on, off, and timer positions.

POWER SUPPLY

The power supply in any electronic device supplies the units life blood. Thus this is the first place to check for proper or improper voltages. Figure 3-3 shows the power supply for the

POWER SUPPLY BLOCK DIAGRAM

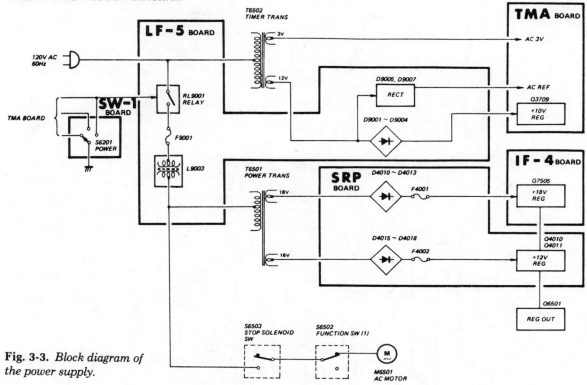

Fig. 3-3. *Block diagram of the power supply.*

KR9000 machine. This diagram shows portions of the power supply that are found on five functional boards: SW-1, LF-5, SRP, TMA, and IF-4, as well as in off-board locations within the machine. Note that Fig. 3-3 is a simplified line drawing and not every lead or wire is shown, but the system connections and the power flow are well indicated.

The 120-volt ac 60 Hz power input goes to and through the LF-5 (line filter) board to the timer transformer, T6502, which is always energized when the line cord is plugged into the ac supply voltage. Two secondary, step-down windings provide 3 volts ac directly to the TMA (timer) board as well as 12 volts ac to the LF-5 board. Diodes D9005 and D9007 (zener) provide a 60 Hz square wave, 6 volts peak-to-peak reference voltage for the timer system (at boards

TMA and TMC). Rectification of the 12-volt ac input to the LF-5 board occurs in a bridge rectifier composed of diodes D9001, D9002, D9003, and D9004, producing an unregulated 10 volts dc output. This is the third voltage connected to timer board TMA. The 10-volt dc supply is regulated by transister Q3709 on the MA board to become the B+ for the timer circuitry.

If the mode selector switch is in the ON position or if the proper time has elapsed to operate the preprogrammed timer system when the switch is in the TIMER position, a connection is made to ground through pin 14 on LF-5 of connector CN9002. Refer to Fig. 3-4. Current then flows from the unregulated 10 volts through the coil of relay RL9001. The relay contacts close, permitting 120 volts ac power to be coupled through choke assembly L9003 and to two other

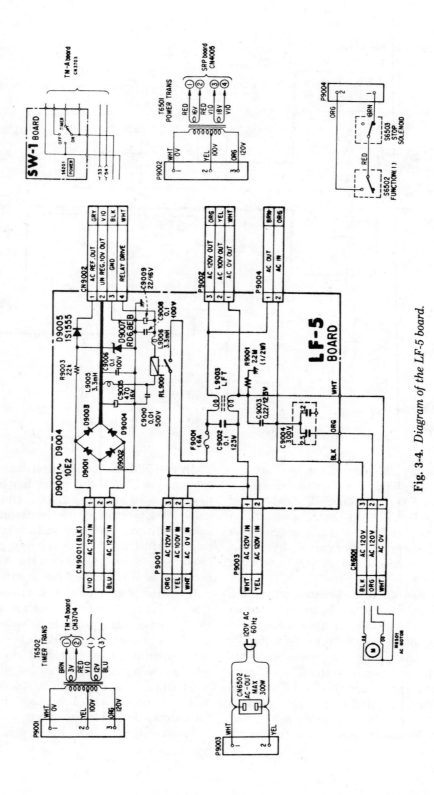

Fig. 3-4. *Diagram of the LF-5 board.*

outputs. One output provides the 120 volts ac to the power transformer T6501 by way of connector plug P9002. The other output (through connector plug P9004, pin 1) produces current through the stop solenoid switch (S6503) which is normally closed, when the function switch is closed (by depressing one of several operating buttons such as record). This current is connected back into the LF-5 board through P9004, pin 2, and out through connector CN6501, pin 3, to the ac motor (M6501). The ac current path from the motor is completed through pin 1 of CN9004.

The two secondary voltage outputs of the power transformer are coupled to the SRP board, entering the latter through connector CN4005 (see Fig. 3-5). These voltages are at 16- and 18-volt ac levels. A bridge rectifier on the SRP board is provided for each ac voltage. Diodes D10, D11, D12, and D13 produce an unregulated 18 volts output via fuse F1 and connector CN4004, pin 9. It then goes to pin 3, connector CN5004, on the CP-3 board, out of CP-3 at pin 1 of CN5003, into the IF-4 board at pin 1 of connector CN7502. Transistor Q7505 on the IF-4 board is the regulator stage of the 18 volt B+ that powers the IF-4 board circuitry as well as the tuners in the VCR machine.

Referring to the SRP board circuit in Fig. 3-5, the output of the other rectifier circuit (D15, D16, D17, and D18) develops +12 volts dc. Regulation of this voltage is accomplished by transistors Q10 and Q11 on the SRP board and externally mounted transistor Q6501. This +12 volt supply is the most used B+ in the KR9000 machine. This voltage is fed to the SRP, IF-4, ARS boards, and the RF modulator.

SYSTEM CONTROL DIAGRAMS

Figures 3-6 through 3-12 show the circuitry involved in the several functions that must be performed in the operation of a VCR recorder such as the Zenith KR9000.

SRP BOARD

Figures 3-6 through 3-10 show the pause, record, play, rewind, and fast-forward systems of the KR9000, respectively. A careful study of these diagrams will increase your understanding of how this machine operates.

SERVO SYSTEM (ARS BOARD)

Figure 3-11 shows a block diagram for that portion of the ARS board that is concerned with the continuous, automatic control of the drum speed. Another function of this portion of the ARS board is the development of the RF switching pulse. This important 30 Hz timing and switching signal is derived from the 30 PG pulses (both A and B) that are generated as the video head disc revolves.

RECORD DRUM SPEED CONTROL

The 30 PG pulse signal from the A head, after processing, becomes one of the two signals in a comparison gate between pins 23 and 24 of IC2502. The frequency of this pulse is a direct measurement of the head speed. The other signal in the comparison gate is developed from the vertical sync pulse, which is separated from the video signal in IC2502, pins 6 and 4. Processing in IC2501 derives a 30 Hz signal that is the reference signal for the comparison gate. Any difference or shifts between the two signals in the comparison gate will cause changes in the current through the drum brake coil, varying the magnetic "drag" on the drum, and correcting the speed.

CONTROL TRACK RECORDING

The control signal is developed during the record mode from the vertical sync pulse after processing and conversion to a 30 Hz signal. The control signal is coupled from pin 19 of IC2501, ARS board and Q2502, and is recorded on the tape by the control head.

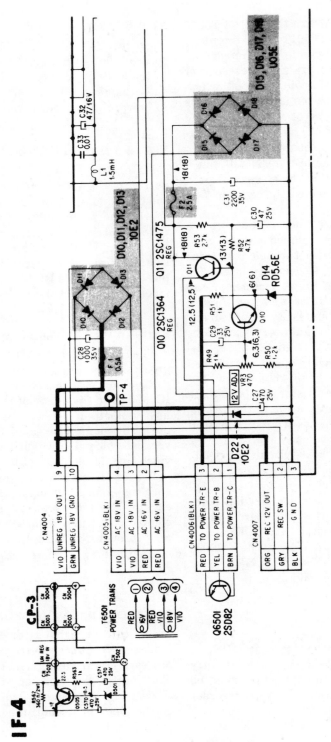

Fig. 3-5. *Power supply circuit diagram.*

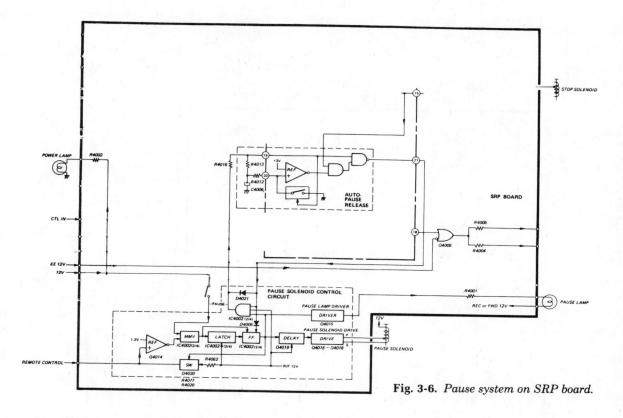

Fig. 3-6. *Pause system on SRP board.*

PLAYBACK DRUM SPEED CONTROL

During playback, the control track signal passes through amplification and modification within IC2502 and becomes the reference signal within the comparison gate. Once again, comparison is made between this signal and the 30 Hz pulse that is directly related to the video head speed. Continuous correction of the drum speed is accomplished in the same way as in the record mode.

TIMER AND AUDIO BLOCK DIAGRAMS

The block diagram (Fig. 3-12) is for the timer system. Included in this system are boards TMA, TMB, TMC, TMD, LF-5, and SW-1.

Shown in Fig. 3-13 is the audio system portion of the ARS board with amplifiers, recording/playback head, and erase heads. A photo of the Zenith KR9000 VCR machine is shown in Fig. 3-14.

CHROMA CIRCUITRY

This section covers the VCR record and playback chroma circuits, plus comb filter circuit operation, and APC and AFC loops. This includes chroma circuit theory of operation and service tips. These chroma circuits are in the Zenith KR9000 and Sony SL-8600 Betamax VCR machines.

THE LUMINANCE CIRCUITS

Before delving into the chroma circuitry, let's cover the luminance circuits (Fig. 3-15) so you will have a better understanding of the total VCR system. As the video heads revolve during the playback mode, they pick up the recorded information from the video tracks on the tape. The combined frequency-modulated luminance (Y) and 688 kHz chroma signals are coupled through the rotating transformer to the ARS

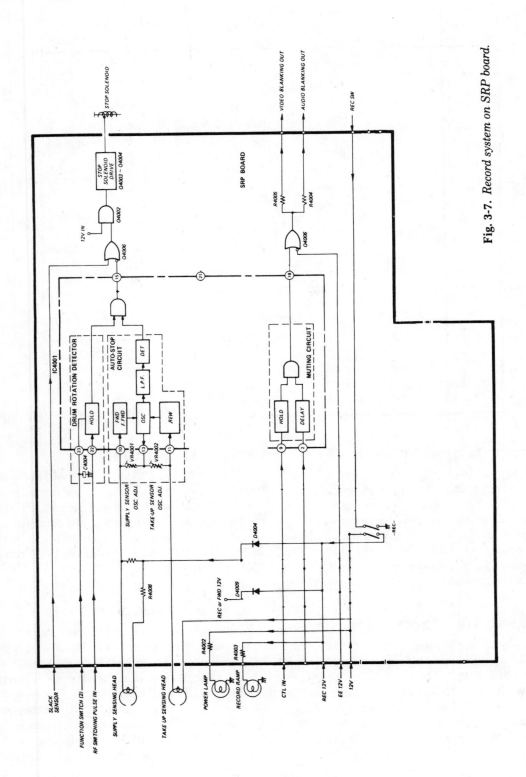

Fig. 3-7. *Record system on SRP board.*

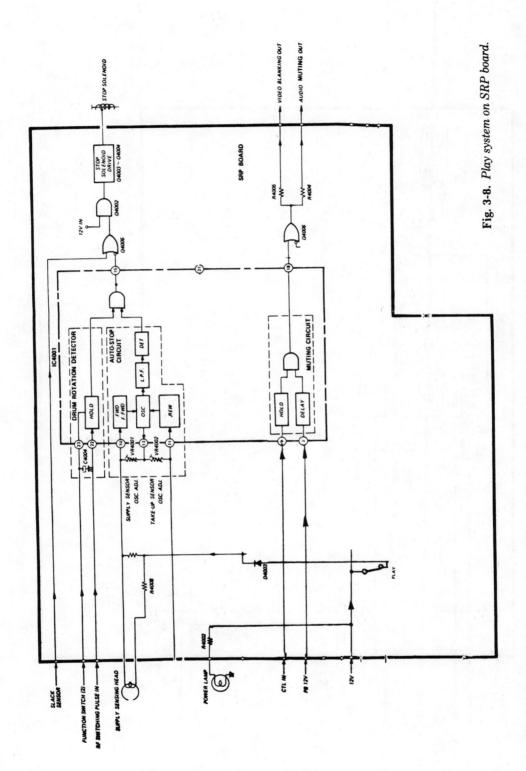

Fig. 3-8. *Play system on SRP board.*

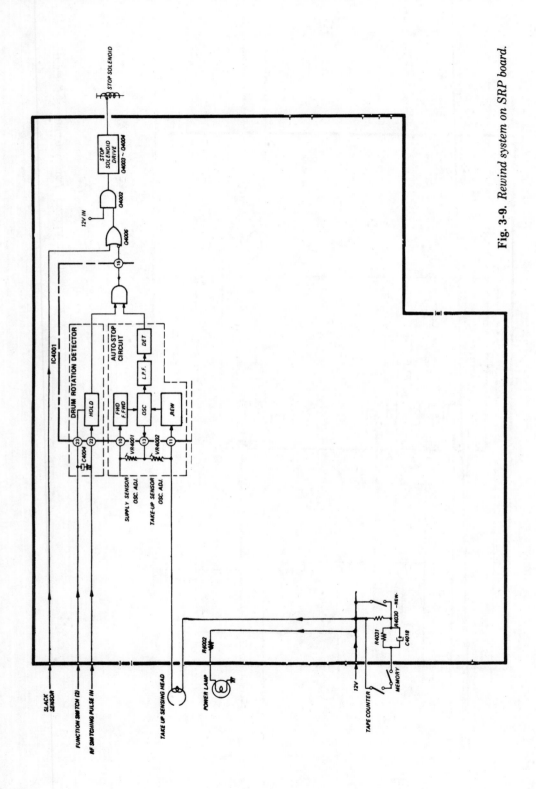

Fig. 3-9. *Rewind system on SRP board.*

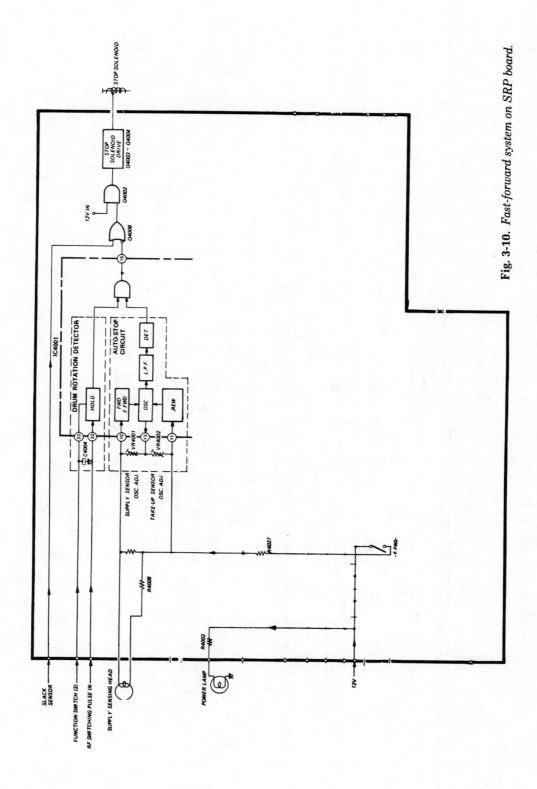

Fig. 3-10. *Fast-forward system on SRP board.*

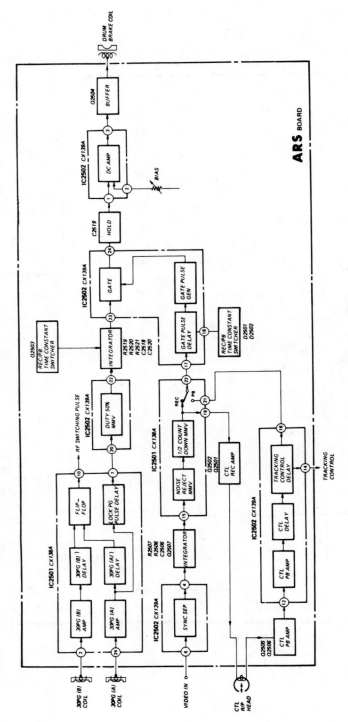

Fig. 3-11. *Servo system on ARS board.*

78

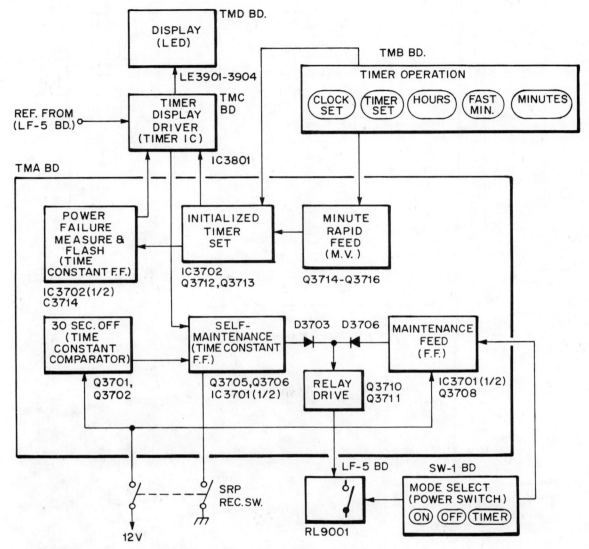

Fig. 3-12. *Block diagram of the clock timer.*

board. There is a "chain" of three functional blocks for the output signal of each of the two video heads. The blocks, mostly inside IC2001 are: the preamplifier, the equalizer amplifier, and the switcher circuits. Inputs to this chain of blocks enter the IC at pins 23 and 24. A 30 Hz signal at pin 18, the RF switching pulse, switches the processed signals to pins 16 and 17 in time with the active tracking of each head.

There are three possible signal outputs from the three center-tapped resistances across pins 16 and 17. These are the monochrome, the luminance portion of the color program, and the chroma (at 688 kHz).

Test Point 5 is the take-off point for a black-and-white signal to pin 14 of IC2001 and internally to the B & W position of a switch. The other switch position is labelled "color" and is

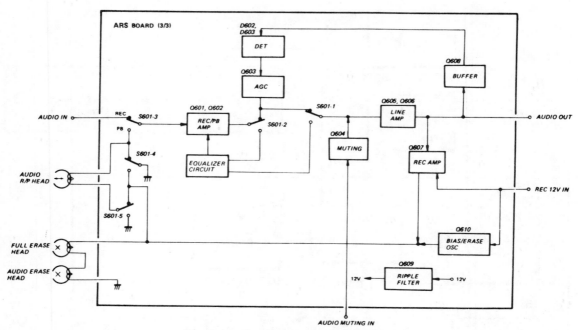

Fig. 3-13. *Audio system block diagram.*

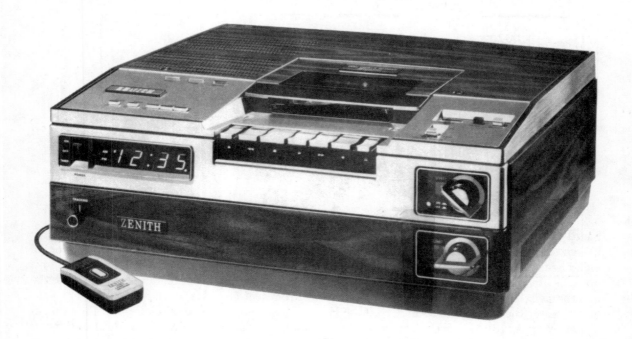

Fig. 3-14. *The Zenith KR9000 VCR.*

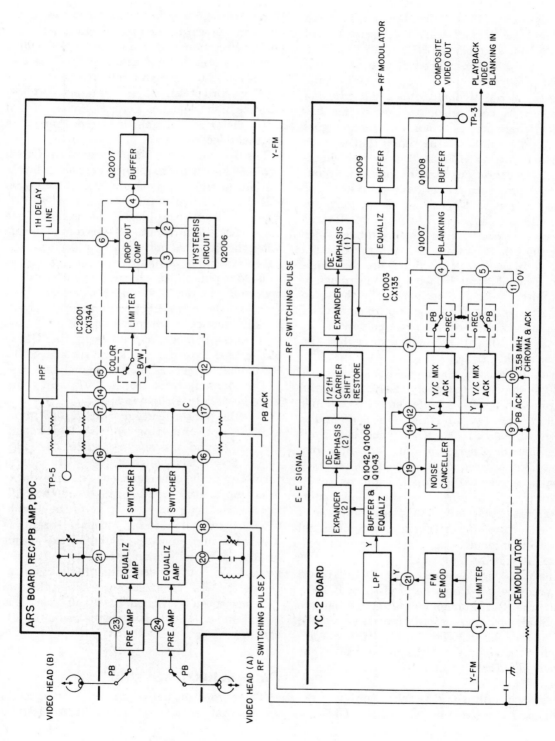

Fig. 3-15. *Block diagram of luminance playback system.*

connected to pin 15. A signal containing both color and B & W can be taken off the top pair of resistors and through the HPF block to pin 15. Only the Y-FM luminance information is connected through the high-pass filter (HPF), because the Y-FM signals are in a high range of frequency by comparison to the chroma frequency; therefore, no color information goes through pin 15. Thus, the signal that is to be processed further in IC2001 is strictly luminance. The condition of the internal switch connected to pins 14 and 15 is determined by the dc voltage on pin 12. This voltage is the playback automatic color killer (PB ACK) signal. If there is no color information in the played-back video tape, the pin 12 voltage is 0 and the internal switch is in the B & W position. Thus the monochrome signal is coupled from pin 14 through to the limiter block. If there is color in the video signal, the pin 12 voltage is approximately +4 volts, the internal switch is in the color position, and the pin 15 signal is coupled through to the limiter block.

In either case, the luminance signal continues through the drop-out compensator block and out pin 4 to Q2007, a buffer stage. The buffer output goes in two directions. One is up and to the left through DL2001, a delay line that delays the signal for a period of one horizontal line, about 63 microseconds. The delayed, fed-back signal is coupled to the dropout compensator by way of pin 6. The purpose of the dropout compensator is to insert the 1H (1 horizontal line) of delayed picture information into the main signal in place of dropout, or no information, on a horizontal line during playback.

The Y-FM signal moves out of the ARS board and into the YC-2 board and pin 1 of IC1003. Inside the IC, the signal is limited to eliminate amplitude variations before it is coupled into the FM demodulator. The demodulator output, which is the original Y-FM signal as well as the recovered, lower frequency luminance signal, appears at pin 21.

A low-pass filter (LPF) selects only the demodulated low-frequency luminance and cou-

ples it through to a buffer and equalizer block and then upward and to the right through a chain of functional blocks. These blocks include expanders and de-emphasis circuits. The emphasis added to the high luminance frequencies for the record now must be counteracted, otherwise the high frequencies of the playback signal will be distorted (of higher amplitude than in the original telecast signal).

The de-emphasized signal returns to IC1003 at pin 19 and inside to the noise canceller block. Out pin 14 and back in to pin 12 goes the signal and then into two similar blocks, labelled Y/C mixers. Within these blocks, the mixing of the played-back luminance and chroma signals takes place, and the outputs are connected to the PB (playback) positions of the internal switches coupled to pins 4 and 5. In the playback mode of operation, the pin 11 voltage is 0 volts, and the switches are in the PB positions. The played-back luminance signal couples out of pin 4, past the blanking block and through the Q1008 buffer, to the video out jack and through to the RF modulator, just as the E-E signal does.

The blanking transistor, Q1007, is in shunt with the signal path between pin 4 of IC1003 and the buffer stage. During normal operation, Q1007 is not conducting, so it is a high impedance to ground and does not effect the video signal flow. Usually when the VCR is first turned on in the play mode, the tape movement is not up to its normal speed. Then the SRP board couples a positive voltage to the base of Q1007, turning it on to saturation. The result is a very low impedance to ground for the output luminance signal, which is thus routed to the ground. Without a video signal into the TV receiver, the CRT is blanked out. When the correct tape speed is reached, the blanking stage is cut off and a normal picture appears on the TV screen.

CHROMA RECORD SYSTEM

The chroma signal is part of the composite video signal that is coupled into pin 13 of IC1001 on the YC-2 board. Refer to the block diagram (Fig. 3-16). The chroma follows the same path as that for the luminance signal, through the AGC block and onto the comb filter. The difference occurs in the output of the comb filter. The chroma portion of the signal is coupled through Q1024, a chroma amplifier, and onto the 3.58 MHz BPF (bandpass filter). The passage of the chroma signal through the comb filter materially reduces the crosstalk that could otherwise produce a noisy picture.

The 3.58 MHz chroma signal is passed by the 3.58 MHz bandpass filter, through to pin 13 of IC1002. All other frequencies that might be in the input signal, such as luminance, do not pass through the filter.

Inside the integrated circuit, the chroma signal enters an amplifier whose gain is controlled by an ACC (automatic color control) voltage. The chroma continues on to the frequency converter block. Another input to this block at pin 16 of IC1002, is a 4.27 MHz cw signal generated within the YC-2 board circuitry. This latter signal beats with the 3.58 MHz signal in the frequency converter.

The frequency converter output at pin 15 includes both the 3.58 MHz and 4.27 MHz signals, as well as the sum and difference of these two frequencies. A low-pass filter (LPF) at pin 15 is designed to pass only the difference frequency, 688 MHz. This downconverted chroma signal is coupled out of the YC-2 board and into the ARS board and transistor stage Q2005, the chroma record amplifier. The output of Q2005 is added to the luminance signal and then coupled through the rotating transformer to the video

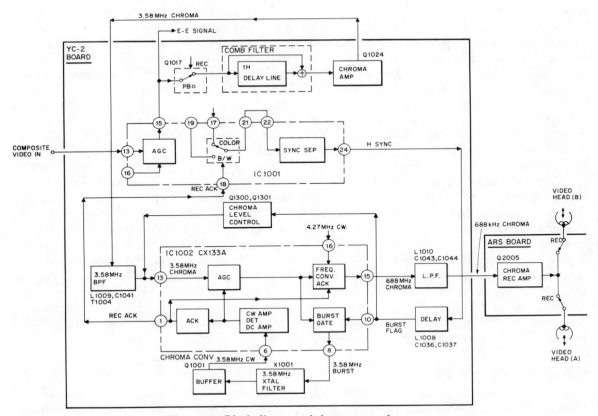

Fig. 3-16. *Block diagram of chroma record system.*

heads, which transfer the video signal to the helical tracks of the video tape.

The development of the ACC voltage that controls the chroma record signal is as follows: The horizontal sync pulse available at pin 24 of IC1001 is connected to a delay block before entering IC1002 at pin 10. The pulse is delayed in time so as to occur during the time of 3.58 MHz color burst on the "back porch" of the horizontal blanking pulse. This delayed pulse is called the "burst flag." It is an input internally in IC1002 to the burst gate block. Another input to this block is the 3.58 MHz chroma signal that includes the color burst. The burst flag permits conduction of the burst gate only during the time of the 3.58 MHz color burst, which then becomes the output signal at pin 8 of IC1001. The level of this color burst is a measure of the level of the chroma signal from which it is derived.

The 3.58 MHz burst is coupled to the 3.58 MHz crystal filter, a tuned circuit that is caused to ring or oscillate. The strength or level of oscillation is determined by the level of color burst. The crystal filter output, connected through a buffer stage, enters IC1002 at pin 6.

Within IC1002, the 3.58 MHz oscillation is amplified and then detected. The resultant dc voltage is amplified to become the ACC voltage. Its level is dependent on the level of the chroma signal and the burst fed back. Thus, this ACC system functions to control the gain of the chroma signal through the IC by effectively sampling the output.

Another useful signal is derived by these elements. That is the ACK or automatic color killer signal, which is available at pin 1. Its value is 0 volts if the video signal being processed is monochrome or about +4 volts if chroma is present in the video. The ACK voltage is used several times in this machine, usually for switching purposes.

CHROMA PLAYBACK SYSTEM

As stated previously, the luminance and chroma signals on playback share several functional blocks on the ARS board, such as the preamplifier, equalizer amplifier, and the switcher blocks in IC2001. The resultant signal across pins 16 and 17 contains both luminance (Y-FM) and chroma (688 kHz) information. These signals are coupled from the tap-off between the resistors that are connected across pins 16 and 17, to the low-pass filter (LPF), that permits only the low 688 kHz signal to move through to pin 15 of IC1004.

The 688 kHz chroma signal enters the IC (see Fig. 3-17), goes through an amplifier whose gain is controlled by the ACC voltage. This voltage is developed in about the same way as in the chroma record process. A playback automatic color killer voltage is also derived from the ACC signal.

The 688 kHz chroma signal leaves the IC at pin 12 but returns at pin 23 and becomes an input to the frequency converter. The other required input, at pin 24, is the same 4.27 MHz cw signal as that used for the record process. Of the outputs of the converter at pin 1, the difference frequency of 3.58 MHz is selected by the external 3.58 MHz bandpass filter and coupled back into the IC at pin 3. The up-converted, recovered 3.58 MHz chroma signal continues through a buffer (emitter follower) stage to pin 4 and the comb filter for further reduction of chroma crosstalk.

The signal then moves via a chroma amplifier (Q1024) into IC1004 at pin 6. A final chroma amplifier delivers the chroma signal, by way of a chroma delay block, to the Y/C mixers in IC1003 at pins 9 and 10. The chroma is mixed with the luminance to recreate the composite video signal that is coupled through the internal switches to pins 4 and 5. These switches are in the PB (playback) positions, because the activating voltage at pin 11 equals 0 volts. The path for the output follows the same route as for the E-E and luminance playback process.

4.27 MHz SIGNAL DEVELOPMENT

The 4.27 MHz cw signal is required for chroma processing, both in record and in playback. In each case, a frequency converter circuit

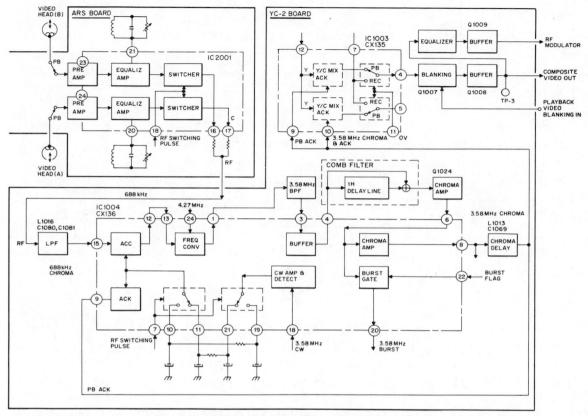

Fig. 3-17. *Block diagram of chroma playback system.*

beats this 4.27 MHz frequency against a chroma frequency to secure the desired chroma signal for further processing. During record, the 4.27 MHz beats with the 3.58 MHz chroma to produce the color-under frequency of 688 kHz. On playback, the 3.58 MHz is recovered by mixing or beating together the 4.27 MHz and the 688 kHz signals. Note block diagram (Fig. 3-18).

The development of the 4.27 MHz cw signal involves the use of another frequency converter, located in the upper left-hand corner of IC1005. One input signal to the converter is 3.57 MHz, delivered by a VXO (voltage-variable crystal oscillator) block within the IC and the external 3.57 MHz crystal, between pins 13 and 17. The second input signal to the converter is 692 kHz, which is equal to 44 times the horizontal frequency (44th) of 15,734 Hz. This 44th signal is

precisely controlled by a phased-locked loop (PLL) that includes functional blocks between pins 5 and 11 of IC1005 and pins 2 and 4 of IC1006.

One of the products of the hetrodyning and the beating of the 3.57 MHz and 692 kHz frequencies within the frequency converter is the desired 4.27 MHz cw signal at pin 15. This signal is coupled through a 4.27 MHz bandpass filter to the primary of a small transformer. T1015, with a center-tapped (at ground) secondary. Two phases of 4.27 MHz are available between the center-tap and either end of the secondary winding, which is connected to pins 8 and 9 of IC1007. These pins connect the in- and out-of-phase signals to the two "contacts" of an internal switch whose output is present at pin 6 of IC1007. The switch is driven at both 30 Hz and

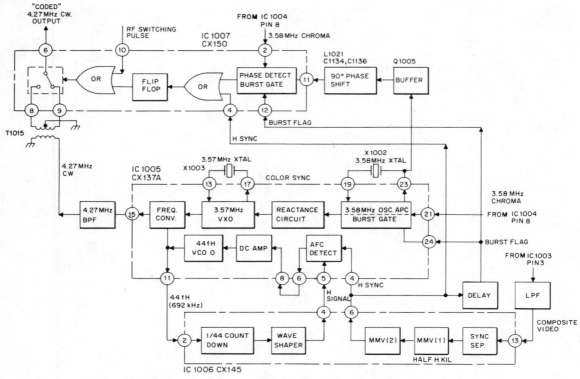

Fig. 3-18. *Block diagram of 4.27 MHz development and APC circuitry.*

15,734 Hz rates by an OR gate and other circuitry within IC1007. The results of this switching is a "coded" 4.27 MHz cw signal, at pin 6, whose phase reverses every horizontal line, for one head only. This "coding" of the 4.27 MHz is part of the crosstalk elimination process in the chroma signal processing. The output at pin 6, IC1007, the 4.27 MHz cw signal, is coupled to the frequency converters in IC1002 (pin 16) and in the IC1004 chip (pin 24).

COMB FILTER OPERATION

The ACC output signal is applied to a frequency converter where it is up-converted back to 3.58 MHz, the difference frequency between the ACC output and the 4.27 MHz carrier frequency. In the frequency conversion in the playback mode, the output frequency and the carrier frequency are close to each other. Any imbalance in the frequency will therefore cause carrier leak. The balance is adjusted by VR1009. The phase of the playback chroma signal is inverted at a 1 H rate when the "A" head is playing back. This is restored to a continuous phase in the frequency conversion process. The restoration is done by inverting the phase of the 4.27 MHz carrier signal every 1 H. Note this relationship as shown in Fig. 3-19. As a result of this process, crosstalk components from the adjacent tracks become phase-inverted every 1 H for both the "A" and "B" head signals. The chroma signal separated in the 3.58 MHz bandpass filter is then sent to a comb filter via an emitter follower. The comb filter is formed by a 1 H delay line and a resistor mix circuit. Utilizing the fact that the signal component is in continuous phase and the phase of the crosstalk component from the adjacent tracks is inverted 180 degrees every 1 H, the comb filter rejects the crosstalk component.

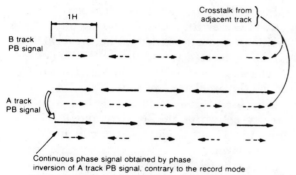

B track
PB signal

1H

Crosstalk from
adjacent track

A track
PB signal

Continuous phase signal obtained by phase
inversion of A track PB signal, contrary to the record mode

Fig. 3-19. *Relationship between playback chroma signal and cross talk signal.*

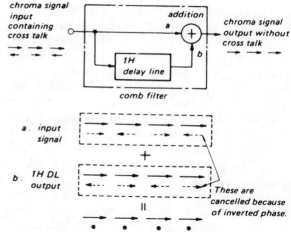

chroma signal
input
containing
cross talk

addition
a

chroma signal
output without
cross talk

1H
delay line

b

comb filter

a. input
signal

+

b. 1H DL
output

These are
cancelled because
of inverted phase.

=

Fig. 3-20. *Comb filter diagram.*

Up to this point in our description of the chroma system it was stated that the chroma signal is continuously in phase when it is observed every 1 H and the crosstalk signal from the adjacent tracks is phase-inverted every 1 H. But originally, the input chroma signal is phase-inverted every 1 H in the NTSC system frequency. The actual phase relationship between the chroma signal and the crosstalk component is that the crosstalk component appears in continuous phase while the chroma signal is phase-inverted every 1 H. The subtraction of the 1 H delayed output and nondelayed output is performed in the comb filter, because both signals flow through R1092 in the opposite direction as shown in Fig. 3-20.

As the crosstalk component is in phase, it is cancelled in subtraction and the output becomes zero. The chroma signals that are phase inverted every 1 H are added, and the chroma output doubles. The chroma signal obtained across R1092 is amplified in the chroma consisting of Q1026 and Q1024 and supplied to pin 6 of IC1004. The chroma signal level is adjusted by VR1010 for correct setting of the color-killer point. The output of the output amplifier is fed to the Y/C mixer of IC1003 from pin 8 via C1223, VR1024, and delay line circuit.

C1069, L1013, R1062, and R1063 form the delay circuit where the delay time difference between the luminance and chroma signals is corrected by delaying the chroma signal by 0.1 mi-

crosecond. A burst gate extracts the burst using a gate signal applied to pin 22. The extracted burst signal is supplied to a crystal filter from pin 20. This is the same crystal filter circuit used in the record system. As in recording, the burst signal is converted into a 3.58 MHz continuous wave (cw) signal in the crystal filter and is fed to pin 18. The output of the crystal filter is at a low level. It is amplified in a cw amplifier to a level high enough to drive the detector circuit.

The 3.58 MHz signal whose amplitude is proportional to burst is picked up in the detector circuit and the detector output is supplied to a hold circuit via switcher 1. Separate hold circuits are provided for each of the two video heads, and switching is done by switchers 1 and 2. Using this method, the ACC loops become independent for each channel. This results in no signal level difference in chroma output for the two combined channels, even if there is a big difference in the chroma output levels of the two heads. The RF switching pulse, applied to pin 7, drives switchers 1 and 2. For A channel, the detector output at pin 21 is held in C1086. R1079, R1090, C1099, and C1100 form an RC loop filter. The output of the filter is supplied to switcher 2 through pin 11. The output of switcher 2 is amplified in a dc amplifier to drive the ACC gain control amplifier.

The filtered detector outputs to pins 10 and 11 are applied to a color-killer circuit via an OR gate. The color killer is set to actuate when the ACC detector output voltage drops below the control range of the loop. The dc voltages at pins 10 and 11 decrease so that the ACC amplifier gain increases when the chroma input signal is low. When the input signal level decreases below the control range of the ACC, the ACC locks out and the voltages at pins 10 and 11 drop to about 1 volt. The color killer functions when either of the two channels locks out and the voltage at pin 10 or 11 drops too low. The output voltage from the color killer is 4 volts in the color mode and zero in monochrome. It is supplied to a color-killer circuit in CX135A, from pin 9.

AFC AND APC CIRCUITS

IC1005, type CX-137A, is involved in both the AFC and APC systems. It contains an AFC, APC, 3.58 MHz and 3.57 MHz crystal controlled oscillator circuits, and frequency converter.

AFC

The AFC circuit obtains a 692 kHz (44 fH) cw signal, which is exactly 44 times the horizontal frequency, both in the record and playback operations. A PLL (phase-locked loop) is utilized for the 44 times multiplication. As the horizontal sync signal is used for the multiplication, the loop is called AFC, as used in some TV receiver circuitry. The AFC circuit is shown in Fig. 3-21.

The divide-by-44 count-down circuit, a part of the AFC loop, is contained in IC1006, type CX145. The waveform at each section is shown in Fig. 3-22. The AFC detection circuit is a sample-and-hold type. Phase detection is performed by sampling the trapezoid waveform applied to pin 5 using the horizontal sync pulse applied to pin 4. The detector output is applied to a dc amplifier through pin 8 from pin 6 via a low-pass filter network. The oscillating frequency of the VCO (voltage-controlled oscillator) is controlled by the dc amplifier output voltage to ob-

tain a 44 fH frequency, phase-locked to H sync. The VCO is an RC multivibrator circuit, and the oscillator frequency is determined by R1111 and C1115 connected to pin 9. VR1013 varies the discharge current of the multivibrator so as to adjust the oscillator frequency.

The oscillator output is supplied to a frequency converter in CX137A and to CX145 from pin 11. A divide-by-44 circuit in CX145 counts down the 44 fH to 1/44 th for obtaining a signal of the fH frequency. The count-down output is a rectangular wave that is shaped into a trapezoidal wave and supplied to the AFC detection circuit in CX137A. The 44 fH VCO output signal is kept phaselocked to the horizontal sync signal by the AFC loop described above. The VCO output signal, supplied to the frequency converter in CX137A, is beat against the output of the 3.57 MHz crystal oscillator to produce a 4.27 MHz output. This 4.27 MHz cw output signal is supplied to the chroma signal frequency converter.

APC

The APC (automatic phase control) functions only in the playback mode. Note the block diagram (Fig. 3-23). The APC circuit consists of a phase detector, reactance circuit, 3.57 MHz (3.58 MHz fH) crystal-controlled oscillator, burst gate circuit which performs phase detection only for the burst signal, and 3.58 MHz crystal-controlled oscillator whose output signal is used as a phase reference. The reactance circuit is a variable capacitance type and is connected to the 3.57 MHz crystal-controlled oscillator to be a part of its oscillating time constant. The reactance circuit and the 3.57 MHz crystal-controlled oscillator are interconnected so that the 3.57 MHz oscillator is a voltage-controlled crystal oscillator (VCXO). In the record mode, the 3.57 MHz oscillator is a stable reference oscillator. In playback, the frequency of the VCXO output is supplied to the frequency converter where it is beat against the output of the 692 kHz (44 fH) VCO to produce 4.27 MHz.

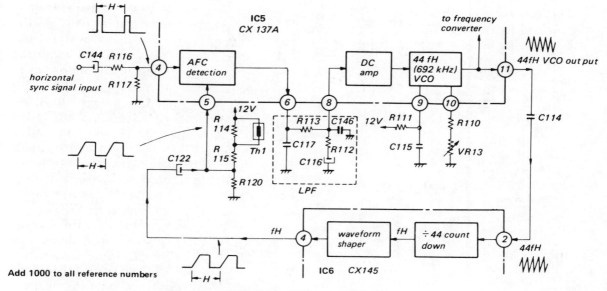

Fig. 3-21. *Block diagram of AFC circuit.*

The 4.27 MHz output is fed to a bandpass filter from pin 15 and fed to the chroma frequency converters of the record and playback systems via CX150. The playback 3.58 MHz chroma signal is fed to the APC phase detector from pin 21. The reference signal is supplied by a 3.58 MHz crystal-controlled oscillator connected between pins 19 and 23. The output signal of this oscillator and the burst portion of the two (record and playback) chroma signals are

horizontal sync input
C137A· ④

trapezoid waveform input
CX137A ⑤

AFC detection input
CX137A· ⑥

1/44 countdown output

44fH
VCO output

Fig. 3-22. *Waveforms of the AFC circuit.*

phase compared. The detector output is filtered in RC filters connected at pins 20 and 22 and supplied to the VCXO to control the 3.57 MHz oscillator frequency. Because the phase comparison is made only with the burst signal, the burst controls the phase detector so that the output of the APC phase detector is supplied only during the burst period.

In the record mode, the output of the APC phase detector is blocked by applying the REC/E-E 12 volts to pin 2 as a control voltage in order to stop the operation of the burst gate. This allows the 3.57 MHz VCXO to function as a fixed-frequency oscillator in the record mode. In playback, the 688 kHz chroma signal reproduced by the video heads is at a frequency which is 43.75 times H sync and contains phase instability caused by mechanical jitter. Note that the AFC loop slightly overcorrects this phase instability because the H sync frequency cannot be multiplied by 43.75. The APC loop detects the small phase variations that result from this over correction and develops a dc control voltage that controls the instantaneous frequency (phase) of the 3.57 MHz VCXO to compensate.

89

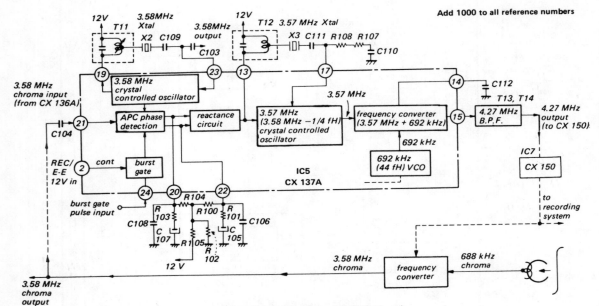

Fig. 3-23. *Block diagram of APC circuit.*

4

Servo and Control Systems

The servo circuitry compares the phase of the 30 PG signal to that of the vertical sync separated from the video input signal in the record mode so as to control the drum rotation and keep its phase constant. A signal is twice as long as the cycle of the VD signal (vertical sync) is fed to the CTL (control) head as the servo reference signal during playback and is recorded onto the tape.

Control functions include autostop at the end of the tape, control of tape threading and unthreading, generation of the audio muting and video blanking signals, control of the pause circuit, and protection operations for such cases as tape slack and rotational failure of the drum.

SERVO TROUBLESHOOTING

VCR servo systems keep the cylinder head, capstan, and video tape moving at the correct speed. The servo must "lock up" in order to record and play a tape properly.

Some of the more common servo problems are

incorrect freerunning speed, defective motor, incorrect voltage to the motor, and loose or worn drive belts. A new set of belts has solved many servo problems. Thus, servo problems can be electronic and/or mechanical. A good way to zero in on a servo problem is to go through all of the adjustment procedures. As an example, all that may be required is to properly set the freerunning speed of the servo motor.

The picture symptom for a servo that is not locked up is the appearance of dark horizontal bands (see Fig. 4-1) that roll vertically up the screen. These bands might come and go, deceiving you into suspecting another type of intermittent problem.

The servo circuits form closed loops to control the speed of motors. Unfortunately, each of the two sets of servo circuits produce nearly the same symptoms when defective, making it difficult to discern which loop is causing the problem. To make matters worse, each set of servos usually contains two loops — one inside the other. The inside loop, the "speed" loop, causes

Fig. 4-1. *Horizontal bands indicate the servo is not locked up.*

the motor to turn at nearly the correct speed. The outside loop, the "phase" loop, provides additional control over the first one. A problem in either loop causes all the signals in both loops to shift away from the normal values as the automatic circuits try to correct for the defective stage. This corrective action causes every component in the loop to appear defective. See the servo block diagram in Fig. 4-2.

Additionally, the servos have both electrical and mechanical components. Mechanical defects produce identical symptoms and measurements as electrical defects. The VA62 Video Analyzer helps break the loops to separate these confusing conditions.

Separating cylinder servo problems from capstan servo problems is the first step in troubleshooting. A simple "look and listen" test tells which servo is causing the problem. What should you look and listen for?

Cylinder Servo

The cylinder motor drives the video heads at the proper speed and position with respect to the tape. Except for the new hi-fi machines that use spinning audio heads, cylinder servo problems

affect only the picture. Refer to the divide-and-conquer VCR block diagram in Fig. 4-3.

One IV frame consists of two fields of video information, or 525 lines. These frames of information are repeated 30 times each second. The VCR upper cylinder contains two video head tabs mounted 180 degrees apart. Each head tab records one field of video information each revolution of the cylinder to form a complete frame. The cylinder motor speed is approximately 1800 rpm or 30 rps. This speed must be maintained even during periods when the tape is not moving around the cylinder. This is the job of the cylinder servo.

Capstan Servo

The capstan motor controls the speed of the tape moving past the heads. Because most VCRs use stationary audio heads, capstan servo problems affect the sound and picture. If the picture is affected and the sound is not, you have a cylinder servo problem. But if the sound is also affected, you have a capstan servo problem.

Before you even open a schematic or reach for a test probe, divide and conquer the problem by listening and looking for the symptom. *Listen:* If the audio sounds too fast or slow, the problem is

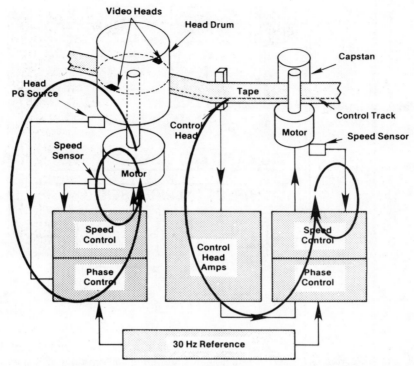

Fig. 4-2. *A pictorial of a VCR head drum and servo drive. Courtesy Sencore.*

in the capstan servo circuits. *Look:* If the sound is okay but the picture is bad, put the VCR into the PAUSE mode. If the picture does not stay stationary and upright, the problem is in the cylinder circuits.

SENCORE SC61 WAVEFORM ANALYZER

Now that you know which VCR servo is at fault, approach the circuits with an oscilloscope. I find that the Sencore SC61 Waveform Analyzer is *the* answer to VCR waveform troubleshooting. It is faster and more accurate than any scope I have ever used. With the SC61, I can conquer servo problems much quicker. Here are just a few of the SC61's features:

- 60 MHz usable to 100 MHz gives you more confidence in waveform analyzing.
- Input protection to 3000 volts protects your

investment and lets you make measurements other scopes can't.
- 100 percent automatic "Auto-Tracking" digital readout lets you make error-free measurements.
- Gives error-free readings of dc Volts, peak-to-peak volts, and frequency at the push of a button. Measures any increment of a waveform in amplitude, time, or frequency and calculates frequency ratio.
- Rock-solid sync circuits latch onto the most elusive signal.
- Faster and more accurate than a conventional oscilloscope.

The block diagram of Fig. 4-4 shows the speed and phase relationships that servos control.

OXIDE BUILDUP

Accurate servo control is only possible in a clean, well-maintained machine. The capstan

93

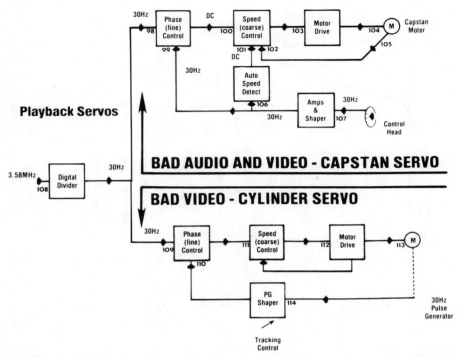

Fig. 4-3. *Block diagram of the servo system Divide and Conquer technique. Courtesy Sencore.*

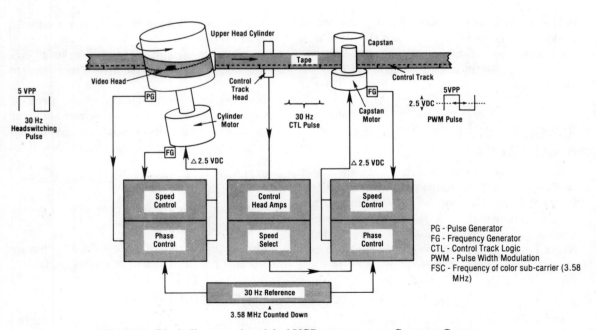

Fig. 4-4. *Block diagram pictorial of VCR servo system. Courtesy Sencore.*

motor is responsible for moving the tape (in record and normal play) from left to right through the tape path. The tape path includes the following critical components:

- Back tension
- Tape path entrance guide
- Full erase head
- Buffer roller
- Cylinder entrance guide
- Video head cylinder
- Cylinder exit guide
- Audio dub erase head
- Audio and CTL heads
- Capstan and pinch roller
- Tape path exit guide

Oxide build-up and tape edge shards deposited on any of these devices can cause some degree of malfunction in the VCR.

The upper cylinder is grooved to provide an air cushion for the tape that wraps around it. Greatly increased drag results when the grooves become clogged. Large oxide deposits on the upper cylinder cause poor tracking and distorted sound. You might think it's a capstan servo problem. Deposits on the audio/CTL head assembly will cause decreased audio or no sound at all, plus the possibility of no CTL pulses reaching the capstan servo circuit.

BASIC OPERATION

The servo system maintains the proper capstan speed during recording and playback and at the same time ensures that the head tabs on the video cylinder retrace the proper tracks during playback that were recorded during the record mode. Refer to Fig. 4-5.

Each video head tab has its gap azimuth set at 6 degrees. One is at plus 6 degrees while the other is at minus 6 degrees. If the video track that was recorded by the plus-6-degree head is played back by the head that is at minus 6 degrees, the result will be noise instead of the desired video information.

Speed through the tape path must be kept constant in spite of increases or decreases in drag caused by dirty surfaces and varying back tension. The motors that operate both the cylinder and the capstan are controlled in speed and phase. Speed control takes care of large rotational changes while phase control takes care of the fine adjustments. The drive to each motor is the result of mixing or adding the speed and phase correction signals from the servo circuit.

SERVO SIGNALS

Let's see what signals are used to control the speed of the capstan and cylinder motors.

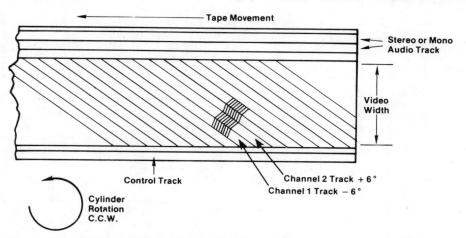

Fig. 4-5. *Using control track signals from the tape, the servo system synchronizes the video heads to play back the proper track. Courtesy Sencore.*

CAPSTAN FREQUENCY GENERATOR (FG) PULSE

Note: Except for the 3.58 MHz and 30 Hz signals, all other servo frequencies might vary from one type of VCR to the next. Those provided in this text are typical for VHS machines.

The capstan motor (Fig. 4-6A) generates FG pulses of a certain frequency determined by the flywheel speed (720 Hz in SP, 360 Hz in LP, or 240 Hz in the EP speed). This is accomplished either by toothed magnetic elements (one fixed, the other moving within it), Hall effect devices, or fixed coils located near the outside diameter of the motor's magnetic flywheel. These frequency generator pulses are used to measure and maintain correct capstan operating speed in both record and playback. They are compared with the divided 3.58 MHz (see point B of Fig. 4-6). The result of this comparison is the speed error signal and goes to the motor drive circuit.

CAPSTAN SPEED AND PHASE CONTROL

Capstan phase control is accomplished during the recording mode by comparing two 30 Hz reference signals. The first is obtained by dividing down the 3.58 MHz that comes from the video/

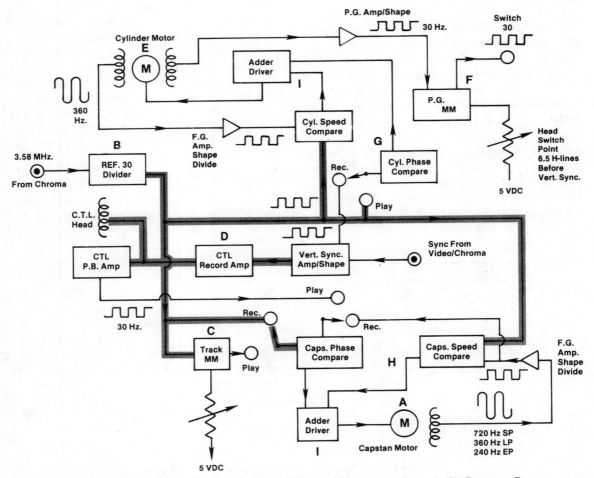

Fig. 4-6. *Block diagram of VCR servo circuits with test points (A) through (I). Courtesy Sencore.*

chroma section of the machine. The second is a divided sample of the 720 Hz capstan FG signal (720 Hz divided by 24 equals 30 Hz) in the SP speed. The result of this comparison becomes the phase error signal that is used to keep the motor turning at a constant speed. In the RECORD mode, a 30 Hz CLT pulse is developed from the vertical sync and is recorded on the control track.

During playback, the machine compares CLT and FG timing in a "speed detect" circuit to automatically select the speed the tape was recorded in. The divided 3.58 MHz reference signal (30 Hz) goes to the tracking control circuit monostable multivibrator, which allows for adjustment when playing tapes that were recorded in other machines (see Fig. 4-6C). The 30 Hz reference is compared with the 30 Hz CTL pulse (Fig. 4-6D) to provide the phase error signal to the capstan motor drive circuit.

CYLINDER MOTOR CONTROL

The cylinder motor (Fig. 4-6E), like the capstan motor, makes use of a frequency generator (FG) device to produce a 360 Hz signal when it is operating at its correct speed. This signal is divided by 12 (to give 30 Hz) and is then compared to a divided vertical sync frequency during record and to a divided 3.58 MHz during playback.

The resulting error signal is sent to the cylinder motor drive circuit to maintain the 30 RPS rate required of the motor in both record and playback functions, regardless of capstan speed.

THE PG PULSE

The PG pulse indicates the position of the heads and is used in head switching. A magnet located on the cylinder motor flywheel corresponds to the position of one on the video head tabs. The magnet produces a pulse in a pickup coil or Hall-effect device with each revolution. This is the PG signal source.

From the PG pulses, another waveform (see Fig. 2-14F) is developed, known as the RF switch

(or head switching) pulse. During formation of the RF switch pulse, the negative PG pulse is ignored and only the positive portion of the pulse is used. The RF switch pulse switches the video head during playback.

When this 33.3 millisecond (30 Hz) waveform is in its negative half cycle (during playback), the track 1 head switches on and the track 2 head is on during the other half cycle. By adjusting the "PG MM" control, the position of the switch point where each head is turned on can be placed at the required 6½ horizontal lines before vertical sync. The RF switch pulse is also used in other areas of the machine for control purposes. Refer to waveform drawings in Fig. 4-7.

In the RECORD mode, cylinder motor phase control is accomplished by comparison of the 30 Hz pulse generator (PG) signal with one-half of the divided vertical sync (30 Hz), which has been separated from the composite sync of the signal being recorded.

In playback, the 30 Hz PG signal is again compared with the 30 Hz reference signal obtained from the divided 3.58 MHz. The result of these comparisons, in both record and play, is the phase error signal output to the cylinder motor drive circuit for phase correction.

Cylinder phase in some machines is accomplished by adjusting it during forward and reverse search modes to lock the noise bar in the picture in one position. This prevents it from entering the vertical interval and provides a stable picture during search. This is done by adjusting the reference 30 Hz signal to 29.49 Hz in forward search and 30.51 Hz in reverse search.

For the cylinder motor, digital counters (Fig. 2-14G), clocked by the divided 3.58 MHz (cylinder speed Fsc/2 and cylinder phase by Fsc/8), develop the error information that is then used to pulse-width-modulate a 1750 Hz carrier. Pulse-width modulation refers to the ability to control the duty cycle or "on" time of a rectangular waveform.

In the case of *capstan* digital speed error development (Fig. 4-6H), the master clock is again a divided 3.58 MHz (see Fig. 4-8). The divide rate

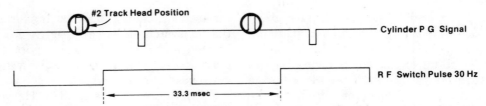

Fig. 4-7. *The RF switch pulse (head-switching pulse) switches the video heads during playback. The positive-going portion of the PG pulse is used to develop the switch pulses. Courtesy Sencore.*

is controlled by the speed select circuit. The digital phase error clock frequency is 447 kHz. The resulting error information is used to pulse-width-modulate a 3.5 kHz carrier for speed control. A 44 Hz carrier is used for phase correction.

These pulse-width-modulated signals, two from the cylinder circuit and two from the capstan circuit, are routed through low-pass filters and become dc levels that are then applied to adder circuits as shown in Fig. 4-6I. The results are the cylinder and capstan motor servo drive voltages.

Mode	Speed Error Development	Resulting Error Information
SP	3.58 MHz divided 4 times	895 kHz
LP	3.58 MHz divided 8 times	447 kHz
EP	3.58 MHz divided 12 times	298 kHz

Fig. 4-8. *Errors that develop for various VCR speed modes. Courtesy Sencore.*

CHECKING SERVO IC INPUTS

All of the circuitry used to perform these functions (with the exception of the low-pass filter) is usually contained in one large scale integrated (LSI) chip. Because so many inputs and outputs are inaccessible within the body of the IC, you must be able to make troubleshooting determinations from the external information available

to you. First, check the control signals from each assembly for proper level at the inputs of this IC.

Because digital counters continue to operate in the absence of reset pulses but provide non-synchronized results, it is very important to confirm the presence of all primary and reference information at the appropriate servo IC inputs. This is best accomplished with a frequency counter and an oscilloscope.

If all necessary primary and reference information is geting to the servo IC and proper adjustment signals are not being produced, the IC itself is defective and must be changed. However, be aware that mechanical malfunctions in the motors themselves can cause erroneous primary information at the servo IC.

Many times, head switching problems are mistaken for servo problems when in fact they are the results of distortion of the RF switch pulse (SW30). The importance of RF switch pulse distortion cannot be stressed enough. This pulse is fed to several areas in the machine including System Control. It can become distorted by component failures that have nothing to do with servo problems and can cause erratic head switching and noise bands in the video output.

SERVO AND PULSE SYSTEMS

The position relationship of the video heads and the 30 PG coils is shown in Fig. 4-9. A block diagram of the servo and pulse system circuits is shown in Fig. 4-10. The servo timing chart is shown in Fig. 4-11.

The drum servo system of this machine is a common magnetic brake servo system. The rotary drum is belt-driven by an ac hysteresis

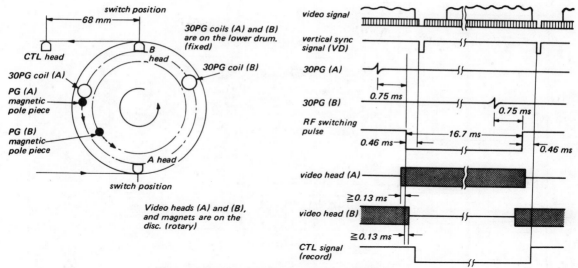

Fig. 4-9. *Relationship between heads and PG coils.*

motor and controlled by a brake coil. The tape is also driven by the ac hysteresis motor.

The rotary head drum assembly is a "stacked" array consisting of the upper drum (fixed), the video head (rotating) and the lower drum (fixed). The two video heads are mounted on the periphery of the head disc. Two magnetic pole pieces, used in conjunction with the two 30 PG coils, are also mounted on the disc. The two PG coils (30 PG coils A and B) are mounted on the lower drum. The 30 PG (A) signal is used in the drum servo system. The 30 PG (A) and (B) signals are used to produce the RF switching pulses.

The drum servo circuit uses two ICs, CX138A and CX139A. The 30 PG(A) pulse triggers the lock PG delay multivibrator MMV(1) to obtain a 30 Hz rectangular wave. The output from MMV(1) toggles a second one-shot, MMV(4), which squares the signal into a 50 percent duty cycle waveform. The output from MMV(4) passes through an integrator network that converts the square wave into a trapezoidal waveform.

The 60 Hz VD signal, separated from the video input signal, is fed to MMV(5), which

eliminates noise. The output from MMV(5) toggles a flip-flop(2), which divides and shapes the signal into a 30 Hz square wave. This signal is used in the record mode as the CTL signal. It is delayed by the gate pulse delay MMV(6) and samples the trailing edge of the trapezoidal waveform in the gate circuit. The result is a voltage corresponding to the time from the PG pulse generation to the VD signal, i.e., the phase relationship between the generated PG pulse and the sampled vertical sync signal.

In the playback mode, the gate pulse delay MMV(6) is triggered by the CTL signal. The sampled voltage is stored in the hold circuit until the next sampling. The stored voltage, amplified by a dc amplifier and driver that also functions as a compensator, controls the drum brake coil. The compensator is used for stabilizing the operation of the servo system.

The correct phase relationship between the drum and the vertical sync signal is obtained by changing the delay time of the gate pulse delay MMV(6). The CTL signal triggers the tracking control MMV(8) and (9) in the playback mode. The output of MMV(9) is applied to the gate pulse delay MMV(6).

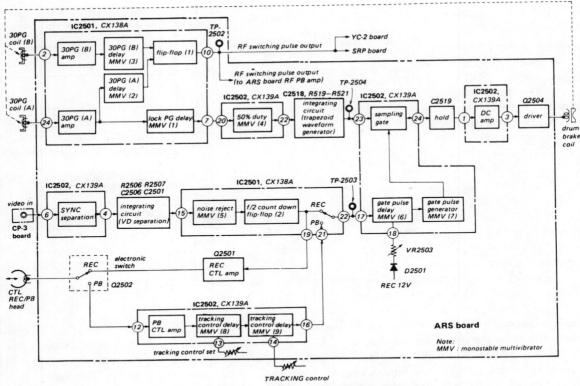

Fig. 4-10. *Block diagram of cervo and pulse circuits.*

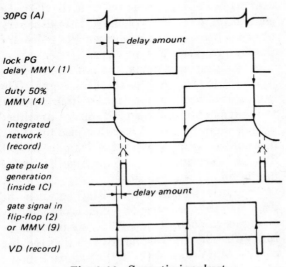

Fig. 4-11. *Servo timing chart.*

Positional errors in the mounting of the 30 PG(A) and (B) units are corrected by the delay MMV(2) and (3). The outputs of the MMV(2) and (3) trigger flip-flop (1) to obtain the 30 Hz RF switching pulse. Only the output from the flip-flop (1) at pin 10 is affected by the delay time of the MMV (2) and (3) chip.

The drum servo circuit uses two ICs that were designed for the drum servo. The CS138A chip contains a block for obtaining various pulse signals from the 30 PG pulses and a block for dividing the vertical sync signal (60 Hz) in half. The phase comparison and sampling circuits and the output drive amplifiers are contained within CX139A. It also contains the sync separator circuit and the CTL playback system block. CX138A has a section for the capstan servo that is unused in this machine.

PG PULSES

The PG coils, mounted on the lower drum, generate a pulse whenever the magnetic poles on the rotary video head disc pass above them (Fig. 4-12). As the magnet approaches a PG coil, a positive-going pulse is generated and when it recedes, a negative-going pulse is generated. There are two sets, (A) and (B), of PG coils and their magnetic pole pieces.

The distance of PG coils (A) and (B) from the center of the drum are different and each generates one pulse for each rotation of the drum. The negative-going pulse of the PG coil output is used to indicate the rotating phase of the drum. The pole pieces, which are strong permanent magnets, pass across the coils and induce a pulse by electromagnetic induction. The small head drum diameter used in this machine requires the use of a permanent magnet system. The pole pieces cut across the coils at a slower relative speed than in the older model VCR machines. A timing chart of the 30 PG pulse circuit is shown in Fig. 4-13 and its block diagram in Fig. 4-14.

Both positive and negative pulses are obtained across R2501 and R2502 by the electric current induced in the PG coils. PG amplifiers amplify and shape the negative-going pulses to the PG coil outputs and trigger PG delay one-shots MMV(2) and MMV(3). The PG amplifier output waveform cannot be observed, because it

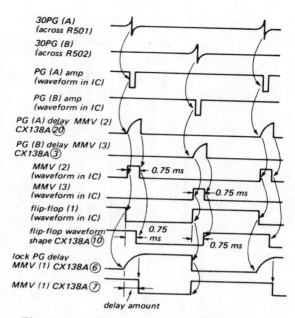

Fig. 4-13. *Timing chart of 30 PG pulse circuit.*

is inside the IC. The output of the PG(A) amplifier directly triggers the lock PG delay one-shot MMV(1). But actually the lock PG delay MMV(1) is triggered by the signal passed through the PB(A) delay MMV(2) and the flip-flop (1). The lock PG delay MMV(1) is triggered by the phase of the PG amplifier output without regarding the delay time of the MMV(2).

The PG pulses (A) and (B) trigger the flip-flop (1) after mounting position errors in the two PG coils are compensated by the MMV(2) and MMV(3) and a 30 Hz square wave is produced. This delay mono multivibrator is formed by a combination of an RC integrated network and a Schmitt trigger circuit. The external circuit connected to pin 3 or pin 20 is the time constant circuit. All other multivibrators in CX138A and CX139A have the same arrangement.

Flip-flop(1) is triggered on the positive-going transitions of the output waveforms of MMV(2) and MMV(3), producing a 30 Hz square wave that is not affected by the delay time of the delay one-shots.

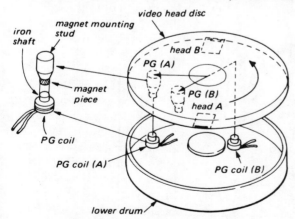

Fig. 4-12. *PG pulse generating mechanism.*

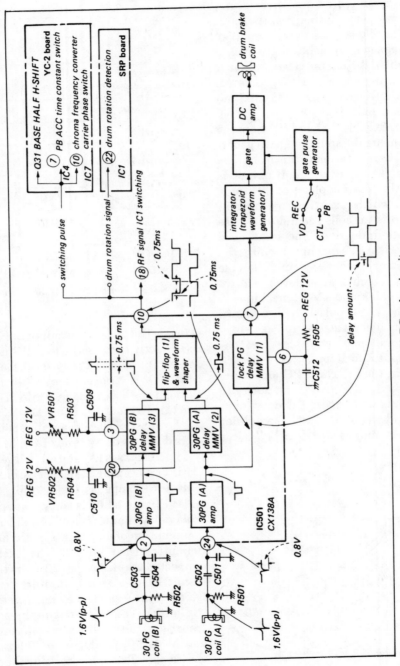

Fig. 4-14. 30 PG pulse circuit.

The output of the flip-flop(1) triggers the lock PG delay MMV(1) to obtain the output whose trailing edge is delayed for a constant width. The time constant circuit for the MMV(1) is connected to pin 6. The output of flip-flop(1) is wave shaped by the outputs of the PG delay MMV(2) and MMV(3) to obtain the same square wave output as the one obtained when the flip-flop is triggered by the delay phase of the MMV(2) and (3). This square wave is the RF switching pulse obtained from pin 10. This pulse goes to the playback RF amplifier, to the drum as a signal for indicating drum rotation, and for various switching signals.

VERTICAL SYNC AND CTL SIGNAL GENERATOR

The block diagram and timing chart of the vertical sync signal and CTL signal systems are shown in Figs. 4-15 and 4-16, respectively.

The vertical sync signal, separated from the video input signal, becomes the servo reference signal in the record mode. The video input connection is provided on the CP-3 board. R5004 and C5001 on the CP-3 board form a low-pass filter. The filter rejects the chroma burst signal and high-frequency noise. The video signal is sync-separated in the circuit between pins 6 and 4 of IC502, CX139A. The sync separation circuit consists of a feedback clamp circuit for sag correction and a switching amplifier. The external circuit connected to pin 5 is the clamp time constant. An RC integrator circuit separates the V sync from the sync separator output and triggers the noise elimination one-shot, MMV(5), via SW(1) of IC2501. SW(1) switches to the pin 15 side for a 0-volt input at pin 16 and to the pin 14 side for 12 volts. The switch is kept permanently switched to the pin 15 side in this unit and is not used as a switch. The noise elimination one-shot, MMV(5), eliminates noise by means of the fact that a one-shot, once toggled, cannot be toggled again until after it has reset itself. The external circuit at pin 17 is the time constant network for MMV(5). The 60 Hz VD vertical sync signal is divided into a 30 Hz square wave in the divide-by-two flip-flop. The 30 Hz square wave passes through SW(3) and appears as the gate signal output at pin 22. The negative-going transition of the flip-flop(2) output becomes the servo reference phase.

The flip-flop(2) output is inverted once in Q501, inverted again and amplified in Q502, and recorded on the tape as the CTL signal. The negative-going transition of the CTL amplifier output is used as the playback servo reference phase. The REC 12 V is supplied as the power supply for the record CTL amplifier.

SW(3) is the servo reference signal switching circuit between the record and playback modes. It is switched to the flip-flop(2) side when pin 23 is 12 volts and to the pin 21 side when pin 23 is 0 volts. In the playback mode, the playback CTL signal is used as the servo reference signal. The positive polarity pulse of the CTL head output becomes the servo reference phase.

The PB 12 volts is fed to the base of Q502 in the playback mode via D503 and R513 and the collector-emitter circuit of Q502 is shorted. The terminal that was in the signal side in the record mode turns to the ground side in the playback mode, and the polarity of the CTL head is inverted. A low-pass filter rejects the high-frequency noise. The playback CTL amplifier amplifies the positive-going input signal in Q502 and Q506 and feeds it to pin 12 of IC502 for amplification and shape so as to obtain the positive pulse output. C533 (connected to pin 9) determines the frequency characteristics of the linear amplifier stage, which is a low-pass filter for 1 kHz, −3dB. C532 (connected to pin 10) is a decoupling capacitor (ac grounded). The tracking control delay MMV(8) and (9) delay the CTL signal by about ⅔ cycle for the manual tracking control in the playback mode. The MMV(9) delays the CTL signal by about ⅓ cycle.

Thus, the output delayed ⅔ cycle is obtained. VR505 adjusts a delay amount of the MMV(8) so that the output phase of the MMV(9) has the relationship of the output phase of the CTL am-

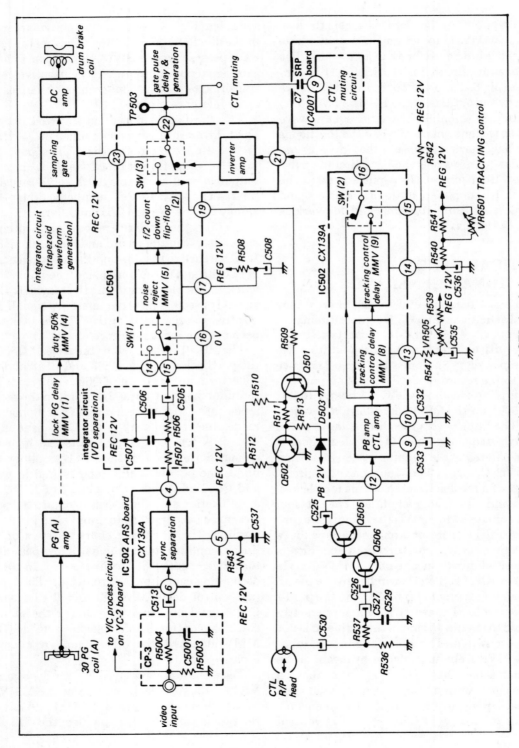

Fig. 4-15. *Vertical sync and CTL signal circuits.*

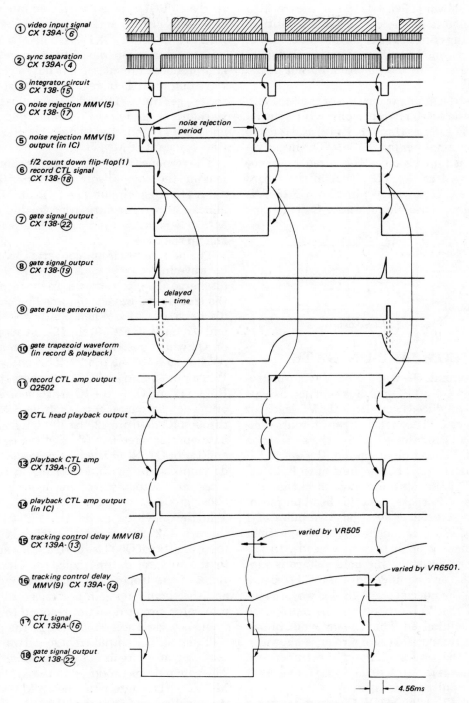

Fig. 4-16. *A & B. Timing chart of vertical sync and CTL signal system.*

plifier (as shown in Fig. 4-17) when the tracking control knob is at the center detent position. The tracking control varies the delay amount of MMV(9) to obtain a delay or advanced CTL signal reference to the playback output of the CTL head.

The positive-going transition of the MMV(9) output is the servo reference phase. The output of the MMV(9) goes through SW(2) and is fed to an inverter amplifier in CX138A. The negative-going transition becomes the servo reference phase as well as the VD signal in the record mode. The CTL signal at pin 22 of CX138A is fed to a gate pulse circuit and a CTL muting circuit.

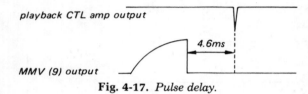

Fig. 4-17. *Pulse delay.*

PHASE COMPARISON GATE

In this circuit, the trapezoidal waveform produced from the 30 PG pulse is sampled using a gate pulse produced from either the VD signal or the playback CTL signal. The gate output is amplified in a dc amplifier to drive the head drum brake coil. The sampling was done by the 30 PG pulse in older units, but in the Zenith KR9000, the VD/CTL signals are used to form the sampling pulses. Because the CTL head output of this video recorder is low (positive pulse of 1 mV), the CTL signal might drop out sometimes due to clogging or dust on the heads. In the CTL/VD gate system, the hold voltage is supplied to the dc amplifier until the CTL signal returns to the normal state. In this way, the influence of the CTL dropout on the reproduced picture is minimized. The schematic diagram of the phase comparison gate circuit is shown in Fig. 4-18, the timing chart of the trapezoidal wave former is in Fig. 4-19, and the timing chart of the gate pulse former is in Fig. 4-20.

The lock PG delay MMV(1) output, triggered

by the 30 PG(A) pulse, is shaped into a 50 percent duty cycle square wave by the duty MMV(4) of IC502, CX139A. The MMV(4) supplies a comparison waveform to a gate circuit and produces the 50 percent duty cycle square wave output without regard to a pulse width of the trigger input. The falling slope of the input waveform is used as the reference phase. R2518 and C2517 (connected to pin 21) determine the time constant of MMV(4).

The response time of the servo is shortened by making the comparison waveform 50 percent duty. The trigger input waveform for MMV(4) is almost 50 percent duty cycle in this unit, but MMV(4) is utilized because it is already available in the chip.

The integrator circuit converts the rectangular wave into a trapezoidal wave by charge and discharge of an RC network. In the record mode, the trapezoidal wave is obtained by charge and discharge of C518. When pin 22 goes high, Q1 and Q2 in the IC are off, the 12-volt supply turns on Q3 which passes current through a 50-ohm resistor to R520 via pin 22. This charges C518. When pin 22 goes low, C518 discharges through R521. Q1 and Q2 in the IC are then on and Q3 is off. The discharge current from C518 feeds through R520 to pin 22, the 50-ohm resistor, the base-emitter circuit of Q2, and the collector to emitter of Q1. R519 is a biasing resistor for the dc amplifier in the later stage and raises the integrator output waveform about 6 volts dc above ground.

In the playback mode, the PB 12 V is fed through R522 to the base of Q503 and the collector to emitter of Q503 is shorted. This adds C520 to the integrator time constant circuit. The slope of the trapezoidal waveform is reduced, and the servo loop gain is reduced so that the integrator circuit does not respond to the high frequency variation contained in the playback CTL signal. The output trapezoidal waveform of the integrator circuit is fed to a sampling gate. The gate pulse is produced in the MMV(6) and MMV(7). The gate pulse delay MMV(6) uses the negative-going phase of the VD/CTL signal

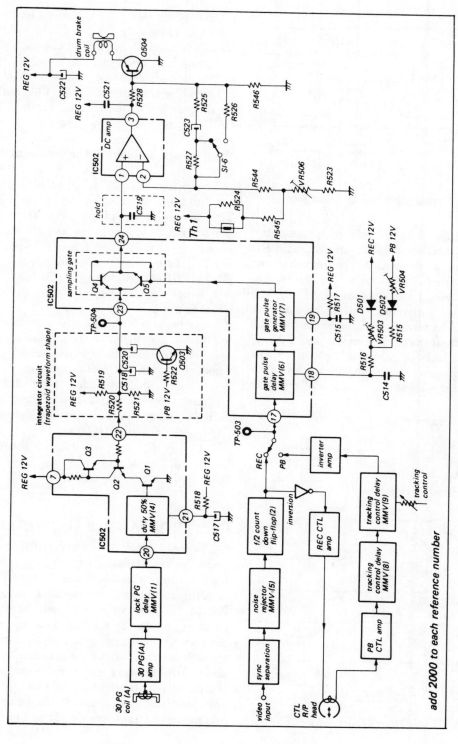

Fig. 4-18. *Phase comparison gate circuit.*

add 2000 to each reference number

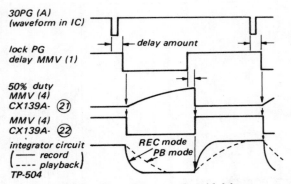

Fig. 4-19. *Timing chart of trapezoidal former.*

supplied to pin 17 as the reference phase and produces a constant delayed output. The RC network at pin 18 determines the time constant of the MMV(6). The time constants are switched in record and playback by D501 and D502. The delay amount is larger in playback than the one in record. This corrects the lock phase, because if the delay times in the MMV(6) are the same in both record and playback, the lock phase becomes incorrect due to the variation of the trapezoidal wave slope of the gate

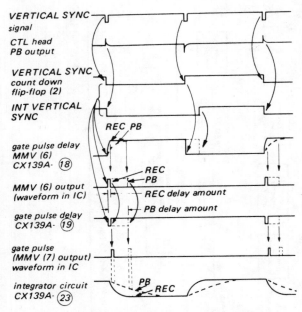

Fig. 4-20. *Timing chart of gate pulse former.*

comparison waveform. The servo lock phase is adjusted by varying the delay time in MMV(6). VR503 is used for adjusting the lock phase in recording and VR504 in playback.

The gate-pulse-former one-shot, MMV(7), is triggered by the output of MMV(6) and generates a gate pulse of constant width. R517 and C515 at pin 19 determine the time constant of the MMV(7) (cannot be observed — part of the chip) and is supplied to the gate-sampling circuit.

The voltage held in C519 is amplified in the dc amplifier and supplied to the brake coil driver. This dc amplifier is a type of operational amplifier (op amp). It amplifies the potential difference between pins 1 and 2, using the potential at pin 2 as the reference. This voltage is applied to pin 2 via R544. VR506 adjusts the bias. The output, which has the same phase as that of the input to pin 1, is obtained at pin 3. This amplifier functions as a phase compensator by inserting an RC network into the feedback loop.

R525 and C523 form a feedback circuit for negative feedback of the output from pin 3 to pin 2. R527 and R526 are added to the circuit in the playback mode to reduce the loop gain. C522 is a high-frequency-bypass filter to prevent the playback RF amplifier from being interfered with by any high-frequency signal flowing to the brake coil. The dc amplifier output is applied to Q504 to drive the brake coil.

CONTROL AND PAUSE CIRCUITS

The main function of the control circuit is to guide the mechanism during tape threading and unthreading operations, together with the subsequent control of the signal system. In the Zenith KR9000 VCR, the switching between function modes is mainly done by manually depressing the function buttons. Therefore, automatic controls of the mechanism by the system control are very few. What *is* controlled automatically by the system control circuit is the auto-stop solenoid. Fig. 4-21 shows the system and pause control block diagram. There are five

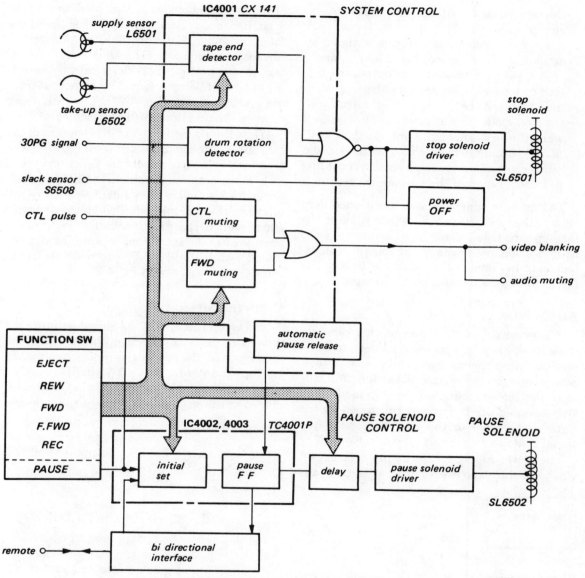

Fig. 4-21. *System control and pause control block diagram.*

major operations provided by the system control circuit:

1. *Tape End Sensor.* The metallic foil attached to both the tape start and end are detected. An electrical circuit is closed, driving the auto-stop circuit.

2. *Head Drum Rotation Detector.* When any of the function buttons are pressed, the system control circuit detects the head drum rotation by detecting the 30 PG pulses. If for any reason the head drum rotation should stop, the auto-stop circuit is energized.

3. *Tape Slack Sensor.* If slack tape is detected in the play or record mode, the tape must be rewound. When the tape slack is detected by

the tape slack sensor element, the auto-stop circuit is energized.

4. *Auto-Stop Circuit.* Auto-stop is initiated by actuation of the devices listed in 1, 2, and 3 above. When auto-stop is initiated, the system control circuit energizes the stop solenoid to place the function mechanism into the stop mode. The ac motor power is cut off when the stop solenoid is energized.

5. *Muting Circuit.* This circuit generates audio muting and video blanking signals according to the operation of the mechanical system.

The take-up sensing coil is positioned very close to the tape in the cassette tape guide on the take-up side. It senses the tape beginning, in the rewind mode, by means of the metallic foil attached to the tape at the tape's beginning. The sensor energizes the auto-stop function.

AUTO-STOP

Whenever any of the detectors that generate the auto-stop operations senses a need to automatically stop the machine, the auto-stop solenoid is energized to release all the function buttons, putting the unit into the stop mode. The stop solenoid drive circuit is controlled by the signals generated in the respective auto-stop signal generator circuit as shown in Fig. 4-22.

The power supply for the ac motor is cut off when the auto-stop is energized. When any one of the function buttons is depressed, the button is latched mechanically by the lock slider of the function button. Release of the function button can occur in two ways: (1) manually by depressing the stop button, or (2) electrically by enegizing of the auto-stop solenoid by an output from the system control (Fig. 4-23). The lock slider releases the function button in each case.

When the auto-stop solenoid is energized, the lock slider is pulled in the direction shown by the arrow in Fig. 4-23 to release the function button lock. If the auto-stop solenoid is still energized, the lock slider will remain pulled even if any one of the buttons is pushed. The ac motor won't start, even if the function button is kept depressed after the auto-stop, because the stop solenoid switch is still on. The auto-stop solenoid drive circuit is shown in Fig. 4-24.

COUNTER MEMORY

The system control of the recorder during the rewind mode is automatically placed into the stop mode by the counter memory circuit when the tape counter reaches a 9999 indication. The counter memory circuit shown in Fig. 4-25 is included in the stop-mode circuit of the tape-end

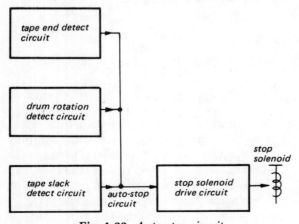

Fig. 4-22. *Auto-stop circuit.*

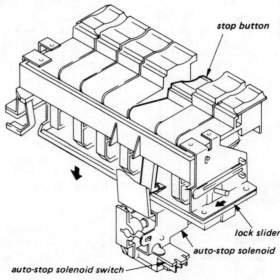

Fig. 4-23. *Drawing of stop solenoid.*

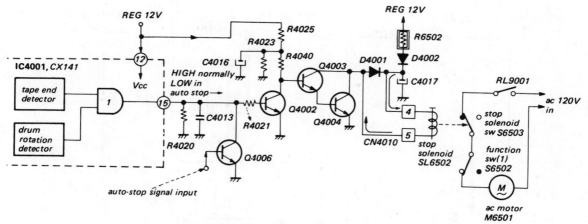

Fig. 4-24. *Stop solenoid drive circuit.*

detector circuit. The bias voltage at pin 11 of the tape-end sensor oscillator circuit is made to increase the supply voltage instantaneously when the tape counter indicates 9999. This stops the oscillation for a moment and the machine is automatically stopped. Capacitor C4018, the counter switch, and the memory on/off switch are connected in parallel with R4030, bias resistor of pin 11. The counter switch located in the tape counter turns on when the tape counter reaches its (0000) position. The oscillation is stopped for the C4018 charging period when the memory on/off switch is on and the counter indication is 9999. If the rewind button is depressed once more after the machine has been automatically put into the stop mode, the rewind operation takes place because C4018 has been charged. Resistor R4601 prevents the machine from being placed in the stop mode automatically when the memory on/off switch is turned on after the tape counter reaches 9999.

DRUM ROTATION DETECTOR

When normal operation cannot be performed due to motor trouble or overload, the recorder must be put into the stop mode. This is also required in the case where the head drum won't rotate due to a broken head drum belt. The drum rotation detector circuit uses the 30 PG signal to

detect drum rotation. When the rotation stops, it activates the stop solenoid circuit. The detector circuit is activated when the ac motor rotates, i.e. when one of the function keys is pressed. The schematic diagram for this circuit is shown in Fig. 4-26.

PAUSE CIRCUITS

The pause circuits include a number of functions including memory, solenoid drive, auto release, and remote control.

Memory

The pause memory circuit is located in IC4003 (Fig. 4-27). The four NOR gates of the IC comprise two latches for the initial set and two toggle flip-flops. The basic operation of the latch is to latch a certain state (tape running or tape stopped) and to lock the gate to ignore a change of the state. The latch operation is utilized for the initial set of the toggle flip-flop connected to the output of this machine. The input of gate 1 of IC4003 is connected to input P (of the pause solenoid), and its output is tied to the input of gate 2.

One of the inputs of each of gate 1 and gate 2 is a low level for time (T). The output of gate 1 becomes P and the output of gate 2 P. The gates 1 and 2 work as inverters. Because one of the

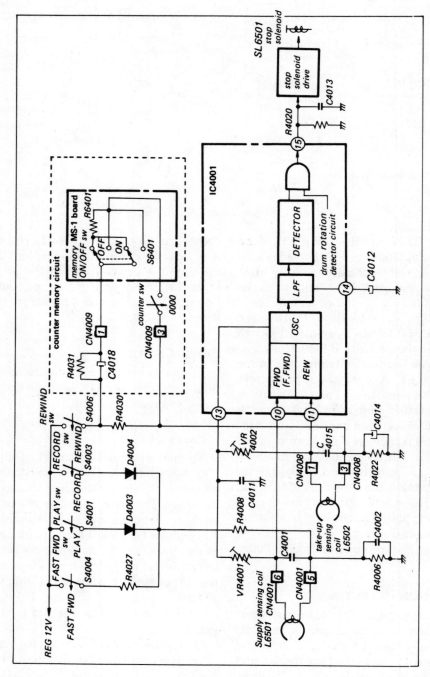

Fig. 4-25. *Tape end detector circuit and memory circuit.*

112

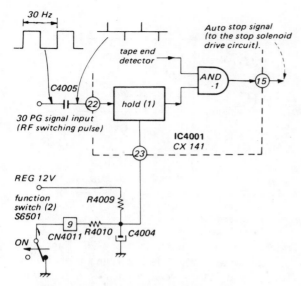

Fig. 4-26. *Drum rotation detector circuit.*

inputs of each gate becomes high when the latch time is (T), the inverter operation does not work, and the following flip-flop serves as an ordinary toggle-type flip-flop. The time is set for about 2 milliseconds by R4066 and C4043.

Solenoid Drive

The pause solenoid drive circuit is formed by Q4016 and Q4017 as shown in Fig. 4-28. The pause solenoid is a two-winding three-terminal type with C (common), P (primary), and S (secondary) terminals. The winding between the C and P terminals serves as an initial pull-in while the one between the C and S terminals works for maintenance so that any temperature increase and the pull-in power requirement of the solenoid is reduced.

Transistor Q4016 turns on in the "run" state, causing current flow through the C-S winding and D4008. This turns on Q4017 and Q4018 (time constant determined by R4078 and C4046). At the same time, a current flows through the C-P winding, and the initial pull-in takes place. When Q4018 turns on, the potential of the P terminal and the current between C and S is cut off. At the same time, Q4017 and Q4018 turn off (by the same time constant determined by R4078 and C4046). Then current flows in the C-S winding again, and the pause solenoid energizing is maintained.

Transistor Q4019 (connected to the base of Q4016) is delayed about 350 ms after the play or record button is pressed by the time constant circuit consisting of R4054, R4055, and C4047, and the run drive signal is bypassed during the delay (Q4019 is on for about 350 ms).

Auto-Release

The pause auto-release circuit is one of the protection circuits for the tape. The pause auto-

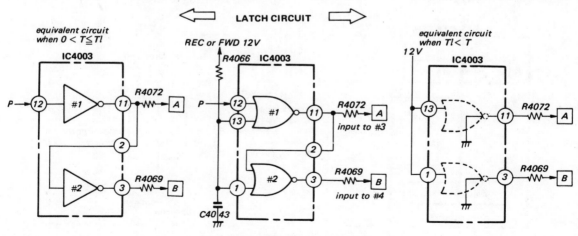

Fig. 4-27. *Pause memory circuit.*

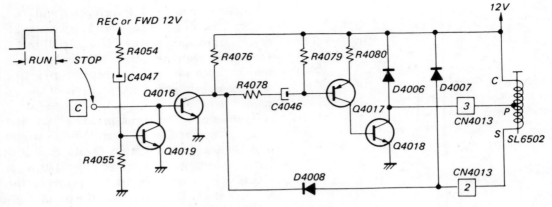

Fig. 4-28. *Pause solenoid drive circuit.*

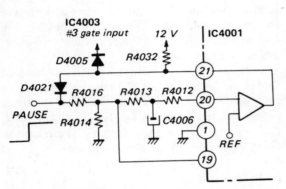

Fig. 4-29. *Pause auto-release circuit.*

release circuit is shown in Fig. 4-29. The pause 12 V becomes a regulated voltage supply by the R4016 and R4014 resistor division. It is processed in the time constant circuit consisting of R4013 and C4006 to be a lamp voltage that increases with time in the pause mode. The lamp voltage is applied to the voltage comparator in IC4001 via pin 20. When it reaches the reference voltage in the IC, pin 21 of IC4001 goes high to set the pause flip-flop of IC4003 to the tape running state and release the pause state.

Resistor R4012 is for protection of the internal circuit of IC4001 and D4021. It serves to lower pin 21 to ground potential when the voltage is not applied to pin 19 in the stop mode or in the tape running state.

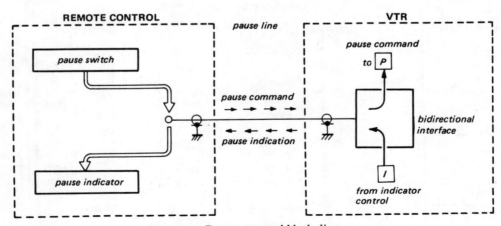

Fig. 4-30. *Remote control block diagram.*

Remote Control Circuit for Pause

A transmission signal and a reception signal are added in one cable in the remote control. The remote control block diagram is shown in Fig. 4-30. The reception signal is the control signal from the remote control to the videocassette recorder and the transmission signal is the pause signal from the recorder to the remote control. Because these two signals are transmitted with one signal line, the reception signal is a negative pulse signal and the pause signal is a positive pulse.

Refer to Fig. 4-31 for the remote control circuit. The received remote control signal is divided by R4061 and R4060 and is applied to the base of Q4014. The threshold voltage of Q4014 is set to about 1.2 volts. When a pulse lower than the voltage comes in, Q4014 turns off and a posi-

tive trigger pulse appears at its collector. This pulse triggers the MMV formed by the two NOR gates (numbers 1 and 2 of IC4002) connected to the following stage. The MMV forms an OR circuit together with the pause switch of the recorder. Normally the output of "D" is a low level.

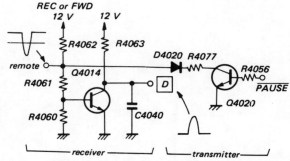

Fig. 4-31. *Remote control circuit.*

5

Sensor Circuits and Electrical Alignment

This chapter explores the various tape detector and sensor circuit operations. Also discussed are the audio record and playback circuits and audio record/play head azimuth alignment and adjustments. Some of the most required electrical field alignment procedures are also covered for the Zenith KR9000 VCR (equivalent to Sony SL-8600) discussed in Chapter 2.

The purposes of the system control and pause circuits are as follows:

- To activate the function selected by the recorder operator.
- To prevent damage to the tape during threading and unthreading or rotor head rotation failure or operation of any other function.
- To generate an audio muting and video blanking signal to keep the TV receiver's audio and CRT inactive until the tape is up to proper speed and the picture is locked in.
- To place the VCR in the pause mode on command. Refer to Fig. 5-1 for block diagram of pause and system control.

AUTO-STOP

To prevent damage to the tape during any problems that might occur, the auto-stop solenoid activates to release the lock slider and the function buttons and stop the VCR. The key to activation of the auto-stop solenoid is the voltage at pin 15 of IC4001 (Fig. 5-2). Whenever the output of pin 15 is high (approximately 12 volts), the auto-stop solenoid remains inactive. When pin 15 goes to low (ground) the auto-stop solenoid activates.

When the VCR is turned on, 12 volts is on pin 12 of IC4001, which puts a high on pin 15. At the same time, C4017 charges to 12 volts through D4002 and R6502. With pin 15 high, Q4002 is forward biased, putting the base of Q4003 at near ground potential, keeping it and Q4004 turned off. The auto-stop solenoid remains in-

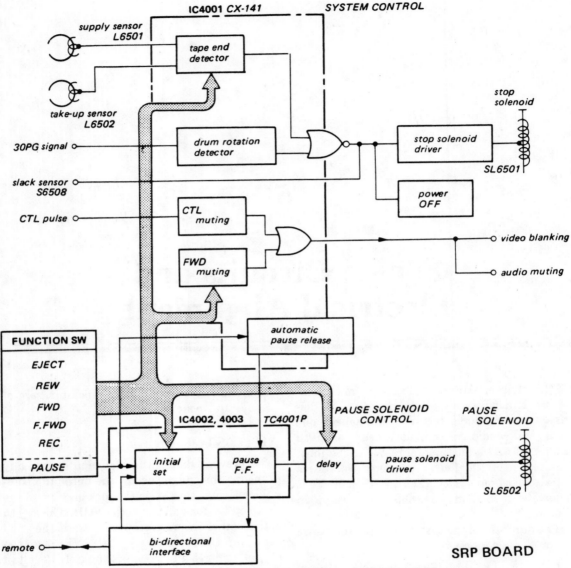

Fig. 5-1. *VCR system control and pause control block diagram.*

active. When pin 15 goes low, Q4002 is turned off. This allows Darlington Q4003 and Q4004 to turn on, providing a path through Q4004 for current to flow from B+ through the auto-stop solenoid. This stops the ac motor, the tape, and the video heads and releases the selected function button.

The solenoid requires 3 amps to pull in, which the VCR power system cannot supply. That is the function of the charge on C4017 — to supply the initial pull-in current surge. After the solenoid pulls in, it only requires 1 amp to hold it in. Resistor R6502 limits the current to 1 amp from the unregulated 12-volt dc supply. When pin 15

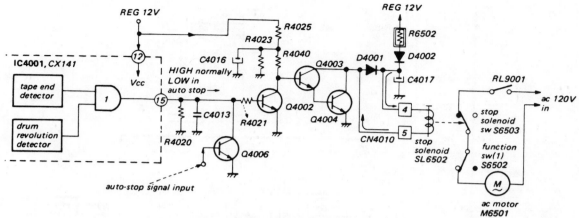

Fig. 5-2. *Stop solenoid drive circuit.*

returns to a high, the circuit action once again returns to the original state, and C4017 charges up again to be ready for the next auto-stop function.

POWER FAILURE

Should the power fail, it is necessary to release the tape. This requires the auto-stop solenoid to activate. When the power fails, pin 15 goes low. The charge on C4016 forward bias's Q4003 and Q4004, allowing the charge on C4017 to bleed off through the stop solenoid, activating it long enough to release the tape and prevent damage.

How pin 15 of IC4001 goes low when an auto stop command occurs is explained as follows:

When either of these two modes is activated, a dc voltage is applied to pin 10 of IC4001 through either one of the switches, through R4008, through the supply sensing coil and to pin 10 of IC4001. This starts the oscillator in IC4001 and creates a feedback loop from pin 13 of the IC through VR4001, which adjusts the amplitude of the oscillations. Then it goes back to the IC pin 10 and through the supply sensing coil (which is part of the oscillator tuned circuit) and through R4006 to ground. This signal is detected in the chip and causes a high on the input of the AND gate. The other high input to the AND gate comes from the drum rotation detector circuit.

This causes pin 15 of the IC to go high and the auto-stop remains inactive.

AT THE SUPPLY TAPE END

Each end of the tape in the cassette has a metallic leader. As this leader passes over the sensing head (during play or fast/fwd), it radically changes the Q of the coil in the sensing head and causes the oscillator to stop. When this occurs, the output of the detector in IC4001 goes low. This causes that input to the AND gate to go low, which in turn causes the output of pin 15 of the AND gate to go low. This activates the auto-stop, causing the VCR to stop.

AT THE TAKE-UP END

In normal rewind operation, a dc voltage passes through the rewind switch (Fig. 5-3) when it is depressed and goes through R4030 and the take-up sensing coil into pin 11 of IC4001. This activates the oscillator circuit and causes a feedback loop out pin 13 of the chip, through VR4002 and the take-up sensing coil (which is part of the oscillator circuit). VR4002 adjusts the amplitude of the oscillations. The signal is passed through the detector in IC4001 and a high is applied to one gate of the AND circuit. The other gate has a high applied to it from the drum rotation detector, causing a high on pin 15,

119

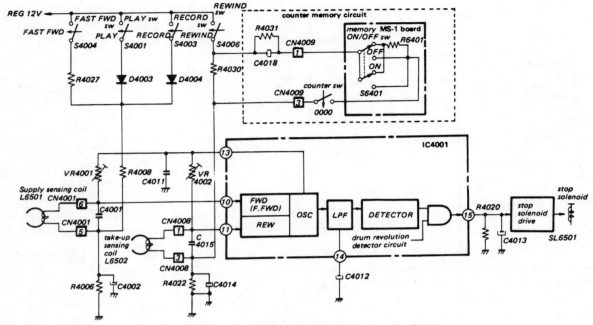

Fig. 5-3. *Tape-end detector and memory circuits.*

which keeps the stop solenoid from activating. The tape in the cassette then rewinds.

When the metallic leader at the end of the tape passes over the take up sensing coil in the rewind mode, it causes a radical change in the "Q" of the take-up sensing coil. This causes the oscillator to stop. The detector output in IC4001 goes low, causing the AND gate output and pin 15 of the chip to go low, activating the auto-stop solenoid. In order to thread the tape and keep the machine out of auto-stop when no function buttons are depressed, the output of the detector in the IC automatically goes high if there is no dc bias on pins 10 or 11.

During the rewind mode, the counter memory can be placed in the circuit by closing the memory switch. This places the circuit in parallel with R4030 and into pin 11 of IC4001. When the counter reaches 0000, the counter switch closes, allowing a large voltage that is almost equal to the supply voltage to be fed to pin 11 of IC4001. This stops the tape end oscillator, thereby initiating the auto-stop mode. The oscillator stops

only for as long as it takes C4018 to charge up. Then the voltage returns to the proper bias for the oscillator to run. If the rewind button is depressed a second time, the tape will rewind until it reaches the end.

DRUM ROTATION DETECTOR

When the head drum stops rotating, it activates the auto-stop circuit (Fig. 5-4). As the drum rotates, it develops the 30 PG pulse that goes to pin 22 of IC4001 via C4005 from CN4003, pin 2. This pulse goes into a hold circuit that is biased on by the regulated B+ through voltage divider R4009 and R4010, out CN4011 pin 1, through S6501, and back in CN4011 pin 3 to ground.

The hold circuit in IC4001 causes a high on its output that goes to the same gate the tape end sensor circuit connects to. This puts a high on pin 15, keeping the auto-stop circuit inactive. If the drum should stop rotating, it will cause the 30 PG pulse to disappear, causing a low at the

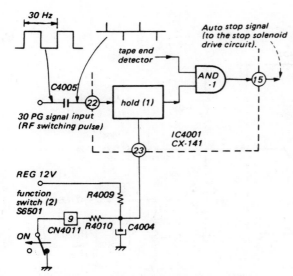

Fig. 5-4. *Drum rotation detector circuit.*

and through voltage divider R4033 and R4034 to ground. The voltage developed at the top of R4033 forward-biases Q4006, turning it on. This causes a near zero resistance from collector to emitter ground, thereby placing the voltage on pin 15 of IC4001 at ground, or low, activating the auto-stop circuit.

output of the hold circuit, and pin 15 will go low, activating the auto-stop mode.

TAPE SLACK SENSOR

Whenever tape slack occurs in the play and record modes only, the auto-stop circuit will be activated. Note this circuit in Fig. 5-5. Whenever slack occurs in the tape, the tape slack lever will move in toward the head drum assembly, closing reed switch S6508 mounted on the TK board. This allows the 12-volt regulator voltage to go from the CP3 board out CN5001 (pin 1), through S6508, back in CN5001 (pin 3), out CN5004 (pin 10), back into the SRP board at CN4004 (pin 3),

BLANKING AND MUTING CIRCUIT

The VCR is permanently video-blanked and audio-muted in all modes, except play, record, and the mode known as "E-E," which stands for "electric to electric." The E-E mode allows the operator to view the picture that is to be recorded (more in Chapter 12). Note the video blanking circuit in Fig. 5-6. With no input to pin 9 or pin 2 of IC4001, the inputs (1) and (2) to the AND gate are high and the output at pin 18 of IC4001 is high, causing the video to blank and the audio to mute. When the play mode is activated, the tape starts moving and the CTL pulse is applied to pin 9 of IC4001, causing a low at pin (1) of the AND gate.

When the play button is pressed, the 12 volts PB voltage is applied to pin 2 of IC4001. However, the input to pin (2) of the AND gate is delayed (remains high) for 3.5 seconds by R4007 and C4003 on pin 24 of IC4001 and the delay circuit inside the IC. This keeps pin 18 high and the video blanking and audio muting on. At the end of the 3.5-second delay, the voltage at pin (2) of the AND gate goes low. This puts pin 18 at a low, thereby releasing the video blanking and audio muting. If at any time either or both the CTL pulse and the PB 12 volts was removed, a high would occur at either or both inputs to the OR gate, causing pin 18 to go high and activating the video blanking and audio muting.

Because no blanking or muting at all is desired when the record mode is activated, E-E 12 volts is applied to the base circuit of Q4005 when the record button is pushed, which forward-biases this transistor, thereby effectively shorting out the voltage on pin 18 of the IC and immediately releasing all blanking and muting.

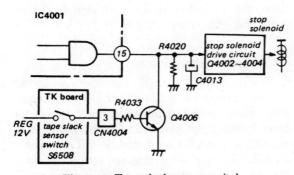

Fig. 5-5. *Tape slack sensor switch.*

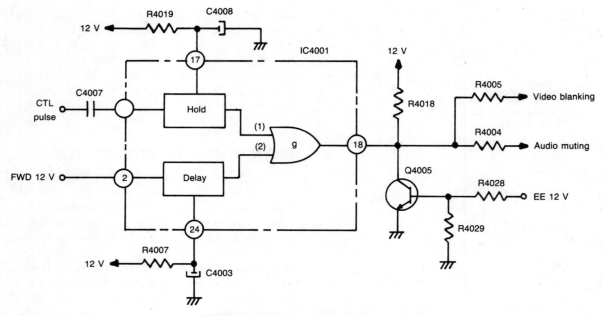

Fig. 5-6. *Blanking and muting circuit.*

ELECTRICAL ALIGNMENT PROCEDURES

The following is information for the electrical alignment for various circuits found in the Zenith KR9000. These adjustments can be performed by the test equipment listed below and a TV signal obtained from a TV receiver.

TEST EQUIPMENT REQUIRED

- Color TV receiver
- Dual-trace oscilloscope with more than 15 MHz and a delay mode
- Frequency counter (more than four digits)
- VTVM or FET meter
- Volt-ohmmeter (20 k ohm/volt)
- Audio signal generator
- Audio attenuator
- Alignment and/or test tape (i.e. Sony KR5-1D)
- Insulated alignment tool

ALIGNMENT SET-UP

Connect antenna or cable to the VHF input terminals on rear of the VCR. Because the signal received through the tuner in the VCR is utilized as the adjustment signal for the alignment of the machine, it is important that the video output signal satisfies these specifications. The VCR should be set to the channel with the best reception. The video signal should be checked with an oscilloscope connected to the Q9 emitter on the YC-2 board. Verify that the sync signal amplitude is approximately 0.3 volt p-p and the video signal amplitude is near 0.7 volt p-p. Adjust the fine tuning while observing the signal and the TV screen so the burst signal amplitude becomes 0.3 volt ± 0.1 volt p-p. Also confirm that there are no spikes at the sync signal portion. Note this video output signal in Fig. 5-7.

VCR ALIGNMENT TOOL

The semi-fixed variable resistors and inductances should be adjusted with an insulated alignment tool. A common screwdriver is too large for adjusting the controls from the conductor side of the PC board, plus it will detune the circuits. The proper alignment tool is shown in Fig. 5-8. The metal blade of the alignment tool is

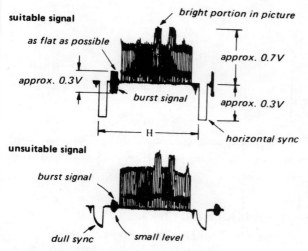

Fig. 5-7. *TV video output signal.*

used for variable resistors and trimmer capacitors, and the plastic tip is used for variable inductances.

SYSTEM CONTROL ALIGNMENT AND CHECKS

These adjustments require the alignment tool. The correct alignment will not result if a common screwdriver is used.

Supply Sensor Oscillator Level Adjustment

1. Insert a video cassette.
2. Connect the scope to TP1.
3. Set up the play mode.
4. Adjust VR1 for a 3.1 ± 0.2 volt (p-p) oscillator output level at TP1.
5. Place the machine into fast-forward and con-

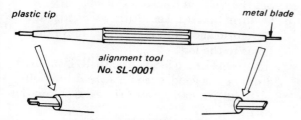

Fig. 5-8. *VCR alignment tool.*

firm that the voltage is almost the same as in the play mode.

Take-up Sensor Oscillator Level Adjustment

1. Insert videocassette.
2. Connect the scope to TP2.
3. Set up the rewind mode.
4. Adjust VR2 for an oscillator output level of 3.1 ± 0.2 volts (p-p) at TP1.

Threading Check

1. Insert the cassette.
2. Check that the threading operation is performed correctly. The threading time should be 2.5 seconds.

Auto-Stop Check

1. Check that the tape is set up at the metal tape portion in the play or fast forward mode.
2. Check to see that the auto-stop operates when the metalized leader on the tape passes the take-up sensing head in the rewind mode.

Counter Memory Check

1. Turn on the memory on/off switch.
2. Check that the auto-stop is set up during the counter indication from 0000 to 9999 in the rewind mode.
3. Check and see that the rewind mode is set up with the rewind button depressed after the auto-stop.
4. Check that the auto-stop is not set up when the memory on/off switch is in the OFF position.

Power Switch Off Check

1. Put the VCR in the record mode. Check that the stop solenoid energizes and the function button is released when the power switch is turned off or switched to OFF from the timer.
2. Check that the function button is released in the play, fast forward, or rewind mode when the power switch is turned off.

Pause Solenoid Check

1. Check that the pause solenoid energizes and the tape runs when the play button is depressed.
2. Ensure the pause lamp lights when the pause button and then the play button is depressed.
3. Connect the remote pause control unit to its terminal. Perform checks 1 and 2 as well as the pause button of the machine.
4. Set up the record mode and perform step 1 through 3 listed above.
5. Put the VCR in the play or record mode and check that the tape moves forward and stops for every pushing of the pause button.
6. Turn off the power switch and depress the play (or record) button to lock it. Check that the play or (record) mode is set up and the tape is taken up when the power switch is turned on.

Eject Mode Check

1. Check that the unthreading operation starts after the ac motor speed increases when the eject button is depressed.
2. Check that the supply reel has stopped, and the cassette is lifted up after the take-up reel has taken up the tape completely. (Verify that the tape is taken up completely though the take-up reel cannot actually be seen.)
3. Check for correct unthreading speed.
4. Eject time should be about 3 seconds after the eject button is depressed.

DRUM SERVO AND PULSE SYSTEM ALIGNMENT (ARS BOARD)

Alignment sequence is as follows (Fig. 5-9):

- Drum free-speed check.
- DC amplifier bias adjustment.
- RF switching position adjustment.
- Record servo lock phase adjustment.
- Playback CTL signal check.

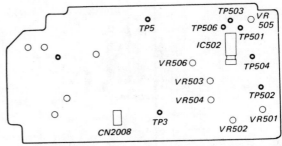

Fig. 5-9. *Test point location on ARS board, component side.*

Drum Free-Speed Check

1. Set VCR tuner to an unused channel.
2. Remove connector CN2008 from the brake coil.
3. Connect a frequency counter to TP502 (Fig. 4-10).
4. Place the machine into the record mode. Verify that the counter reading is 30.68 to 30.53 Hz.
5. If not, check for stretched or worn drum drive belt. When there is no problem with the drum drive belt, change the drum pulley with a different diameter pulley in order to obtain the right specification.
6. Re-connect the CN2008 connector removed in step 1 after these checks.

If a frequency counter is not available:

1. Use a station signal from the VCR's tuner.
2. Insert a video cassette.
3. Disconnect plug from the brake coil.
4. Connect a scope to TP502, set the horizontal axis knob of the scope to 2 ms/cm, and trigger the scope from TP503.
5. Set the VCR in the record mode.
6. Confirm that it takes 20 to 27 seconds for the waveform shown in Fig. 5-10 to move fifteen times to the left.
7. If not, check for a faulty drive belt. If the drive drum belt is good, exchange the drum pulley with a different diameter pulley to obtain the correct speed.

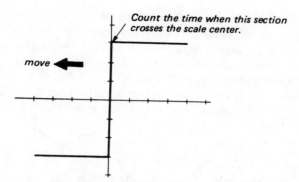

Count the time when this section crosses the scale center.

move ←

Fig. 5-10. *Drum free-speed check.*

8. Connect the brake coil plug disconnected in step (3) after this check.

DC Amplifier Bias Adjustment

1. Tune in a TV station with the VCR tuner.
2. Insert a cassette and set up the record mode of VCR.
3. Connect the scope to pin 1 of IC502 and set it to the dc range. Trigger the scope externally from TP502.
4. Adjust VR506 for a dc level of 3.5 Vdc. Note waveform shown in Fig. 5-11.

RF Switching Position Adjustment

1. Play back the color bar portion of the alignment tape.
2. Connect the scope to TP5 and trigger it externally from TP502.
3. Set the tracking control for maximum amplitude of the RF signal at TP5.
4. Adjust the vertical hold control of the TV monitor so that the vertical blanking portion appears on the screen.

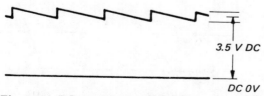

3.5 V DC

DC 0V

Fig. 5-11. *DC amplifier bias adjustment waveform.*

5. Adjust the contrast control so that the sync portion of the blanking period is observed. The blanking period is arranged with front porch (3H), sync (3H), and back porch (13 to 15 H) from the top as shown in Fig. 5-12.
6. Adjust VR501 and VR502 so that the interval between the switching position and the blanking is the same as the width of the front porch. VR501 is for the adjustment of the switching position of the 30 PG (B) and VR 502 for the 30 PG (A) switching position. When both of the switching positions are adjusted correctly, the positions are superimposed on the screen.

Record Servo Lock Phase Adjustment

1. Receive a TV station via the VCR tuner.
2. Insert the cassette and place the machine into the record mode.
3. Connect the scope to TP502 and trigger it externally from TP502.
4. Set the scope horizontal time base generator to 2 ms/cm, the horizontal sweep to ×10 (or ×5 MAG), and the scope trigger slope to (+). Adjust the scope horizontal position control until the negative-going edge of the waveshape is positioned at the exact center of the scale. See Fig. 5-13A.
5. Remove the scope probe from TP502 and connect it to IC502 pin 6. Adjust VR503 so that there are 7 horizontal sync pulses (±2 H) between the scope scale center and the beginning of the vertical sync as shown in Fig. 5-13B.

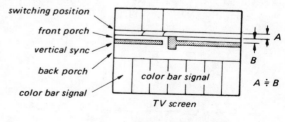

switching position
front porch
vertical sync
back porch
color bar signal
color bar signal
A
B
$A \doteqdot B$
TV screen

V.HOLD knob is moved.

Fig. 5-12. *RF switching position adjustment.*

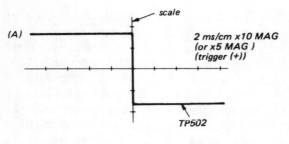

(A)

scale

2 ms/cm x10 MAG
(or x5 MAG)
(trigger (+))

TP502

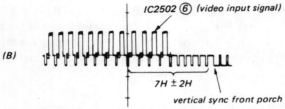

(B)

IC2502 ⑥ (video input signal)

7H ± 2H

vertical sync front porch

Fig. 5-13. *Record servo lock phase adjustment.*

Playback CTL Signal Check

1. Tune in a TV station through the VCR tuner.
2. Insert the cassette and perform a recording for a few minutes.
3. Rewind and play back the portion on which the recording was made in step 2.
4. Connect the scope to TP506 and confirm that the CTL signal shown in Fig. 5-14 is obtained.

Setting Tracking Control

1. Tune in a TV station with the VCR tuner.
2. Insert the cassette and make a recording for several minutes.
3. Play back the recorded position. Set the tracking control knob to the center detent position.
4. Connect the scope to TP501 and trigger it externally from TP503.

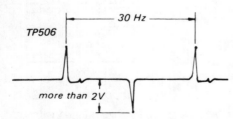

TP506

30 Hz

more than 2V

Fig. 5-14. *Correct playback CTL signal.*

5. Set the scope trigger slope to (−), the horizontal time base to 0.5 ms/cm, and adjust the horizontal position knob so that the beginning of the waveform is positioned at the left end of the scale. See waveform in Fig. 5-15.
6. Adjust VR505 so that the rising portion of the waveform is at the position shown in Fig. 4-16.

Play Servo Lock Phase Adjustment

1. Use station signal from the VCR tuner.
2. Insert a cassette and perform a recording for several minutes.
3. Play back the recorded portion. Set the tracking control to its center detent position.
4. Connect a scope to TP501 and trigger it externally from test point TP502.
5. Set the scope trigger slope to (−) and the horizontal time base to 0.5 ms/cm. Adjust horizontal position control so that the beginning of the waveform is positioned at the left end of the scale.

PLAYBACK SYSTEM ALIGNMENT

For the playback rf amplifier frequency response adjustment (ARS board), the CH-A and CH-B amplifiers require independent frequency response adjustments. The adjustments for CH-B is indicated by parentheses.

1. Play back the RF sweep portion of the alignment tape.
2. Connect the scope to TP5 and trigger it externally from TP502.

TP501 waveform

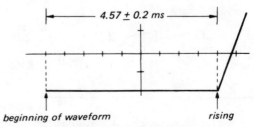

4.57 ± 0.2 ms

beginning of waveform

rising

Fig. 5-15. *Tracking control set.*

3. Adjust the tracking control for the largest RF output.

4. Set the horizontal time base of the scope to 2 ms/cm. Select negative (−) trigger slope to observe the CH-A signal and positive (+) trigger slope to observe the CH-B signal. The RF sweep portion has frequency markers at 1 MHz, 2 MHz, 3.58 MHz, 4.5 MHz, and 5.1 MHz from left to right.

5. Set VR2 (VR1) to fully ccw position as viewed from the component side and VR7 (VR6) to the fully cw position.

6. Adjust VC2 and (VC1) so that the waveshape peak amplitude (tuned frequency) is located at the 5.1 MHz ± 0.1 MHz. Refer to waveform in Fig. 5-16.

7. Adjust VR2 (VR1) and VR7 (VR6) so that the waveform between the 2 MHz and 4.8 MHz markers is as flat (±1 dB) as possible. Note the waveform in Fig. 5-17.

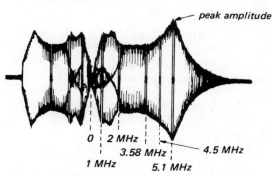

Fig. 5-16. *Playback amplifier frequency response waveform.*

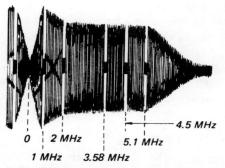

Fig. 5-17. *Playback amplifier frequency response.*

8. Set the scope time base generator to 5 ms/cm and adjust VR5 to equalize the levels of CH-A and CH-B at 3.58 MHz.

Dropout Compensator Threshold Level Adjustment (ARS Board)

1. Play back any pre-recorded tape known to have many dropouts.

2. Turn VR8 fully clockwise cw as viewed from the circuit board side so that dropouts appear on the screen.

3. Turn VR8 counterclockwise slowly until the dropouts disappear. Set VR8 to that point.

4. Rewind and play back the tape. Confirm that the dropouts are compensated for at the section where they appeared.

Playback Video Output Level Adjustment (YC-2 Board)

1. Play back the color bar segment of the alignment tape.

2. Connect the slope to the emitter of Q25. Trigger the scope externally from pin 12 of IC7.

3. Set VR21 for 1.3 volt p-p ± 0.1 volt. Note waveform shown in Fig. 5-18.

4. Connect the scope to the emitter of Q9 and adjust VR8 for a 1 ± 0.1 volt reading. Refer to waveform in Fig. 5-19.

VCO Oscillator Frequency Adjustment (44 fH)

This adjustment is located on the YC-2 board.

1. Use an off-the-air station signal and set up the E-E mode.

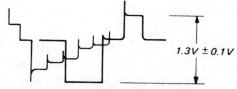

Fig. 5-18. *Playback video output level adjustment.*

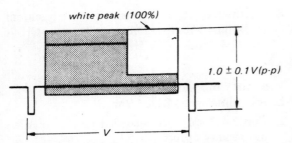

white peak (100%)

$1.0 \pm 0.1 V(p\text{-}p)$

V

Fig. 5-19. *Playback video output level waveform adjustment.*

2. Short pins 5 and 6 of IC5 on the YC-2 board with a jumper lead.
3. Connect a frequency counter to pin 11 of IC5.
4. Adjust VR13 for a counter reading of 692.308 kHz $\pm$ 2 kHz.
5. Remove the jumper lead connected in step 2 and confirm with a counter reading of 692.308 kHz.

3.57 MHz VXO Free-Running Frequency Adjustment (YC-2 Board)

1. Tune in a TV station signal and set up the E-E mode.
2. Connect the frequency counter for the junction of R109 and T14.
3. Turn the core of oscillator transformer T12 until the counter reads the frequency of 4,267,919 $\pm$ 5 Hz. T12 can be adjusted from the component side with the alignment tool

3.58 MHz Osc Frequency Adjustment (YC-2 Board)

1. Use an off-the-air TV station signal and set up the VCR in the E-E mode.
2. Connect the frequency counter to the emitter of Q5.
3. Turn the core of T11 until the counter reads 3.57545 kHz $\pm$ 5 Hz.

AUDIO SYSTEM ADJUSTMENTS

Record. The audio information that is available for *recording* comes from either a separate outside source (through the audio-in jack) or from another board. The latter case is the most used mode of operation. The audio signal portion of a TV program is recovered from the IF signal within the IF-4 board, as in the block diagram of Fig. 5-20. Whatever the audio signal source, it is coupled to and through the CP-3 board to the ARS board.

Audio amplification circuits process the signal and connect it to the audio head. The information is transferred to the tape as a longitudinally recorded track.

Playback. On playback, the audio head picks up the recorded audio information from the tape's audio track and routes it to the ARS board (Fig.4-20). Amplification is accomplished and the audio signal is connected, out of the ARS board, to and through the CP-3 connection board.

The audio information is available at the audio out jack at the rear of the VCR machine (for connection to a TV monitor). Audio is also coupled from the CP-3 board to the RF modulator for development of the VHF OUT signal.

The connections of the equipment for VCR audio adjustments are shown in Fig. 5-21. The adjustment sequence is

- Azimuth adjustment.
- Playback frequency characteristic adjustment.
- Playback output level adjustment.
- Bias oscillator check.
- Bias trap adjustment.
- Record bias adjustment.
- Record current adjustment (not covered in this section).
- Overall frequency characteristic check (not covered in this section).
- AGC operation check (not covered in this section).
- S/N Ratio check (not covered in this section).
- Distortion check (not covered in this section).

Audio Head Azimuth Adjustment

1. Terminate the audio out terminal with a 100 kilohm resistor and connect an ac VTVM.

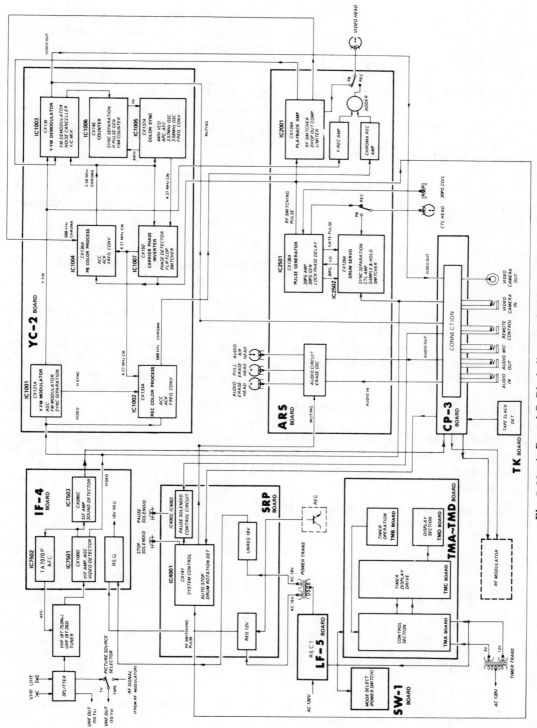

Fig. 5-20. *A, B, and C. Block diagram showing audio circuits.*

129

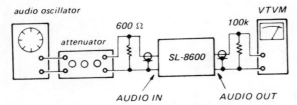

Fig. 5-21. *Connections for audio test equipment.*

2. Play back the 5 kHz portion of the alignment tape.
3. Adjust the azimuth adjustment screw on the audio head for a maximum VTVM reading. See adjustment location in Fig. 5-22.

Playback Audio Frequency Characteristic Adjustment

1. Terminate the audio line output terminal with a 100-kilohm resistor and connect an ac VTVM. Note test instrument set-up and adjustment locations in Fig. 5-23.
2. Play back the 333 Hz portion of the alignment tape and measure the output level with the VTVM. Make a note of the measured value.
3. Play back the 5 kHz portion of the alignment tape.
4. Adjust VR601 so that the playback output level of the 5 kHz audio signal is zero ±1 dB in reference to the measured level of the 333 Hz in step 2 above.

Example: If the measured value in step (2) is −25 dB, adjust the 5 kHz signal level for

−25 dB with VR601. Note: Each level of the 5 kHz and 333 Hz audio level recorded on the alignment tape is at −25 dB.

Playback Output Level Adjustment (ARS Board)

1. Terminate the audio out terminal with a 100-kilohm resistor and connect a VTVM.
2. Play back the 333 Hz audio signal portion of the alignment tape.
3. Adjust VR602 for a VTVM reading of −25 dB. Note equipment setup and adjustment location in Fig. 5-24.

Bias Oscillator Check (ARS Board)

1. Set the audio oscillator to zero output and the attenuator to maximum.
2. Insert the cassette and set up for record mode.
3. Connect a frequency counter to test point TP603. Confirm that the oscillating frequency is 65 kHz ± 6.5 kHz.

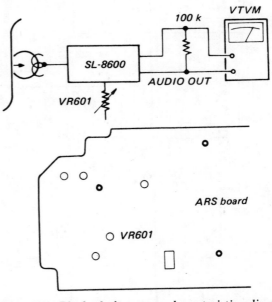

Fig. 5-23. *Playback frequency characteristic adjustment.*

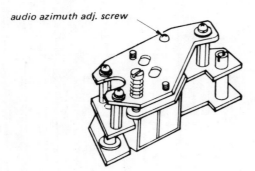

Fig. 5-22. *Audio head azimuth adjustment.*

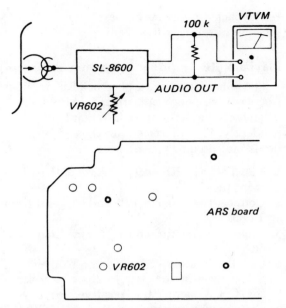

Fig. 5-24. *Test set-up and adjustment locations for playback output level.*

Bias Trap Adjustment (ARS Board)

1. Set the audio test oscillator output to zero and the attenuator to maximum.
2. Insert the cassette and set VCR in the record mode.

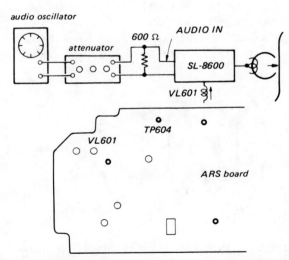

Fig. 5-25. *Bias trap adjustment test set-up and locations.*

3. Connect the scope to TP604.
4. Adjust VL601 for minimum bias leak value. Minimum level below 0.5 volt p-p. See Fig. 5-25 for test setup and adjustment locations.

Record Bias Adjustment (ARS Board)

Make sure that the playback frequency characteristic adjustment has been completed before making these checks.

1. Ground the TP601 with a jumper to turn off the AGC operation.
2. Connect the 333 Hz audio signal to the audio in.
3. Terminate the audio out terminal with a 100 kilohm resistor and connect to a VTVM.
4. Insert a cassette and set up VCR in the E-E mode.
5. Adjust the audio out terminal level with the attenuator for − 25 dB.
6. Depress the record button and make a recording for about five digits on the tape counter.
7. Change the audio signal to 5 kHz at − 25 dB and make a recording for about five digits on the tape counter.
8. Play back the portion of the tape recorded in steps 6 and 7. Measure the playback output levels of the 333 Hz and 5 kHz signals.
9. Confirm that the playback output level of the 5 kHz is zero dB ± 0.5 dB in reference to the 333 Hz playback output level. If not, repeat steps 2 to 8 and adjust the bias current with VC601.
10. After the adjustment, remove the jumper connected to TP601.

VIDEO SYSTEM ADJUSTMENTS

The playback system of the video stages is aligned first with the alignment tape. After the playback system is confirmed to operate properly, the record system is aligned. The alignment sequence is shown below. The "Y" signal and chroma signal alignments are performed for each of the playback and record systems.

ALIGNMENT SETUP

—Connect the color bar signal of 1 volt (p-p) to the video/camera in.

—Connect the color monitor TV to the video/camera out. This is for terminating the video output with 75 ohms.

—Set the tracking control to the center detent position unless otherwise required.

—The YC-2 board can be checked when the machine is placed on the tuner block side. Refer to Fig. 5-26. Set up the record or play mode in that state. When the YC-2 board is removed from the machine, place the machine horizontally for the alignment.

PLAYBACK SYSTEM ALIGNMENT

• Playback RF amplifier frequency response.
• Dropout compensator threshold level.
• Playback video output level.
• 44 fH VCO free-running frequency.
• 3.57 MHz VXO free-running frequency.
• 3.58 MHz oscillator frequency.
• 4.27 MHz carrier phase alternate check.
• 4.27 MHz carrier leak.
• Color-killer threshold level.

• Playback chroma output level.
• Comb filter fine adjustment.

Playback RF Amplifier Frequency Response

The following is for the playback RF amplifier frequency response adjustment (ARS board). Align both the CH-A and CH-B amplifiers independently. The adjustment for the CH-B amplifier is indicated by parentheses.

1. Play back the RF sweep portion of the alignment tape.
2. Connect the scope to TP5 and trigger it from TP502.
3. Adjust the tracking control for maximum RF output.
4. Set the time base of the scope to 2 ms/cm. Select negative (−) trigger slope to observe the CH-A signal and positive (+) trigger slope to observe the CH-B signal. The RF sweep portion has frequency markers at 1 MHz, 2 MHz, 3.58 MHz, 4.5 MHz, and 5.1 MHz from the left to right. Note scope waveform in Fig. 4-27.
5. Turn VR2 (VR1) fully counterclockwise as viewed from the component side. Turn VR7 (VR6) fully clockwise. Adjust VC2 (VC1) so that the waveshape peak amplitude (tuned frequency) is located at 5.1 MHz ± 0.1 MHz, as shown in Fig. 5-27.
6. Adjust VR2 (VR1) and VR7 (VR6) so that the waveform between the 2 MHz and 4.8 MHz

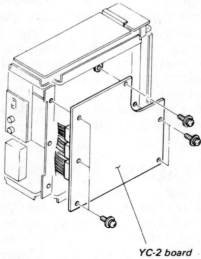

YC-2 board

Fig. 5-26. *Location of YC-2 board.*

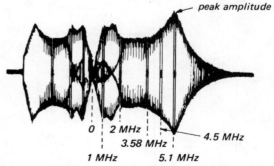

peak amplitude

0   2 MHz   4.5 MHz
3.58 MHz
1 MHz   5.1 MHz

Fig. 5-27. *High-frequency range tuned frequency adjustment.*

markers is as flat (±1 dB) as possible. Note waveform in Fig. 5-28.

7. Set the scope time base to 2 msec/cm. Adjust VR5 to equalize the levels of the CH-A and CH-B at 3.58 MHz.

Dropout Compensator
Threshold Level (ARS Board)

1. Play back any pre-recorded tape known to have many dropouts on it.
2. Set VR8 in fully clockwise position as viewed from the conductor side of the board. The dropouts can now be viewed on the monitor screen.
3. Turn VR8 counterclockwise slowly until the dropouts disappear. Set VR8 to that point.
4. Rewind the tape. Play back the tape again and confirm that the dropouts are compensated for at the tape portion where the dropouts were observed in step 2.

Playback Video Output Level (YC-2 Board)

1. Play back the color bar portion of the alignment tape.
2. Connect the scope to the emitter of Q25 and adjust VR21 for 1.3 ± 0.1 volt p-p. Note scope trace in Fig. 5-29. The scope is triggered externally from pin 12 of IC7.
3. Connect the scope to the emitter of Q9 and adjust VR8 for 1 ± 0.1 volt p-p. Note waveform in Fig. 5-30.

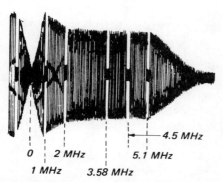

Fig. 5-28. *Playback amplifier frequency characteristic adjustment.*

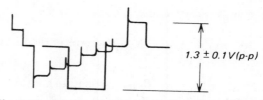

Fig. 5-29. *Playback video output level adjustment.*

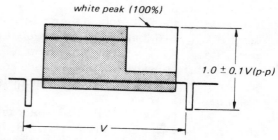

Fig. 5-30. *Playback video output level adjustment.*

44 fH VCO Free Running
Frequency (YC-2 Board)

1. Supply the color bar signal to the video/camera in terminal.
2. Short pins 5 and 6 of IC5 with a jumper.
3. Connect the frequency counter to IC5, pin 11.
4. Adjust VR13 for a counter reading of 692.308 kHz ± 2 kHz.
5. After the adjustment, remove the short connected in step 2 above.

3.57 MHz VXO
Free-Running Frequency (YC-2 Board)

1. Feed the color bar signal into the video/camera in terminal.
2. Connect the frequency counter to the junction of R109 and T14.
3. Adjust the core of T12 to obtain a frequency counter indication of 4,267,919 ± 5 Hz. Utilize the alignment tool for this adjustment.

3.58 MHz Oscillator Frequency (YC-2 Board)

1. Feed a color bar signal into the video/camera in terminal.

2. Connect the frequency counter to the emitter of Q5.
3. Adjust the core of T11 for a counter reading of 3,579,545 Hz ± 5 Hz.

4.27 MHz Carrier Phase Alternate Check (YC-2 Board)

1. Feed a color bar signal into the video/camera-in terminal.
2. Connect the channel 1 probe of the dual-trace scope to IC7, pin 6, and the channel 2 probe to pin 8. Trigger the scope externally from IC7, pin 10. Set the time base to 5 ms/cm and adjust so that the amplitudes of both channels are equal on the scope.
3. Set the scope to add (A + B) mode and confirm that the waveform shown in Fig. 5-31 is obtained.

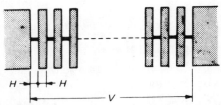

Fig. 5-31. *4.27 MHz carrier phase alternate check waveform.*

4.27 MHz Carrier Leak (YC-2 Board)

1. Play back the color bar portion of the alignment tape.
2. Connect the scope to IC4, pin 8.
3. Adjust VR9 for minimum leak of 4.27 MHz. Refer to proper waveform shown in Fig. 5-32.

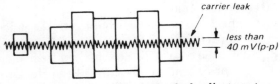

Fig. 5-32. *4.27 mHz carrier leak adjustment.*

Color Killer Threshold Level (YC-2 and ARS Boards)

1. Play back the color bar portion of the alignment tape.

2. Connect the scope, adjusted to 5 ms/cm time base, to IC4 pin 15 on the YC-2 board and trigger it externally from IC4, pin 7.
3. Turn VR006 (channel balance) on the ARS board either clockwise or counterclockwise to obtain 0.03 volt p-p of chroma level in either channel.
4. Turn VR10 on the YC-2 board fully counterclockwise as viewed from the conductor side. The picture displayed on the color TV monitor will be in black and white on the screen because the color killer is working.
5. Turn VR10 on the YC-2 board slowly clockwise and set it at the point where the color picture appears on the TV monitor screen.
6. The channel balance must be reset because the balance in this state is incorrect.
7. Connect the scope to TP5 on the ARS board and play back the RF sweep signal of the alignment tape.
8. Adjust VR5 on the ARS board so that CH-A and CH-B have equal amplitude at the point of 3.58 MHz.

Refer to Fig. 5-33 for adjustments and component locations for the YC-2 board. Refer to Fig. 5-34 for adjustments and component locations for the ARS board. Refer to Fig. 5-35 for adjustments and component locations for the SRP board.

RECORD SYSTEM ALIGNMENT (NOT COVERED IN THIS SECTION)

- White clip adjustment.
- Sync tip carrier frequency adjustment, carrier balance, and dark clip adjustment.
- 3.58 MHz trap.
- FM modulation deviation.
- E-E output level.
- 3.58 MHz crystal filter adjustment.
- Record ACC level.
- Half H shift and card clamp adjustment.
- Y-FM record current adjustment and chroma record current check.

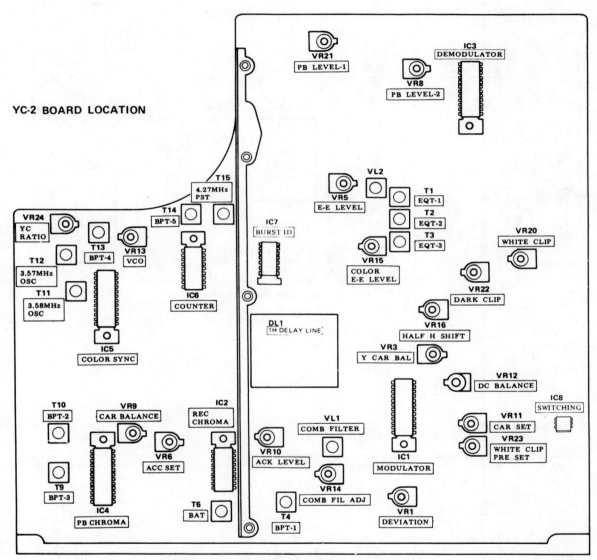

YC-2 BOARD LOCATION

Fig. 5-33. *Component locations on YC-2 board.*

ARS BOARD LOCATION

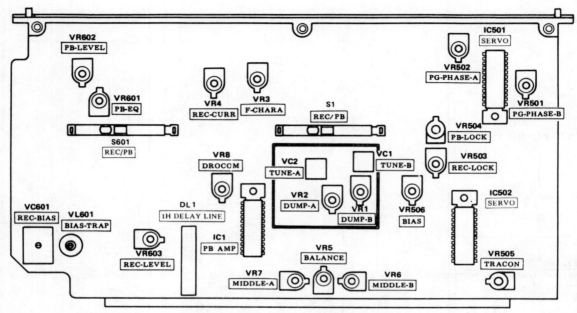

Fig. 5-34. *Component locations on ARS board.*

SRP BOARD LOCATION

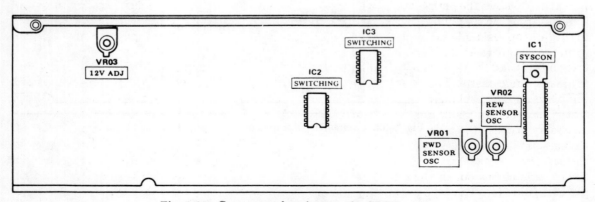

Fig. 5-35. *Component locations on the SRP board.*

6

Mechanical VCR Operation and Drum Head Adjustments

This chapter covers mechanical VCR operation plus drum head change-out and adjustments for the Zenith model KR9000 (Sony SL-8600) machine. All numbers in parentheses are keyed to the specific part in the illustration that is under discussion.

TAPE PATH

Figure 6-1 shows the path taken by the tape as it moves from the supply reel table assembly (1) to the take-up reel table assembly (32). The tape from the supply reel table assembly first passes the guide (2) and the tape slack preventing plate (3) inside the cassette. The tape slack preventing plate applies backing pressure to the tape when slack occurs inside the cassette. This is necessary because the tape will be damaged by the cassette lid (5) if the slack tape comes out of the cassette.

The tape comes out of the cassette through the exit guide (4). The tape passes the sensing head (6) mounted on the tension arm assembly

(7). When all the tape has wound onto the take-up reel in the play mode, a trailer tape, attached to the end of the video tape, appears after the normal tape passes the sensing head. The sensing head detects the trailer tape and the auto-stop is activated.

The tape passes the tension arm assembly (7), which prevents tape slack between the exit guide (4) and the supply tension regulator arm assembly (8). It also acts to keep a constant clearance between the tape and sensing head. There are two (upper and lower) tape guide bosses on the tension arm assembly. These are to regulate the height of the tape wound by the supply reel in the rewind mode.

The tape, having passed the tension arm assembly, passes the supply tension regulator arm assembly. The supply tension regulator arm assembly changes the forward direction of the tape toward the drum (13). At the same time, it senses tape tension in the play operation and serves to control the braking pressure to the supply reel table assembly via the brake band assembly (9). This keeps a uniform holdback

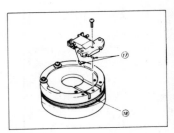

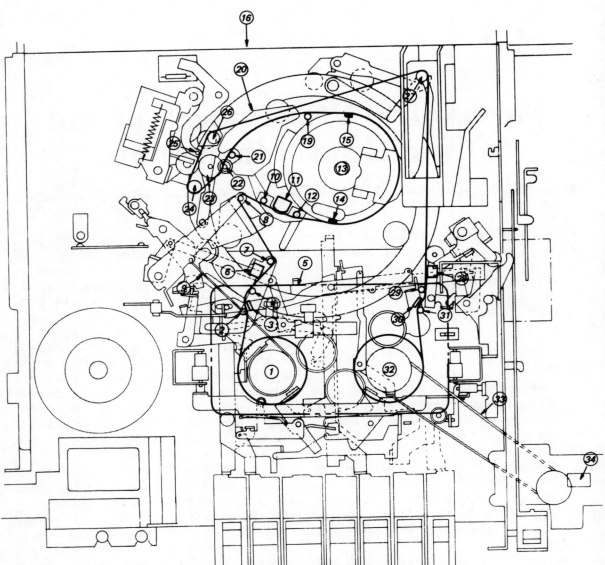

Fig. 6-1. *Illustration of the path taken by the tape threading and drum head assembly.*

tension on the tape. The right angle of the pin on the supply tension regulator arm assembly (8) allows the best tracking around the entrance of the drum. The tape then passes the rotary tape guide A (10), the full-width erase head (11), and the tape guide B (12).

The full width erase head erases the tape during the record mode. The rotary tape guide A and the tape guide B regulate the tape on its upper and lower edges in order to afford optimum tracking from the first contact point of the tape to the drum to the middle point. The tape guide B serves to provide a constant overlap.

The tape runs approximately 180 degrees around the drum head where the rotary video heads (14) and (15) make contact with it during record and playback. The tape path is designed to be parallel with the supply and take-up reel tables at the entrance side of the drum and to be parallel with the loading board (16) at its exit. The tape retaining springs (17) at the middle section of the drum press down the tape running along the circumference of the drum. The tape is pressed by the springs on the top of the drum in order to keep the best tracking at the middle section.

The tape, having passed the drum, passes the tape guide C (19), the ACE (Audio, CTL and Erase) head and the tape guide P. The tape guides C and P regulate the tape position from the top in order to keep the best tracking from the middle section of the drum to the exit point of the tape. The tape guide C serves to keep a constant overlap. The ACE head performs the erasure, record, and playback of the audio signal, and the record and playback of the control signal. The ACE head is an assembly and the position of the core of the head is adjusted with respect to the upper flanges of the tape guide C (19) and the tape guide P (21). It is so designed that no height adjustment of the audio head in the ACE head assembly is required if the tape runs along the upper flanges of the tape guides C and P.

The tape is squeezed between the capstan flywheel assembly and the pinch roller assembly.

The capstan flywheel assembly (22) rotates at a constant speed, advancing the tape at a fixed rate. The angle of the capstan assembly against the tape is very precisely set. The pinch roller (23) is pressed against the tape at an angle of 90 degrees with respect to the forward direction of the tape with automatic alignment in order to make the tape run in a stable manner. The tape passes the guide roller assembly, and the forward direction of the tape turns 180 degrees in order to reverse the running direction of the tape. The guide roller assembly (24) is designed to turn freely so that the friction between the guide roller assembly and the tape is minimized for the least tape wear.

The tape, having passed the guide roller assembly, passes the tape slack sensing lever assembly (25), which senses any tape slack resulting from the reduction of tape tension when in the record or play mode. The slack sensing lever assembly terminates the record or play mode and sets up the stop mode. The tape passes the fixed tape guide. The space between the upper and lower flanges on the fixed tape guide is narrower than the spaces of other flanges because it is designed so that the tape height is optimum in the fast forward (or rewind) mode or at the completion of the threading operation.

In normal operation, the tape runs smoothly at the beginning of the play or record mode. The tape passes the loading arm assembly (27), the forward direction of the tape turns right about 60 degrees, and it is positioned at the same height as that of the take-up reel table assembly. The tape then passes the take-up sensing head (28). When all the tape is wound onto the supply reel (in the rewind mode), the leader tape appears at the head of the video tape. The sensing head senses the leader tape and the rewind mode stops. The tape, having passed the sensing head, enters into the cassette, passes the exit guide, the tape slack prevention plate, the guide, and finally is wound by the take-up reel. As noted above, the tape advances from the supply reel to the take-up reel. Because the take-up reel rotates in the play, record, fast forward, or rewind

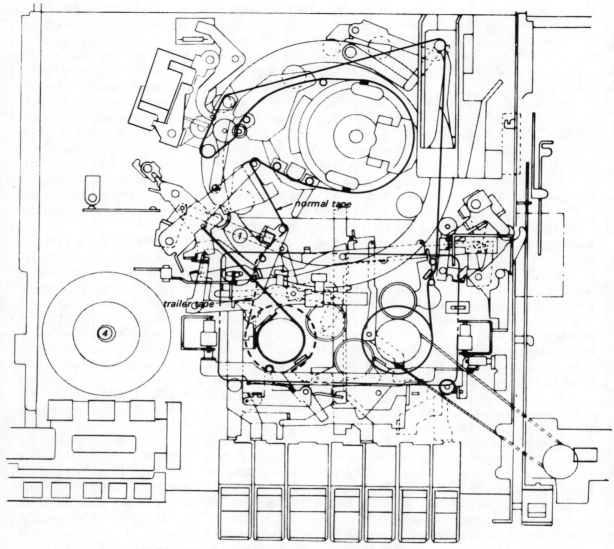

Fig. 6-2. *Tape sensing mechanism and machine shut-off.*

mode, the counter belt is connected to it and also moves to turn the counter.

SAFETY MECHANISM

In the record, play, or fast forward modes, the trailer tape appears at the end of the video tape on the supply reel. The sensing head senses the trailer tape when it passes the sensing head, signalling that all the video tape has been taken up. Refer to Fig. 6-2.

When the sensing head senses the trailer tape, the stop solenoid (2) on the button block assembly is energized and a function button is released. At the same time, the microswitch (3) is pushed by the stop solenoid (2) when the stop solenoid is energized and the microswitch (3) is turned-off, and the ac motor stops. Refer to Fig. 6-3. The stop of the ac motor makes the rotation of the take-up reel stop and the tape movement stops without being wound further. If the play or the fast forward button is depressed in this state, the safety mechanism functions in such a manner that the tape is not wound because the sensing head senses the trailer tape immediately, the stop solenoid (2) is energized, the microswitch (3) turns off, and the ac motor (4) does not rotate. See Figs. 5-2 and 5-3.

TAPE-END SENSING MECHANISM

In the rewind mode, the leader tape appears at the beginning of the normal video tape when all the video tape on the take-up reel is rewound to the supply reel. The sensing head senses the leader tape when it passes the sensing head. Refer to Fig. 6-4.

When the sensing head senses the leader tape, the stop solenoid on the button block assembly is energized, and the rewind button is then released. At the same time, the microswitch (3) is pushed by the stop solenoid (2) when the stop solenoid is energized, the ac motor stops. Refer to Fig. 6-5. The supply reel stops its rotation and the tape is not rewound further. If the rewind button is depressed in this state, the safety mechanism functions in such manner that the tape is not rewound because the sensing head senses the leader tape, the stop solenoid is energized, the microswitch is turned off, and the ac motor stops.

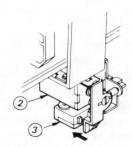

Fig. 6-3. *Microswitch assembly.*

TAPE-SLACK SENSING MECHANISM

The tape is squeezed by the capstan and pinch roller and fed through at a fixed rate. The tape is taken up with a certain specified torque by the take-up reel assembly. If the torque is lost due to breakage of the belt or other causes, the tape is fed by the capstan housing assembly and the pinch roller but won't be wound up by the take-up reel. To prevent the tape from winding around the capstan housing assembly and the pinch roller, the following actions take place: when the take-up torque decreases, the tape slack sensing lever (1) shown in Fig. 6-6 moves in the direction shown by the arrow, thus tripping the reed switch.

When the reed switch (2) acts, the stop solenoid on the function block is energized and the play or record button is released. See Fig. 6-7. The pinch roller is then disengaged from the capstan housing. At the same time, the microswitch (4) is pushed by the stop solenoid to be turned off when the stop solenoid is energized, and the ac motor stops. The capstan housing and the take-up reel stop their rotation, and the tape is not taken up. If the play or the record button is depressed in this state, the safety mechanism functions in such a manner that the tape is not wound because the tape slack sensing lever assembly (1) (Fig. 5-6) senses the slack, a function button is released, and the ac motor and the drum do not rotate.

POWER-OFF SENSING MECHANISM

Rotation of the ac motor stops if the ac power is cut off for any reason in the operating modes or in the threading and dethreading operations. In this case, the reel table must be braked at once. If the reel goes on due to inertia, the tape slackens and will be damaged. For this reason, whenever the ac power is cut off, the safety mechanism functions as follows to prevent the tape from becoming slack and damaged: The

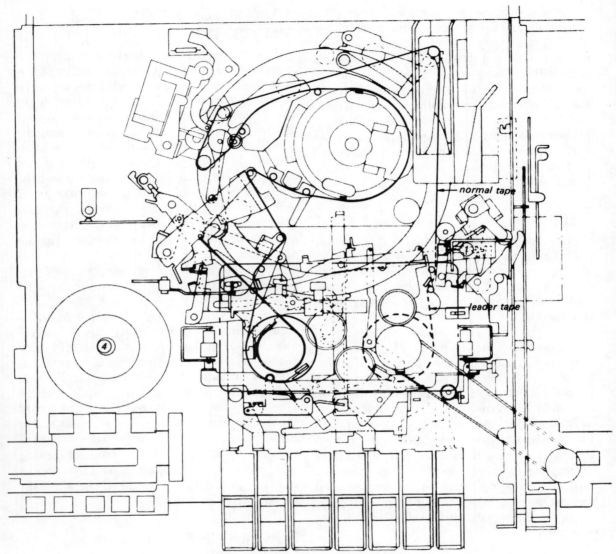

Fig. 6-4. *Tape-end sensing machine diagram — rewind mode.*

stop solenoid on the function block is energized, the function button is released, and the brake assemblies actuate to apply the brakes on the reel tables. This is shown in Figs. 6-8 and 6-9.

FUNCTION MODES

Consider the eject button operation. The operation of the other buttons is basically the same. When the eject button is depressed, section A of the button pushes down the lock and it is locked by it. There is a safety mechanism to prevent other buttons except the stop button from being pushed. When the eject button is pushed, the stopper B assembly and the stopper A assembly move in that direction, and buttons other than the stop button cannot be depressed because they are blocked by the stopper assem-

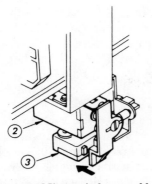

Fig. 6-5. *Microswitch assembly.*

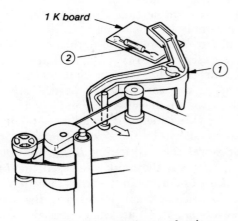

Fig. 6-6. *Slack sensing mechanism.*

1 K board

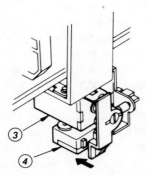

Fig. 6-7. *Stop solenoid assembly.*

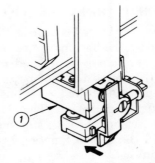

Fig. 6-8. *Drawing of stop solenoid.*

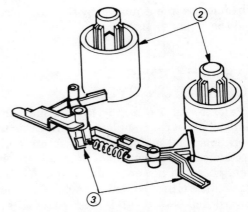

Fig. 6-9. *Brake drum assembly.*

blies. The depressed eject button pushes the microswitch lever, which actuates the microswitch, and the ac motor starts.

OPERATION OF RECORD AND PLAY BUTTONS

When the record button is locked, the stopper B assembly and the stopper A assembly move in the opposite direction so that any button other than the stop button cannot be pressed because the stopper B and A assemblies block the depressed button and the button cannot be depressed. When the play button is depressed and locked, the stopper B assembly and the stopper A assembly moves in the same direction and any button other than the stop and pause buttons cannot be depressed.

OPERATION OF THE STOP BUTTON

When the stop button is depressed, the lock is pushed and all the buttons are released. When the eject operation is completed in the eject mode and the cassette-lift compartment moves upward, section A of the lock is pushed by the cassette-lift compartment. This is the same state as the set-up when the stop button is depressed and the locked function button is released. In any case where the safety mechanism actuates, the stop solenoid turns on, the lock is pulled, and the locked function button is released, just as when the stop button is depressed. The microswitch is pushed by the stop solenoid and the ac motor stops.

CASSETTE INSERTION AND COMPLETION CYCLE

Refer to Figs. 6-10 and 6-11 for the mechanical components at the completion of the cassette-in operation. The motor does not rotate and no mechanical component moves in this machine even when the power is turned on. When the eject button is pushed, the cassette lift (1) moves upward. The cassette (2) is inserted into the cassette lift. Section B of the cassette moves in the direction of the arrow with Section A of the cassette lift. The lock mechanism of the cassette lid is released and the lid can be opened or closed freely. In this condition, if the cassette lift is pushed down, the cassette is held by the cassette guide (4) and it does not come out.

When the cassette lift is pushed down, section C of the cassette contacts the lid opener (3) and moves in the direction of the arrow. Section D of the cassette moves in the direction indicated by the arrow away from the take-up reel and supply reel. The take-up and supply reels rotate. As the cassette-lift compartment goes down further, section F of the take-up and supply reels fits into section E of the reel table assembly. Thus, the power of the reel table can be transmitted to the take-up and supply reels. When the cassette-lift compartment is all the way down, the lock re-lease lever (9), the positioning adjust lever (11), the cassette fastening lever, the adjusting plate, the right cassette retainer assembly (14), the connection link (15) and the left cassette retainer assembly (16) hold the cassette-lift compartment (1) by the force of the lock-lever returning spring.

When the cassette drops into place, the cassette sensor goes down, moving the cassette switch lever and making the switch link free. The force of the spring moves the release switch link to push the microswitch (21), and the ac motor starts to run.

The supply sensor relay lever is pushed down by the cassette, the threading gear base assembly moves in the arrow direction and the gear A contacts the tire section of the intermediate pulley assembly (26), starting rotation. The rotational force is transmitted to the entire threading ring assembly (31) via gear D, the gear F, and the gear G. With the rotation of the entire threading ring assembly, the positioning lever (10) moves in the direction indicated by the arrow, moving the position adjust lever in the direction shown by the arrow. The positioning lever moves the positioning limiter (32) in the arrow direction, the crank is released, the spring moves the brake link in the arrow direction. The take-up brake assembly (36) and the supply brake assembly (37) are then released from the take-up and the supply reel table assemblies.

Because the entire threading ring assembly turns in the direction shown by the arrow, the guide roller assembly (40) moves in the same direction with that of the threading ring assembly, and the tape in the cassette is withdrawn.

The threading arm assembly (41) and the arm lock assembly (42) move in the direction indicated by the arrow with the turning of the entire threading ring assembly. When the entire threading ring assembly reaches the position just before the completion of threading, the loading brake lever assembly moves in the direction shown by the arrow by the entire threading ring assembly because of the cam shape of the ring and the end brake link (44) is moved in the

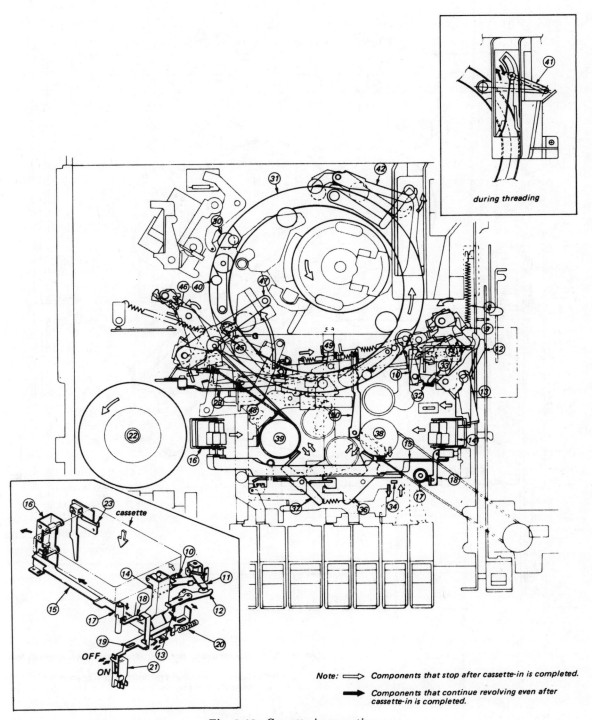

during threading

Note: ⟹ Components that stop after cassette-in is completed.

➡ Components that continue revolving even after cassette-in is completed.

Fig. 6-10. *Cassette-in operation.*

145

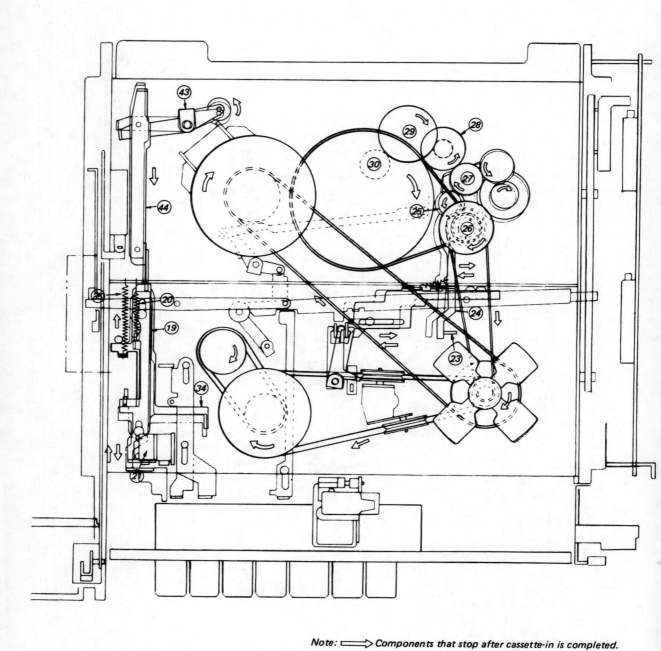

Note: ⟹ Components that stop after cassette-in is completed.

➡ Components that continue revolving even after cassette-in is completed.

Fig. 6-11. *Diagram of complete threading operation.*

arrow direction. The switch link is moved in the arrow direction, turning off the microswitch. At the same time, the switch link (19) moves the brake link (34), and the take-up brake assembly (36) and the supply brake assembly (37) apply the brakes on the take-up reel table and the supply reel table.

When the entire threading ring assembly stops, the positioning roller (45) drops into section G of the positioning lever (46), and the supply tension regulator arm assembly (47) and the brake band assembly (48) move in the direction shown by the arrows.

Because the eject stopper assembly (49) moves in the direction indicated by the arrows when the positioning lever moves in the direction shown by the arrow, the mechanical lock is released so that the fast forward, play, or rewind button can be depressed. As the eject stopper assembly (49) moves, the loading brake assembly (50) moves in the direction shown by the arrow and the take-up reel table assembly brake is released.

PLAY MODE

Figures 6-12 and 6-13 show the position of the mechanical parts at the completion of the play mode. The play button, when depressed, pushes the microswitch on the SRP board in the direction shown by the arrow and the switch turns on. The play button pushes the play link and section A of the play link moves the forward link in the direction indicated by the arrow. Section B of the forward link releases the supply brake assembly (5) and the take-up brake assembly (6), and the supply reel table and the take-up reel table can then rotate.

The microswitch (10) is pushed by the switch lever (9) of the function block when the play button is depressed and turns on. As the ac motor starts rotation in the arrow direction, the capstan pulley (11), motor pulley assembly (12), fast forward idler belt, intermediate pulley assembly (14), fast forward assembly (15), fast forward idler assembly, forward belt, forward as-

sembly, relay belt, relay pulley assembly (20), capstan belt, capstan flywheel assembly (22), drum belt, drum pulley, drum flywheel, and the video head disc assembly (26) rotate, each in the directions indicated by the arrows.

When the play button is depressed, the play link is pushed in the direction shown by the arrow, the forward link and the ARS link assembly (27) are pushed in the directions shown by the arrows, and the sub-lever assembly (27) moves in the arrow direction. As the ARS link assembly moves, the ARS slide plate assembly (29) moves in the direction indicated by the arrow, switching the slide switch to the PB position.

When the microswitch turns on, the pause solenoid operates and the pressure spring (33) is pulled. As the forward link moves in the arrow direction, the C section of the forward link moves the pinch roller pressure link assembly (31) and the pause solenoid assembly block (34) in the direction shown by the arrow, making the pinch roller assembly (35) contact the capstan flywheel assembly (22). The pinch roller assembly (35) starts rotation and the tape is squeezed between the pinch roller and the capstan flywheel assembly so that it is fed forward at a constant speed.

Because the pause solenoid assembly block moves in the arrow direction, the tape slack detecting lever assembly (37) contacts the tape by the spring (38). Thus the assembly is ready to detect tape slack.

When the forward link (4) is pushed, section F of the forward link moves the play link assembly (39), tension regulator release link assembly (4) and the supply tension regulator arm assembly (42) each in the direction shown by the arrow. The supply tension regulator arm assembly detects tape tension, making the brake band assembly (43) move in the direction indicated by the arrow so as to apply the brake on the supply reel table assembly (7) for a constant tape tension.

Because section E of the play link assembly (39) moves the idler slide plate (44) in the arrow

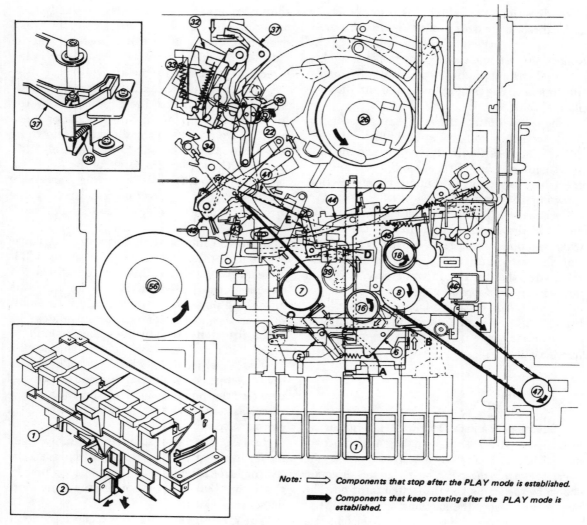

Note: ⟹ Components that stop after the PLAY mode is established.

➡ Components that keep rotating after the PLAY mode is established.

Fig. 6-12. *The play mode.*

direction when the play link assembly moves in the direction shown by the arrow, the forward assembly is moved by the spring and contacts the take-up reel table assembly. The take-up reel table assembly starts to rotate in the arrow direction and the tape is taken up.

With the rotation of the take-up reel table, the tape counter is turned by the counter belt in the direction shown by the arrow and the count is indicated. Because the ARS link assembly moves when the play button is depressed even if the lever knob of the picture source selector is in the TV position, the antenna auto lever moves in the direction indicated by the arrow, the antenna lock arm and the antenna slide plate are moved by the spring in the direction shown by the arrow, and the antenna slide plate moves in the direction shown by the arrow so as to switch the antenna select switch to the tape position. Thus, the antenna select switch is automatically connected to the tape position when the play button is depressed.

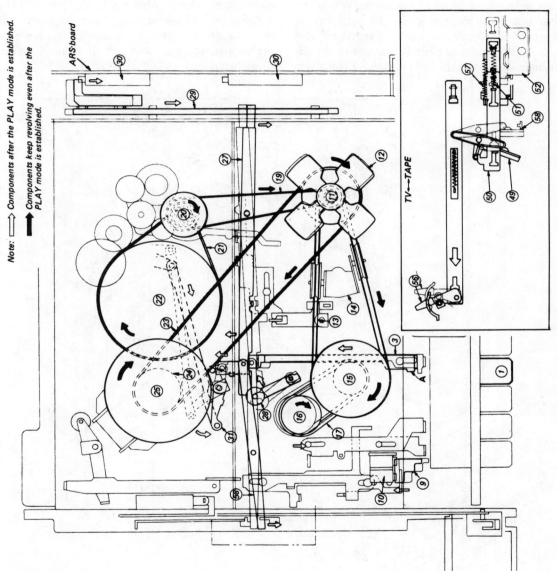

ARS·board

TV←→TAPE

Fig. 6-13. *Position of parts in the play mode.*

149

PAUSE MODE

Figure 6-14 illustrates the operation of the mechanical parts when the pause mode is set up with the machine in the play mode. When the pause button is depressed in the play or the record mode, the pause solenoid turns off and the pressure arm assembly (3) moves in the direction indicated by the arrow, making the pinch roller assembly (4) disengage the capstan flywheel assembly. At the same time, the pause brake rubber (6) presses the tape on the pinch roller assembly (4) and the tape movement stops. Note: the play assembly (7) contacts the take-up reel table assembly (8) in this state, but the take-up reel table assembly (8) is not rotating because the tape does not move and the friction clutch acts.

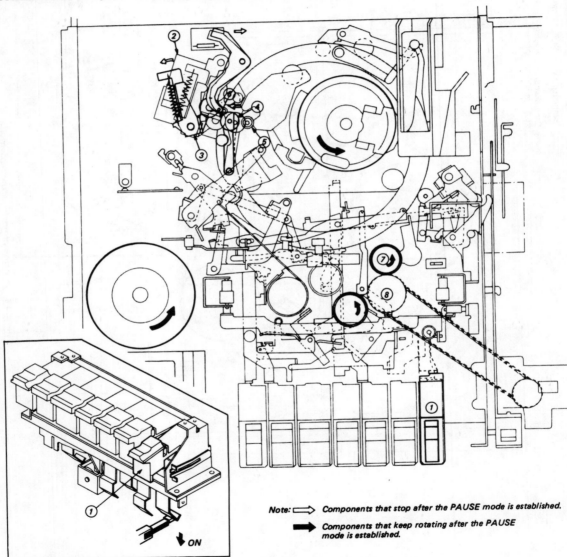

Note: ⟹ Components that stop after the PAUSE mode is established.

⟹ Components that keep rotating after the PAUSE mode is established.

Fig. 6-14. *Mechanical parts placement in the pause mode.*

RECORD MODE

Figure 6-15 shows the operation of the mechanical parts when the record button is depressed. The record button cannot be depressed if the record safety tab in the cassette is broken out, because the cassette cannot push down the probe (3) and section A of the record slide plate (4) contacts the probe. Thus the record button cannot be depressed. If the record safety tab is not removed, the cassette pushes the probe (3), the record slide plate (4) can move, and the record button can be depressed.

When the record button is depressed, the

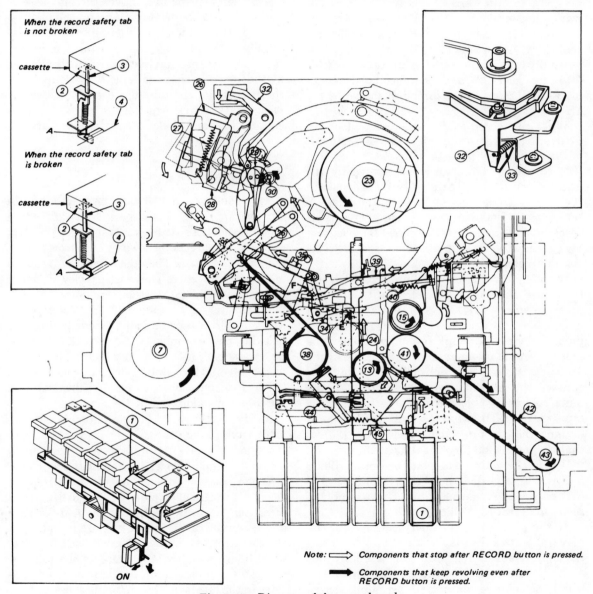

Note: ⟹ *Components that stop after RECORD button is pressed.*

➡ *Components that keep revolving even after RECORD button is pressed.*

Fig. 6-15. *Diagram of the record mode.*

switch lever (5) of the function block moves in the direction shown by the arrow, turning on the microswitch (6). The ac motor then starts its rotation.

As the ac motor starts to rotate, the capstan pulley (8), motor pulley assembly (9), fast-forward idler belt (10), intermediate pulley assembly (11), fast-forward assembly (12), fast-forward idler assembly (13), play belt (14), play assembly (15), relay belt (16), relay pulley assembly (17), capstan belt (18), capstan flywheel assembly (19), drum belt (20), drum pulley (21), drum flywheel (22), and video head disc assembly (23) all move in the direction shown by the arrows.

When the record button is depressed (see Fig. 6-16) and the record slide plate moves, section B of the record slide plate (4) pushes the play link (24) to move in the direction shown by the arrow. Section C of the play link (24) releases the supply brake assembly (44) and take-up brake assembly (45) so that the supply reel table assembly (38) and the take-up reel table assembly (38) and the take-up reel table assembly (41) can now rotate.

With the turning-on of the microswitch (6), the pause solenoid (26) energizes to pull the pressure spring (27). When the play link (24) moves, section F of the play link moves the pinch roller pressure link assembly (25) and the pause solenoid assembly block (28) each in the arrow direction. The pinch roller assembly (29) contacts the capstan flywheel assembly (30), the pinch roller assembly starts its rotation, and the tape is squeezed between the pinch roller assembly and the capstan flywheel assembly (30) so that the tape is fed forward at a constant speed.

When the pause solenoid assembly block moves, the tape slack detection lever assembly (32) moves by its spring. Thus, the assembly is ready to detect tape slack.

As the play link is pushed, section E of the play link moves the play link assembly (35) and the supply tension regulator arm assembly (36) each in the direction shown by the arrow. The supply tension regulator arm assembly senses

tape tension and makes brake band assembly (37) move in the direction shown by the arrow so as to apply braking on the supply reel table assembly (38) to keep the tape tension constant.

As the play link assembly moves, section F of the play link assembly (34) moves the idler slide plate (39) in the direction shown by the arrow, the pressure spring (40) moves the play assembly (15) in the arrow direction and the play assembly (15) contacts the take-up reel table assembly (41). The take-up reel table assembly (41) starts its rotation in the direction indicated by the arrow so that the tape is taken up. With the rotation of the take-up reel table assembly (41), the counter belt (42) rotates the tape counter (43) and the count is then indicated.

FAST-FORWARD MODE

Figures 6-17 and 6-18 show the operation of the fast-forward mode. When the fast-forward button is depressed, the switch lever of the function block moves in the direction shown by the arrow to turn the microswitch on. Then the ac motor (4) rotates in the arrow direction. The capstan pulley (5), motor pulley assembly (6), fast-forward idler belt (7), intermediate pulley assembly (8), fast-forward assembly (9), fast-forward idler assembly (10), play belt (11), play assembly (12), relay belt (13), relay pulley assembly (14), capstan belt (15), capstan flywheel assembly (16), drum belt (17), drum flywheel (18) and video head disc assembly (19) all move in the directions indicated by the arrows.

As the fast-forward button is depressed, the leaf switch (20) on the SRP board turns on and the sensing head (21) initiates its operation to set up the stop mode when sensing the tape end.

When the fast-forward button is depressed, the fast-forward slide plate (22) is pushed in the direction shown by the arrow, the fast-forward arm assembly (23) is moved by the spring (24) in the direction shown by the arrow, the fast-forward idler assembly (10) contacts the take-up reel table assembly (25), and this assembly rotates in the direction shown by the arrow to take

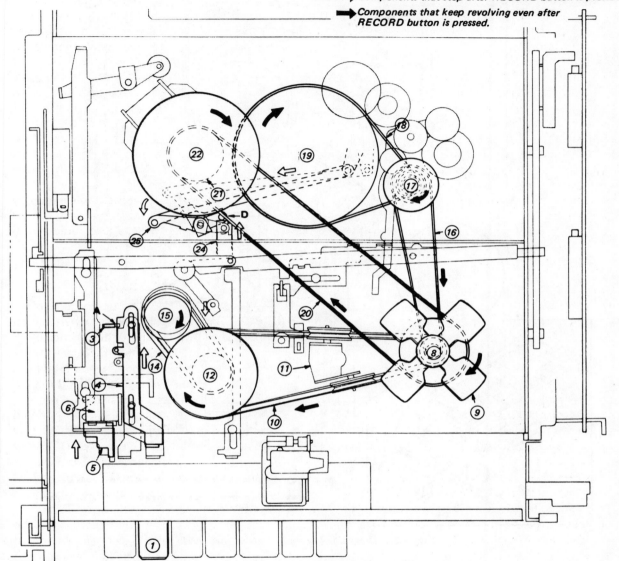

Fig. 6-16. *Mechanical parts placement in the record mode.*

up the tape. With the rotation of the take-up reel table assembly, its rotational force is then transmitted to the counter belt.

As the fast-forward slide plate is pushed, section A of the plate moves the RF slide plate (28) in the direction shown by the arrow, section B of the RF slide plate moves the supply brake as-

sembly (29) and the take-up brake assembly (30) in the direction shown by the arrows, and the brake assemblies disengage from the supply reel and the take-up reel tables. Now the reel tables can rotate freely.

As the RF slide plate moves, section D of the RF slide plate moves the RF link (33) and the

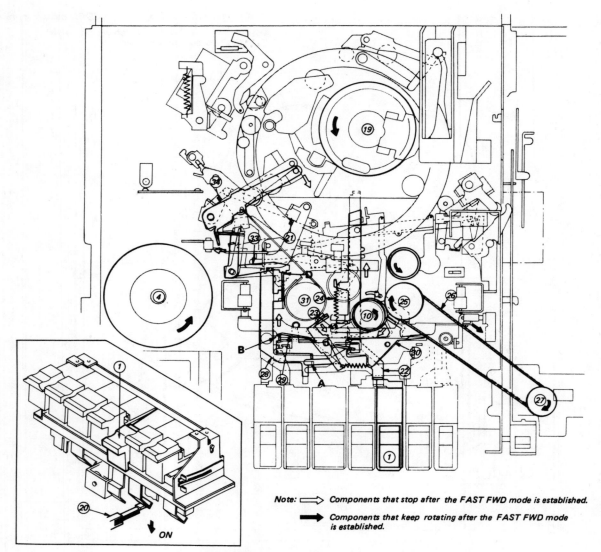

Note: ⟹ Components that stop after the FAST FWD mode is established.

➡ Components that keep rotating after the FAST FWD mode is established.

Fig. 6-17. *Diagram of parts in fast-forward mode.*

supply tension arm assembly (34) in the direction shown by the arrows for decreasing the friction caused by the tape running.

REWIND MODE

Figures 6-19 and 6-10 shows operation of the mechanical components at the completion of the rewind mode. When the rewind button is depressed, the switch lever of the function block moves in the direction indicated by the arrow to turn on the microswitch. The ac motor will then rotate in the direction shown by the arrow. The capstan pulley (5), motor pulley assembly (6), fast forward idler belt, intermediate pulley assembly (8), fast forward assembly (9), fast-forward idler assembly (10), play belt (11), play assembly (12), relay belt, relay pulley assembly, capstan belt (15), capstan flywheel assembly (16), drum belt (17), drum flywheel (18) and

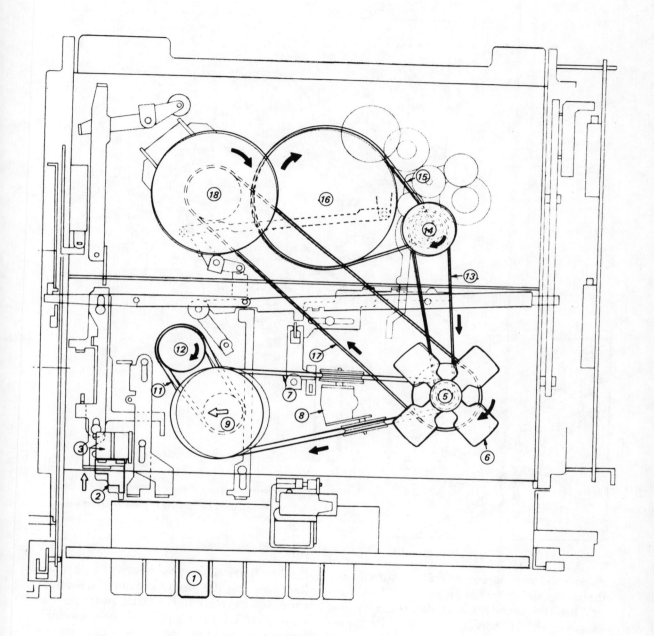

Note: ⟹ Components that stop after FAST FWD mode is established.

➡ Components that keep revolving even after the FAST FWD mode is established.

Fig. 6-18. *Machine at completion of fast-forward mode.*

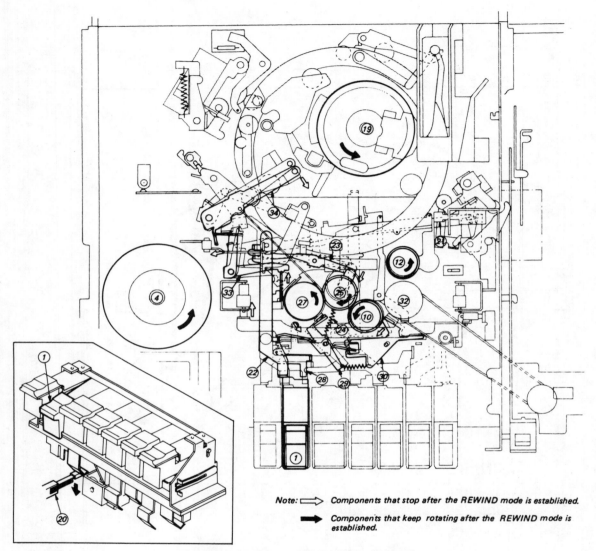

Note: ▷ *Components that stop after the REWIND mode is established.*

▶ *Components that keep rotating after the REWIND mode is established.*

Fig. 6-19. *Machine operation in rewind mode.*

video head disc assembly (19) all turn in the directions shown by the arrows.

When the rewind button is depressed, the leaf switch on the SRP board turns on and the sensing head (21) actuates to set up the stop mode when sensing the tape end.

The rewind slide plate (22) is pushed in the direction indicated by the arrow when the rewind button is depressed. The rewind link assembly (23) moves by the force of the spring in the direction indicated by the arrow. The rewind idler assembly (25) on the rewind link assembly transmits the force to the fast forward idler assembly (10) to the supply reel table assembly. The supply reel table assembly rotates and the tape is rewound.

The RF slide plate is also pushed in the same direction as the rewind slide plate (22) moves. The supply brake assembly (29) and take-up brake assembly (30) disengage from the supply

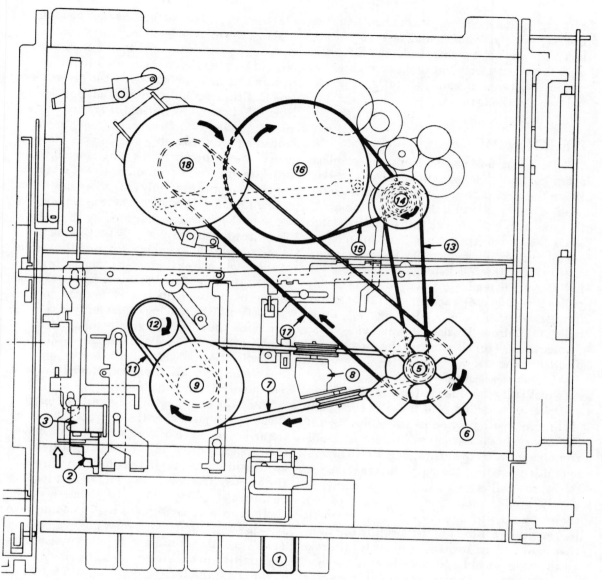

Fig. 6-20. *Parts location at completion of rewind mode.*

reel table (27) and the take-up reel table assembly, allowing the reel table assemblies to turn freely. As the RF slide plate moves, the RF link (33) and the supply tension arm assembly (34) move in the directions shown by the arrows so as to decrease the friction caused by the tape running.

STOP MODE

If the stop button is depressed in any mode, all the links are released and return to the same state as that which occurs when the power is turned on and the cassette is inserted. Because the ac drive motor and the capstan motor are not rotating in this mode, any components such as the reel table assemblies and capstan housing assembly are not moving. Only the ST brake assembly operates, and the brake band assembly and loading brake assembly are off. All the solenoids are in the OFF state.

EJECTION MODE

Refer to Figs. 6-21 and 6-22 for the mechanical operation at the completion of the eject mode. When the eject button is depressed, the switch lever of the function block moves in the direction shown by the arrow to turn on microswitch (3). The ac motor then starts. The capstan pulley (5), motor pulley assembly (6), fast-forward idler belt (7), intermediate pulley assembly (8), fast-forward assembly (9), relay belt (13), relay pulley assembly (14), capstan belt (15), capstan flywheel assembly (16), drum belt (17), drum pulley (18), drum flywheel (19) and video head disc assembly (20) all rotate in the direction indicated by the arrows.

When the eject button is depressed, the eject slide plate (21) is pushed in the direction shown by the arrow. Section A of the eject slide plate (21) moves the supply brake assembly (22) and the take-up brake assembly (23) each in the direction shown by the arrow. The brake assemblies disengage from the supply and take-up reel tables, thus allowing them to rotate freely. As the eject slide plate moves in the direction shown by the arrow, the eject brake assembly (65) moves by the force of the spring in the direction shown by the arrow so as to apply the brake on the supply reel table. Section B of the eject slide plate makes the eject link turn in the arrow direction. Section C of the eject link moves the eject plate B (27) in the direction indicated by the arrow. Section D of the eject slide plate B moves the idler slide plate (28), spring (29) and play assembly (12) each in the direction shown by the arrow. The play assembly contacts the take-up reel assembly, the take-up reel table assembly (25) rotates to take-up the tape. With the rotation of the take-up reel, the counter belt

(31) then moves to operate the tape counter.

When the eject slide plate B moves in the direction shown by the arrow, the lock release lever (33) at the lock release sub-lever (34) each turn in the direction shown by the arrow. Section E of the eject link turns the left positioning lever (35) in the direction of the arrow. The positioning roller (36) disengages from the entire threading ring assembly (37) and a state in which the entire threading ring assembly can turn is set up. As the positioning lever (35) turns, the eject stopper assembly (38), threading brake spring (39), and threading brake assembly (40) move in the direction indicated by the arrows. The threading brake assembly (40) contacts the take-up reel table assembly. Because the torque of the take-up reel is greater than the braking force of the threading brake assembly (40), the torque of the take-up reel table does not decrease. With the moving of the left positioning lever (35), the supply tension regulator arm assembly (41) and brake band assembly (42) move in the direction indicated by the arrows, and the brake band assembly disengages from the supply reel table, releasing the brake.

When the eject button is depressed, the eject slide plate is pushed in the direction shown by the arrow, section F of the eject slide plate moves the threading drive link (43) and the threading switch side plate (44) in the direction shown by the arrows. The spring (45) pulls the threading gear base assembly (46) in the direction shown by the arrow so as to press the threading gear base assembly (46) on the relay pulley assembly (14). The rotation of the relay pulley assembly is transmitted to the entire threading ring assembly (37) via the gear B, gear C, gear D, gear E, gear F, gear G, and the entire threading ring assembly turns in the direction indicated by the arrow.

As the entire threading ring assembly rotates, the arm clock assembly (55) is pushed in the direction shown by the arrow because of the cam shape of the entire threading ring assembly (37) so as to release the hold state of the threading arm assembly (56). The threading arm assembly

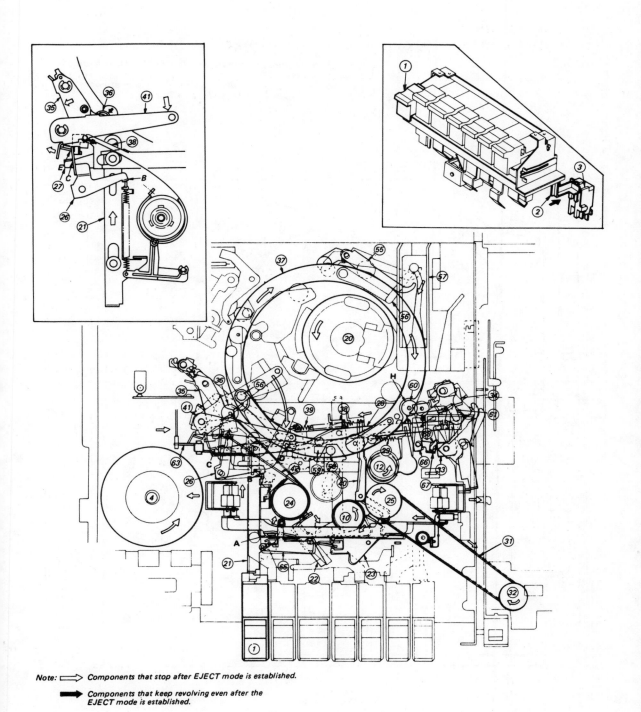

Note: ▷ *Components that stop after EJECT mode is established.*

▶ *Components that keep revolving even after the EJECT mode is established.*

Fig. 6-21. *Machine operation in the eject mode.*

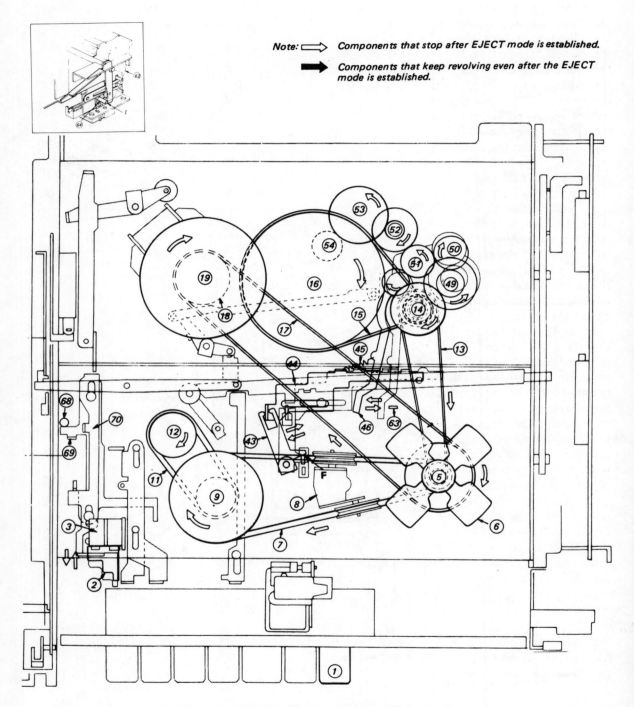

Note: ⇨ Components that stop after EJECT mode is established.

➡ Components that keep revolving even after the EJECT mode is established.

Fig. 6-22. *Parts location at completion of eject mode.*

moves along the threading arm guide (57) in the direction indicated by the arrow. The positioning roller (36) and the left positioning lever (35) are pushed further from the positions to which they were pushed by section F of the eject link (26) when the eject button was depressed. With the moving of the positioning lever (35), the supply tension regulator arm assembly (41) and the eject stopper assembly (38) are pushed further in the directions shown by the arrows from the position to which they were moved when the eject button was depressed.

When the entire threading ring assembly turns and reaches the position of unthreading completion, the guide roller assembly (58) contacts the ring stopper (59), the positioning roller (60) drops into the recess "H," and the limiter assembly D (61) turns in the direction shown by the arrow. The cassette lift assembly (62) is unlatched and moves upward so that the cassette can be taken out.

When the positioning roller (60) drops into the recess "H," the positioning lever (66) moves, the positioning limiter (67) moves in the arrow direction, and the brake link (69) is moved by the crank (68) in the arrow direction. With the popping-up of the cassette lift assembly and the movement of the brake link, the switch link (70) moves in the direction shown by the arrow.

When the cassette-lift assembly moves upward, the supply sensor relay lever goes upward so that the threading gear assembly (46) disengages the relay pulley assembly (14). When the cassette lift assembly goes up, section I of the cassette-lift assembly moves in the direction indicated by the arrow to push the lock plate (64), releasing the eject button. The switch lever (2) moves in the direction shown by the arrow to turn off the microswitch (3). The ac motor stops rotation and the eject operation is completed.

MECHANICAL ADJUSTMENT PROCEDURES

The following adjustments include the video head disc eccentricity, the video head dihedral, the tape path, the RF envelope/tracking, audio azimuth, and CTL head position adjustments.

VIDEO HEAD DISC ECCENTRICITY ADJUSTMENT

1. Remove the threading arm guide (1).
2. Install eccentricity gauge with the thumbscrew (II) so that its probe contacts the head disc circumference about 2 mm below the top edge of the video head disc assembly (3). Refer to Fig. 6-23.
3. Rotate the ac motor slowly counterclockwise. Adjust the video head disc assembly (3) eccentricity so that the gauge reading deflection is within 3 microns by very gently tapping the inner circle edge of the video head disc assembly (3) with the blade of a screwdriver. Refer to Fig. 6-24.

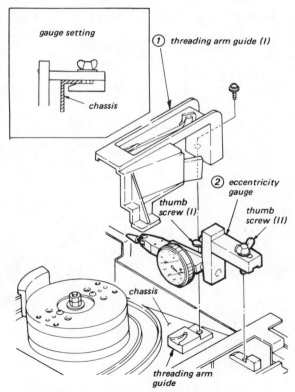

Fig. 6-23. *Drum head eccentricity adjustments.*

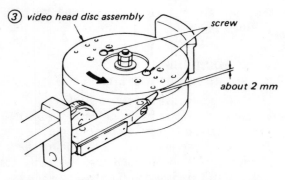

Fig. 6-24. *Drum eccentricity measurements.*

4. When the eccentricity is less than 3 microns, finger tighten the two mounting screws alternately and finally tighten them fully with a screwdriver. (Tightening torque must be more than 10 kg/cm).

5. Make a final test of eccentricity using the gauge after the screws are fully tightened. Solder the leads of the video head to the solder terminal (3) as shown in Fig. 6-25.

Note: Bend the video head leads flat on the top surface of the video head disc assembly (4) so that the slack of the leads does not touch the upper drum assembly (1). Set the upper drum assembly (1) to the drum support. Tighten the mounting screws with an Allen wrench while holding the height-determining screws with your fingers. Refer to Fig. 6-25.

VIDEO HEAD DIHEDRAL ADJUSTMENT

Note: This adjustment might not be required (Fig. 6-25).

1. Play back the monoscope signal segment of the alignment tape and check for dihedral error at the top of the picture (or at the bottom, at the switching point.)

2. If dehedral error is observed, install four tapered screws (5) in each of the four holes on video head disc assembly (4) through the slots in the upper drum assembly (1).

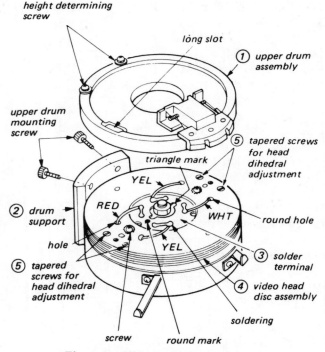

Fig. 5-25. *Video head disc assembly.*

3. Turn one of the four screws until the screw touches the head base. Give this screw one additional quarter turn.
4. Play back the monoscope signal of the alignment tape and check again for dihedral distortion.
5. If the symptom is worse, unscrew the screw identified in the step above, and turn down the screw on the opposite side of the same video head. Repeat, in quarter turn steps, to eliminate dihedral error. Remove all 4 screws after check.
6. Perform electrical adjustment.

TAPE PATH ADJUSTMENT

Note: This adjustment requires high skill. Practice it fully before performing the actual adjustment. The unit is in play mode.

Tape and Fixtures Required for Tape Path Adjustment

- Alignment tape (Sony part number KR5-1D).
- Eccentric screwdriver (for CTL head position adjustment).
- Inspection mirror (for tape running check).
- Tension regulator bending fixture.

Guides and Screws to be Adjusted for Tape Path Adjustment

- Entrance side tape guide nuts (1) and (2).
- Exit side tape guide nuts (3) and (4).
- Azimuth adjusting screw (5).
- CTL head position adjusting screws (6).

Tape Path Adjustment

Adjustment sequence (see Fig. 6-26).

1. Coarse adjustment.
2. RF envelope/tracking adjustment.
3. Audio azimuth adjustment.
4. CTL Head position adjustment.

Coarse Adjustment

Insert a cassette. Press the play button to run the tape in the play mode. Adjust the screws (1),

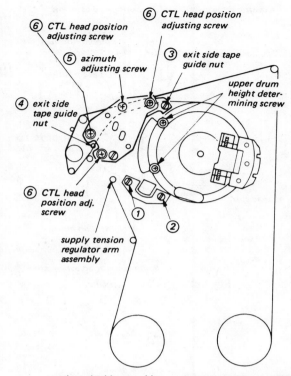

supply real table assembly   take-up reel table assembly

Fig. 6-26. *Tape path adjustment.*

(2), (3), and (4) so that the tape runs smoothly and any tape curl is removed while observing tape at the entrance tape guide flange and exit tape guide flange. Any fold-out or curl of the tape should be removed at the capstan and at the entrance and exit of the revolving head disc assembly. The tape should run so that the top edge of the tape maintains stable contact along the upper flange of the tape guides, and any tape curl and wrinkles should be removed. Use an inspection mirror to observe tape-edge-to-flange guide contact during this adjustment. Refer to Figs. 6-27 and 6-28

RF ENVELOPE/TRACKING ADJUSTMENT

Connect the scope to TP 5 on the ARS board. Trigger the scope from TP 501. Play back the monoscope segment of the alignment tape. Refer to Fig. 6-29. While observing the RF enve-

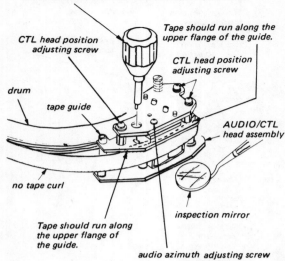

Fig. 6-27. *Audio/control head.*

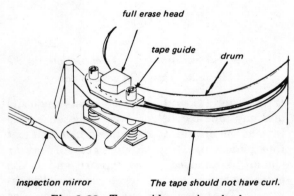

Fig. 6-28. *Tape guide running check.*

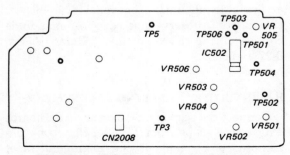

Fig. 6-29. *ARS board test point locations.*

lope on the scope, adjust the tracking control for best results. Connect an ac VTVM to the audio out jack in order to measure the audio signal level fluctuations, as shown in Fig. 6-30.

Adjust tape guide nuts (1), (2), (3), and (4) according to the following adjustment procedures so that the rf envelope displayed on the scope screen meets the specification required. The audio signal level fluctuation should be within 2 dB. See Figs. 5-27 and 6-30.

Entrance Side Adjustment

For the entrance side adjustment, play back the monoscope segment of the alignment tape. Make an attempt of screwing either in or out, slightly, the entrance side tape guide nuts, as shown in Fig. 6-26 as (1) and (2), to optimize the center and right half portion of the RF waveshape.

It is allowable that the tape edge contacts either the upper or lower flange of the entrance side tape guide, but it should only contact the upper flange of the entrance side tape guide nut.

The supply tension regulator arm may be bent slightly if necessary to remove a curl at the entrance guide, but only after *all* adjustments have been made for an optimum flat envelope. It is most common that the RF envelope amplitude at the point shown at the right side of Fig. 6-31, the entrance, keeps fluctuating a little.

Exit Side Adjustment

Play back the monoscope segment of the alignment tape. Make an attempt of either screwing in or out the exit side tape guide nuts in order to obtain the most flat envelope at the left

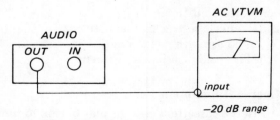

Fig. 6-30. *Audio output level check.*

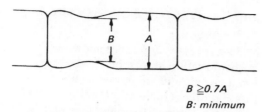

B ≥ 0.7A

B: minimum

Fig. 6-31. *RF envelope waveform.*

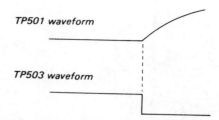

Fig. 6-32. *Dual-trace waveform shift adjustment.*

half of the waveshape. Adjust first the exit side tape guide nuts so that the upper edge of the tape runs along the upper flange of the audio/control head assembly. This adjustment should be done while observing the tape-to-guide contact using an inspection mirror.

Screw in the exit side tape guide nuts gradually so that the envelope of the waveform becomes flat while tape curl is kept at a minimum. Check that audio level variation is within 2 dB. If level variation is great, check for curls on the tape, again using the inspection mirror.

AUDIO AZIMUTH ADJUSTMENT

Adjust the audio azimuth adjustment screw for maximum audio level. Refer to Fig. 6-27. Note: the height of the audio head can be set by running the tape along the flanges of the exit-side tape guide nuts. It is not necessary to adjust audio head height independently.

CTL HEAD POSITION ADJUSTMENT

Use the following procedure to change a shift state to a no-shift state.

1. Connect the channel-1 probe of a dual-trace scope to TP501 and the channel-2 probe to TP503. Set the scope to chop and trigger it externally from TP503.
2. Adjust VR505 on the ARS board so that the phase of the positive-going portion of channel-1 waveform matches that of the negative-going portion of the channel-2 waveform. Note these two waveforms in Fig. 6-32. Set the tracking control knob to its center detent position.

3. Play back the monoscope signal of the alignment tape.
4. Loosen the three CTL adjusting screws and adjust the position of the CTL head with an eccentric screwdriver for maximum rf output at TP5 on the ARS board.
5. Check to see that the audio/control head is positioned in the approximate center of the cut-out hole of the guide base plate.
6. Tighten the three CTL adjusting screws.

DEGAUSSING VIDEO HEADS AND OTHER PARTS

If the video head is magnetized, the S/N ratio deteriorates and slant beat and noise appear on the picture. The video head and other parts must then be demagnetized. To degauss, bring the tip of the demagnetizer as close as possible to the head tip without actually contacting the head tip. Withdraw the demagnetizer very slowly and turn off power of the demagnetizer when it is at least 7 feet away from the VCR deck (refer to Fig. 6-33).

VIDEO AND AUDIO HEAD REPLACEMENT

The following are procedures for removing and replacing the heads.

VIDEO DRUM HEAD DISC REPLACEMENT

1. Remove the upper drum mounting screws shown in Fig. 6-25 with an Allen wrench and

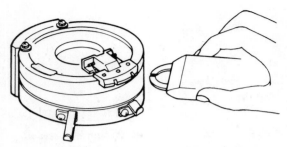

Fig. 6-33. *Degaussing the video heads.*

1. Move harness clamp (3) as shown in Fig. 6-34 in the direction shown by the arrow.
2. Remove exit end adjustment nut (2).
3. Lift audio/control head assembly (1) and unsolder the leads of the head (see Fig. 6-34).
4. Replace audio/control head assembly (1) and solder the lead connections.
5. Perform the tape path (play) adjustment.

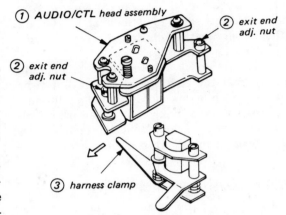

Fig. 6-34. *Audio/control head assembly.*

remove the upper drum assembly (1) from the drum support (2).

2. Remove the four wires of the video head disc from the solder terminal (3).
3. Remove the two screws and video head disc assembly (4).
4. Clean the bottom and flange surfaces of the replacement video head disc assembly with a piece of alcohol dampened soft cloth.
5. Place the replacement video head disc assembly (4) so that the red lead of the disc assembly is close to the small black mark on the solder terminal (3) and secure the disc assembly temporarily with the two screws.

7

VCR Tuner and IF Circuits

This chapter covers the tuner and IF circuit board used in the Zenith/Sony KR9000 VCR machine. This includes tuner and IF circuit board alignment data. Then there is a brief look at this machine's system control circuits. The chapter concludes with information about the built-in clock timer and its circuit description.

RF TUNER

The VHF tuner has a compact turret switching mechanism. Its electrical performance is equal or superior to that of a large-scale VHF tuner. Dual-gate MOSFETs with a high transmission characteristic are used in the RF amplifier stage as the transistor amplifier. This MOSFET enables the tuner to provide excellent characteristics for cross modulation, noise figure, and automatic gain control.

An IC (CX-097A) is utilized in the mixer stage and local oscillator circuits for improving operational reliability by reduction of the parts re-

quired. The mixer stage has been designed to give a good channel 6 chroma beat characteristic. This is important in the reception of the channel 6 station signal. A constant impedance in the output circuitry eliminates the need of adjustment for optimum coupling to the IF circuitry stages.

IF CIRCUIT BOARD

The developed output signals of the IF board circuitry are the video and audio signals to be recorded by the VCR tape machine. The TV station signal coupled into the tuner terminals of the VCR are fed to each tuner where the selected signal is converted to the IF signal with 45.75 MHz for the video carrier and 41.25 MHz for the audio carrier. The IF signal is connected via the input filter group to a trap for the required attenuations of unwanted signals.

The video IF response of the input stage is adjusted in the circuit consisting of T506, T507, and CV501 trimmers. The signals are amplified

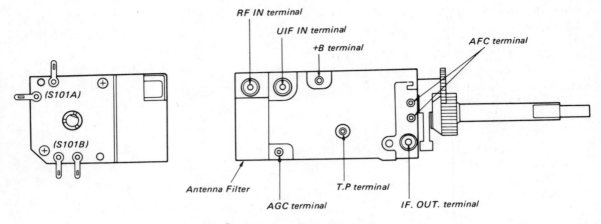

Fig. 7-1. *Tuner adjustment locations.*

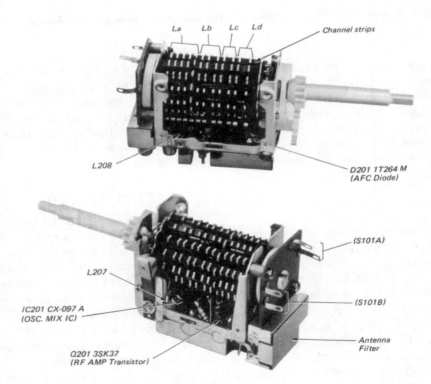

Fig. 7-2. *Tuner pictured with covers removed (drum type).*

while maintaining the required bandpass and are detected within IC501. Each signal from IC501 is fed to the AFC circuit and audio circuit via Q501 (IF buffer). The AFC integrated circuit, IC502, amplifies and detects the IF signal and develops the AFC voltage that controls the local oscillators of the VHF/UHF tuners for stable operation.

The audio circuit in IC503 amplifies and detects the 4.5 MHz audio IF signal. The AGC output of IC501 gain-controls the IF amplification and the RF stage of the VHF tuner for stable reception.

TUNER ADJUSTMENTS

The VHF tuner adjustment locations are shown in Fig. 7-1. A photo of the VHF tuner with cover removed is shown in Fig. 7-2. The VHF and UHF wiring information is shown in Fig. 7-3.

EQUIPMENT REQUIRED FOR TUNER ALIGNMENT

- Sweep generator
- Marker generator
- Oscilloscope
- Regulated dc power supply
- Detector (see Fig. 7-4)

TUNER ADJUSTMENT PROCEDURE

1. Make all the connections for the test equipment set-up. Refer to block drawings in Figs. 7-5 and 7-6.
2. Scope — set vertical sensitivity range at 1 to 2 mV.

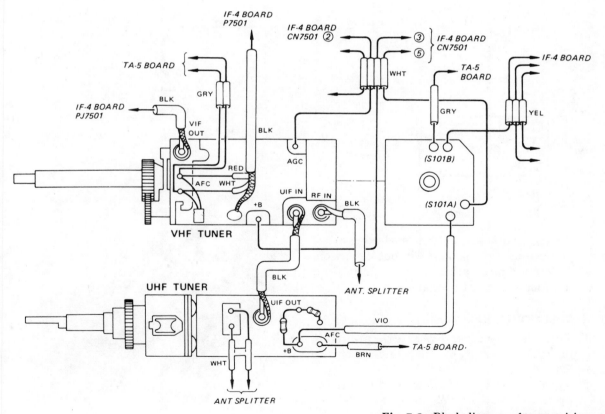

Fig. 7-3. *Block diagram of tuner wiring.*

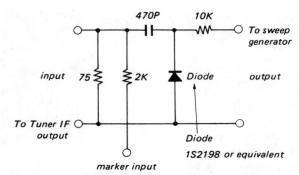

Fig. 7-4. *Detector alignment circuit.*

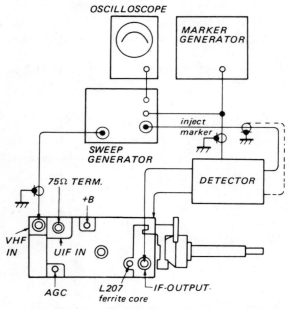

Fig. 7-5. *Tuner alignment set-up.*

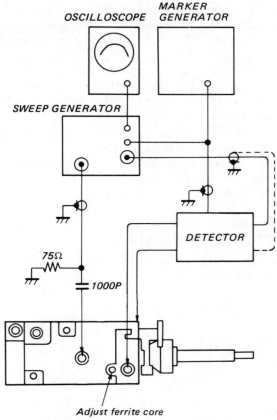

Fig. 7-6. *Tuner alignment test equipment set-up.*

3. Sweep generator — output level adjusted to produce usable response without distortion. Note: short pins 7 and 8 together of AFC prior to tuner alignment procedure.

TUNER COVER REMOVAL

1. Push the tuner cover to release the projection on the cover from the long slot. See Fig. 7-7.

2. Turn the tuner upside down and release the protection on the cover from the long slot in the same manner as above.

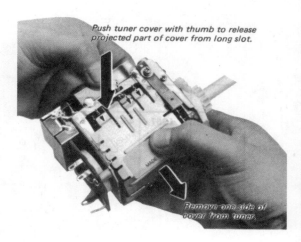

3. Remove the tuner cover as shown in **Fig. 7-8**. Note: be sure to remove the cover by following the above procedure sequence. Reverse the removal procedure for the cover attachment.

4. To remove and install the tuner channel strips refer to **Fig. 7-9**. The complete tuner alignment procedure is shown in steps one, two, and three of **Fig. 7-10**.

IF BOARD ALIGNMENT

Equipment required for IF alignment of VCR is as follows:

- VIF sweep generator
- Standard signal generator
- Monitor scope
- Volt/ohm meter
- Resistor divider pad
- Insulated alignment tool
- 50-ohm to 75-ohm converter
- 3 dB attenuator

Preset CV501, T501, and T509 as follows, prior to the IF alignment procedure. CV501 preset: Set CV 501 (trimmer) as shown in **Fig. 7-11**. Note the position for the maximum capacity of CV501 (the position shown where the MAX mark aligns to the other mark) and then set CV 501 to the required preset position. Trap coil

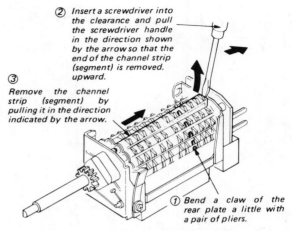

② Insert a screwdriver into the clearance and pull the screwdriver handle in the direction shown by the arrow so that the end of the channel strip (segment) is removed upward.

③ Remove the channel strip (segment) by pulling it in the direction indicated by the arrow.

① Bend a claw of the rear plate a little with a pair of pliers.

Fig. 7-9. *Channel strip removal.*

preset: Turn the cores of T501 (VIFT-T1) and T509 (VIFT-T4) about two turns clockwise as viewed from the component side of the tuning coil. Set the VIF sweep generator output to -22 dBm. Refer to **Fig. 7-12**.

Maintain the waveform amplitude on the monitorscope at 3 volts peak-to-peak by adjusting the MGC VR (note **Fig. 7-13**). Adjust the cores of T508 and T510 for maximum output at 44.00 MHz. See waveform in **Fig. 7-14**. Adjust the cores of T506 and T507 so that the 45.75 MHz marker (point P) is positioned at the -7 dB point. Refer to **Fig. 7-15**.

4.5 MHz Trap Adjustment

Mix the VIF sweep generator output and the 41.25 MHz, -30 dBm output from the SSG and apply the mixed signal to the IF-4 board. Refer to set-up in **Fig. 7-16**. Adjust T513 (4.5 MHz trap) to minimize the 4.5 MHz signal leak on the monitor scope screen. Note waveform in **Fig. 7-17**. Amplitude should be less than 50 mV.

Trap Coil Adjustment

This trap coil adjustment procedure is with the sweep generator.

1. Connect the equipment as shown in **Fig. 7-12**.
2. Set the generator output to -15 dBm.

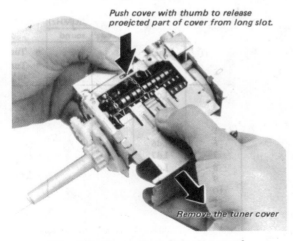

Push cover with thumb to release proejcted part of cover from long slot.

Remove the tuner cover

Fig. 7-8. *Tuner cover being removed.*

Step	Channel	Adjust	Marker point (MHz)		Adjustment procedure	Equipment setup
			Video	sound		
1	UIF	L207	45.75	41.25	• Adjust L207 (turn ferrite core) so that the peak of response curve is as shown in Fig. 3-10. 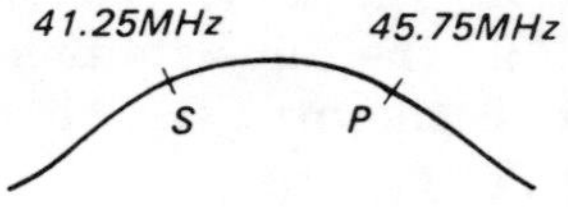 41.25MHz 45.75MHz S P Fig. 3-10.	See Fig. 3-6.

Step	Channel	Adjust	Marker point (MHz)		Adjustment procedure	Equipment setup
			Video	sound		
2	2-13	Fine tuning shaft			• Turn AFC switch to "OFF". • Turn the fine tuning shaft for correct tuning of each channel.	See Fig. 3-5.
	2	Ld2			If correct tuning is not obtained with above adjustment, adjust as follows. • Adjust Ldn (2-13) for fine tuning of each channel.	
	3	Ld3				
	4	Ld4				
	5	Ld5				
	6	Ld6				
	7	Ld7				
	8	Ld8				
	9	Ld9				
	10	Ld10				
	11	Ld11				
	12	Ld12				
	13	Ld13			• Turn AFC switch to "ON".	

Step 3	Channel	Adjust		Marker point (MHz)		Adjustment procedure	Equipment setup
				Video	sound		
3	2	Lb2	Lc2	55.25	59.75	• Adjust Lbn (2-13,) and Lcn (2-13,) of each channel so that the two markers of each channel are within the peak (75-100%) of the response curve. (See 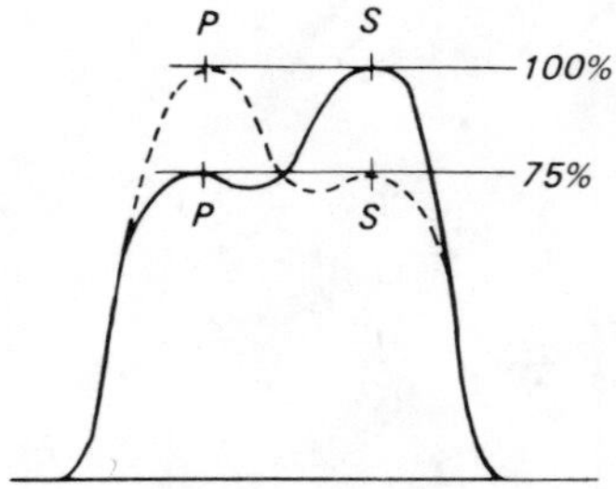 Fig. 3-11.)	See Fig. 3-5.
	3	Lb3	Lc3	61.25	65.75		
	4	Lb4	Lc4	67.25	71.75		
	5	Lb5	Lc5	77.25	81.75		
	6	Lb6	Lc6	83.25	87.75		
	7	Lb7	Lc7	175.25	179.75		
	8	Lb8	Lc8	181.25	185.75		
	9	Lb9	Lc9	187.25	191.75		
	10	Lb10	Lc10	193.25	197.75		
	11	Lb11	Lc11	199.25	203.75		
	12	Lb12	Lc12	205.25	209.75		
	13	Lb13	Lc13	211.25	215.75		

Fig. 3-11.

Fig. 7-10. *A & B. Complete tuner alignment charts.*

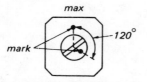

Fig. 7-11. *CV501 trimmer adjustment.*

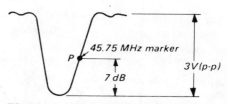

Fig. 7-15. *T506 and T507 adjustments.*

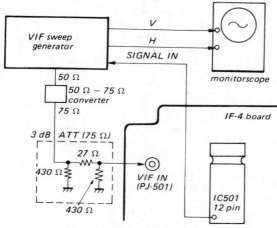

Fig. 7-12. *VIF circuit adjustment.*

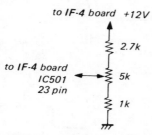

Fig. 7-13. *Resistor divider network.*

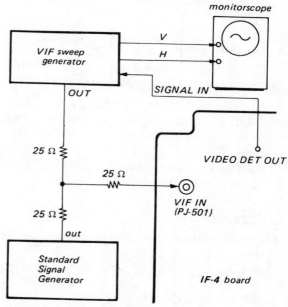

Fig. 7-16. *4.5 MHz trap adjustment.*

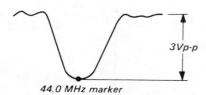

Fig. 7-14. *T508 and T501 adjustments.*

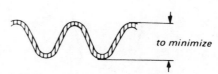

Fig. 7-17. *Waveform required for T513 adjustment.*

3. Adjust the MGR VR (see Fig. 7-13) so that each trap section on the monitor scope screen is easily observed.
4. Adjust T501 so that the 41.25 MHz marker point is positioned at the peak. Refer to Fig. 7-18.
5. Adjust T504 so that the 39.75 MHz marker point is in coincident with the top of the solid line as shown in Fig. 7-19.
6. Adjust T505 so that the 47.25 MHz marker point coincides with the peak of the solid line in Fig. 7-20.
7. Adjust T502 so that the 49 MHz marker point coincides with the top of the solid line shown in Fig. 7-21

Equipment Setup

This is the adjustment procedure when you are using the standard signal generator.

1. Amplitude modulate the SSG output with 400 Hz, 40 percent signal and set its output level for the best observation on the monitor scope. Refer to Fig. 7-22.
2. Adjust each trap coil for minimum amplitude of the trap frequency.

AFC Preadjustment

1. Complete the connections for the equipment as shown in Fig. 7-12.
2. Set the VIF sweep generator output to −22 dBm.
3. Adjust the MGC VR for 3 volts (p-p) waveform on the monitor scope. Refer to Fig. 7-14.

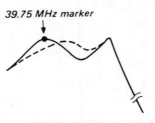

Fig. 7-19. *T504 adjustment waveform.*

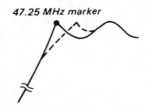

Fig. 7-20. *T505 adjustment.*

Fig. 7-21. *49 MHz trap adjustment.*

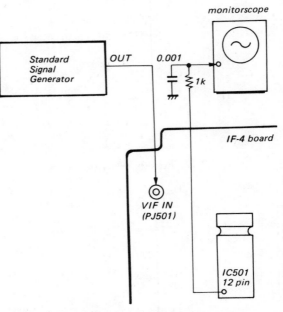

Fig. 7-22. *Trap coil adjustment set-up.*

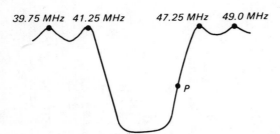

Fig. 7-18. *T501 adjustment waveform.*

4. Adjust T515 (AFC-3) for maximum displacement of the 45.25 MHz marker point (P) in the direction shown by the arrow in Fig. 7-23.

VIF Overall Adjustment

1. Make the equipment set-up as shown in Fig. 7-12.
2. Set the VIF sweep generator output to −22 dBm.
3. Maintain the monitor scope waveform with the MGC VR.

Adjust T506, T507, and CV501 so that the VIF response is as shown in Fig. 7-24. The VIF response around 44 MHz should be as flat as possible.

AFC Adjustment

1. Complete the connections shown in Fig. 7-12 and set the VIF sweep generator output to −22 dBm.
2. Ensure that the waveform on the monitor scope is 3 volts (p-p). If not, adjust the MGC VR for 3 volts (p-p).
3. Change the test setup to the one shown in Fig. 7-25.
4. Adjust VL509 and VL510 to obtain the monitor scope waveform shown in Fig. 7-26.
5. Set the VIF sweep generator output to −26 dBm.
6. Adjust VL510 so that the output waveform becomes as symmetrical as possible. Refer to waveform in Fig. 7-27. If a satisfactory result is not attained, adjust T515. Confirm that the output waveform is effected by the limiter.

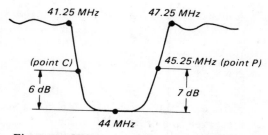

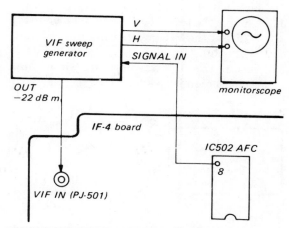

Fig. 7-24. *VIF overall adjustment waveform.*

Fig. 7-25. *Waveform adjustment for AFC.*

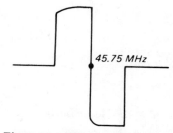

Fig. 7-26. *AFC output waveform.*

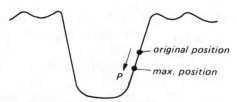

Fig. 7-23. *AFC-3 adjustment waveform.*

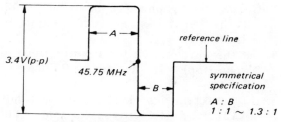

Fig. 7-27. *AFC output waveform.*

7. Adjust VL509 so that the 45.75 MHz marker coincides with the reference line shown in Fig. 7-27.
8. Change the test setup to that shown in Fig. 7-28.
9. Set the SSG output to 45.75 MHz, −25 dBm, and adjust the MGC VR for a voltmeter reading of 3.8 ±0.3 volts.
10. Connect the voltmeter to pins 7 and 8. Adjust VL509 so that the voltmeter reading is less than 0.5 volts dc.
11. Change the test set-up to that shown in Fig. 7-25. Set the VIF sweep generator output to −22 dBm.
12. Ensure that the waveform on the monitor scope screen is as shown in Fig. 7-26.

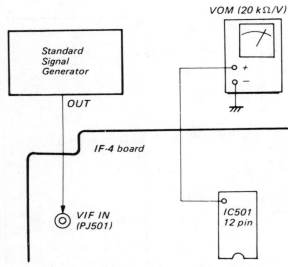

Fig. 7-28. *AFC test adjustment set-up.*

BAND CORRECTION CIRCUIT CHECK

1. Connect the equipment as shown in Fig. 7-12.
2. Set the VIF sweep generator output to −22 dBm.
3. Set the waveform on the monitor scope to 3 volts (p-p) with the MGC VR. Refer to Fig. 7-14.
4. Disconnect the signal in terminal from IC501 pin 12 and connect it to the video out (X) of the IF-4 board. Refer to Fig. 7-29.

5. Confirm that the waveform on the monitor scope is the same as the one shown in Fig. 7-30.

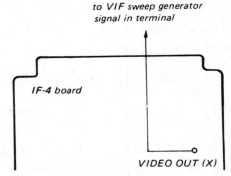

Fig. 7-29. *Band correction circuit check.*

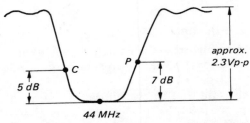

Fig. 7-30. *VIF overall waveform characteristic.*

SWEEP IF (SIF) ADJUSTMENT

1. Set the standard sweep generator output to 5 dBm (4.5 MHz at 25 kHz FM).
2. Adjust T517 for the maximum waveform deflection on the monitor scope screen (see Fig. 7-31).
3. Set the SIF generator output to 5 dBm.
4. Confirm that the linearity of the S-shape waveform is good and the 4.5 MHz marker is on the baseline.

TUNER AGC ADJUSTMENT

1. Check that the 18 volts is applied to the B+ terminal of the VHF tuner.
2. Set the tuner control to channel 12 and tune the local adjustment finely.
3. Set the SSG to 205.25 MHz, an available output level of −53 dBm ± 2 dB and 1 kHz 40

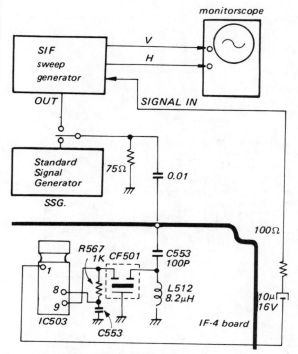

Fig. 7-31. *SIF adjustment set-up.*

percent AM. Note: the peak level of the wave-form is -50 dBm $\pm$ dB.

4. Adjust VR501 for a voltmeter reading of approximately 11.5 volts. Refer to Fig. 7-32 for the tuner adjustment setup. The VHF/UHF tuner circuits are shown in Fig. 7-33.

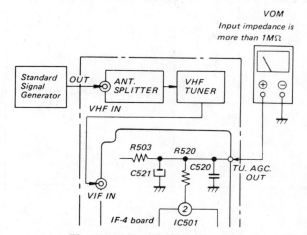

Fig. 7-32. *Tuner AGC adjustment.*

SYSTEM CONTROL AND PAUSE CONTROL CIRCUITS

The main function of the system control circuits is to control the mechanism during the tape threading and unthreading periods, together with the subsequent control of the signal systems. In this VCR, the switching between function modes is done mainly by manually depressing the function buttons. Therefore, automatic controls of the mechanism by the system are very few. What *is* controlled automatically by the system control circuit is the autostop solenoid. There are five major operations provided by the system control circuit.

- Tape-end sensor — the metallic foil attached to both ends are detected. An electrical circuit is closed, driving the auto-stop circuit.
- Head drum rotation detector — when any of the function buttons are depressed, the system control circuit detects the head drum rotation by detecting the 30 PG pulses. If for any reason the head drum rotation stops, the auto-stop circuit is energized.
- Tape slack sensor — if tape slack is detected in the play or record mode, the tape must be re-wound. When the tape slack is detected by the tape slack sensor element, the auto-stop circuit is energized.
- Auto stop circuit — auto-stop is initiated by actuation of the devices listed above. When auto-stop is initiated, the system control circuit energizes the stop solenoid to place the function mechanism into the stop mode. The ac motor power is cut off when the stop solenoid is energized.
- Muting circuit — this circuit generates audio muting and video blanking signals according to the operation of the mechanical system.

RECORDER MECHANISM COMPONENT LOCATIONS

Location of the switches and solenoids related to the system control is shown in Fig. 7-34. The item numbers in the following description are

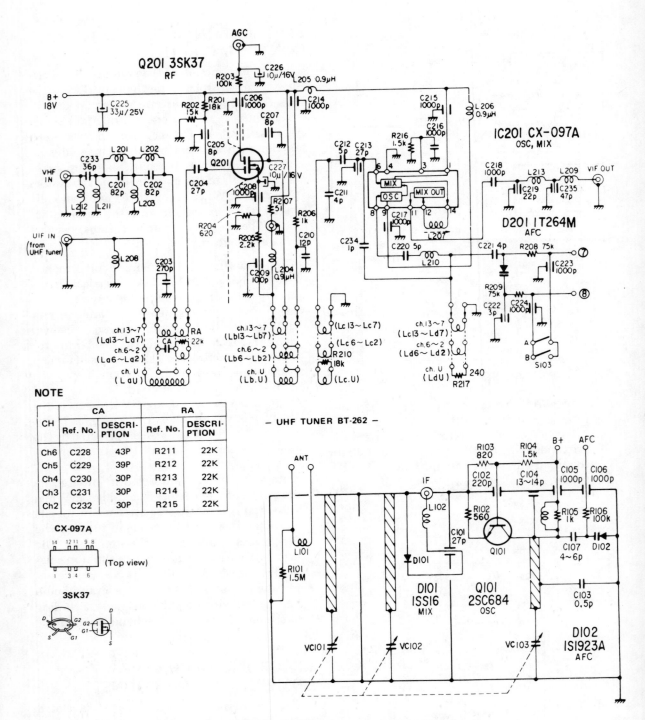

Fig. 7-33. *A & B. Circuit diagram for VHF/UHF tuners.*

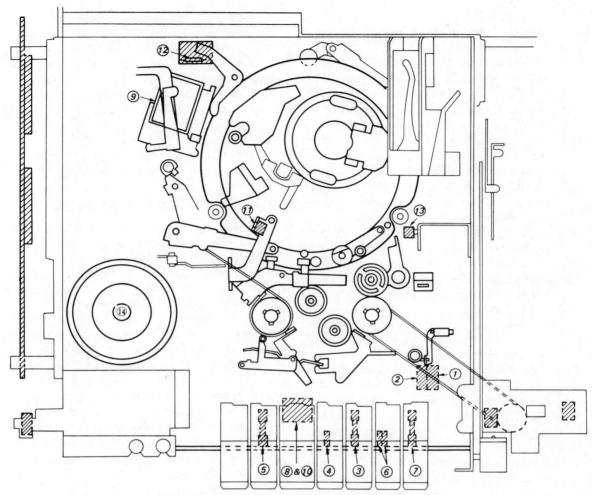

Fig. 7-34. *Switch locations on system control circuit.*

the reference numbers of the parts in this drawing.

(1) Function switch — This switch is located on the opposite side of the chassis (as viewed from the top) and actuated by the function button mechanism. This function switch turns on when any one of the mode buttons is pressed. When the switch is on, power is supplied to the ac motor.

(2) Function switch — This switch is operated the same way as switch (1). The drum rotation detector protection circuit is operated by this switch. When this switch is on, the protection circuit functions and vice versa.

(3) Fast-forward switch — This is a leaf switch on the SRP board. It is actuated by the fast-forward function button (turns on when the button is depressed).

(4) Play switch — This is a microswitch on the SRP board. It is actuated by the play button. It is turned on as the play button is pushed.

(5) Rewind switch — This is a leaf switch on the SRP board. It is actuated by the rewind function button and turned on as the button is depressed.

(6) Record switch — This is a microswitch on the SRP board. It is actuated by the record function button.

(7) Pause switch — This is a leaf switch on the SRP board. It is actuated by the pause function button and turned on when depressed. Note: all the switches described above are released by either pressing the stop function button manually or energizing the stop solenoid electrically.

(8) Stop solenoid — The stop solenoid is located beneath the function button block and is energized in the auto-stop mode. All the function buttons and switches are released by the solenoid action. The solenoid action also actuates to turn off the stop solenoid switch.

(9) Pause solenoid — The pause solenoid is located on the upper section of the threading chassis and is energized in the play, record, and pause modes. It serves to make the pinch roller press against the capstan pulley tape movement.

(10) Stop solenoid switch — This switch is positioned beneath the stop solenoid and is actuated directly by the mechanical action of the stop solenoid. As the stop solenoid is powered on, this switch is actuated to turn off the ac motor power.

(11) Supply sensing coil — This coil is positioned on the tension arm, very close to the tape but not in contact with it. This coil senses the tape end during the play, record, and fast forward modes by means of a metallic foil on the tape ends. The sensor energizes the auto-stop function.

(12) Tape slack detection switch — This switch is located on the TK board and is actuated magnetically by the tape slack detection lever. This slack detection lever can move in the play and record modes. If tape slack occurs, this switch is turned on, energizing the auto-stop circuit.

(13) Take-up sensing coil — This coil is positioned very close to the tape in the cassette tape guide on the take-up side. It senses the tape beginning (in the rewind mode). The sensor energizes the auto-stop function.

(14) AC motor — This is a hysteresis synchronous motor, powered by the ac power line. Its stability is a function of the line frequency stability. When any of the function buttons is depressed, the ac motor is powered and drives the capstan, forward idler, fast-forward idler, rewind idler, gear (A), gear (B), and rotary video head disc assembly, via the rubber drive belts.

VCR CLOCK TIMER SYSTEM

The block diagram for the system control (on the SRP board) is shown in Fig. 7-35.

The timer clock device has a standard LED readout clock, capable of running on either 50 or 60 Hz frequencies. Note timer system block diagram in Fig. 7-36. This timer circuit can be programmed to start the VCR at a preset time.

Both the clock and the timer can be set by the buttons on top of the machine, above the clock at the left front corner. The clock is set by depressing the clock time and hours buttons and watching the AM/PM light. The minutes can be set by depressing the minute or fast minute button. Note: the hours do not advance as the minutes pass the 60 mark. Each must be set independently. The timer start time can be set by depressing the turn-on time button instead of the clock time button and following the above procedure.

The clock also has the ability to reset itself in case of a momentary power interruption (2 seconds or less) and will give a one second display flashing indication in case of a longer power outage. Resetting the clock will return it to normal operation.

Timer Circuit Description

The timer comprises IC801 on the TMC board and functions normally as a 12-hour clock synchronized with a power supply frequency. It is possible to reserve a starting time of the VCR

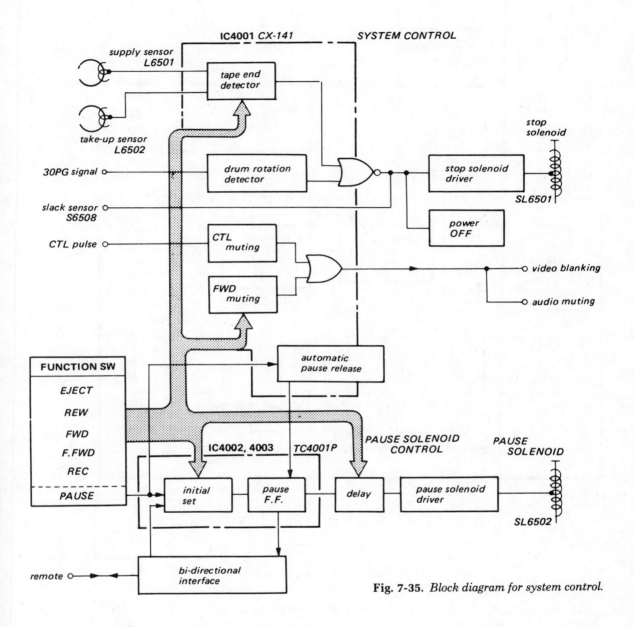

Fig. 7-35. *Block diagram for system control.*

timed by the minute once within 24 hours from the current time.

The clock function must determine whether the power frequency is 50 Hz or 60 Hz, because the timer synchronizes with the power supply. The line frequency is determined at the moment when the power is first connected to the IC. A reference signal, supplied from the LF-5 board where an ac 12 volts is wave-shaped, is applied to pin 27. Counting for driving the timer is performed in the IC by comparison of the ac 12 volts with the OSC 1 (pin 26) level.

The timer operates in the following manner. A start time can be reserved in the form of a reser-

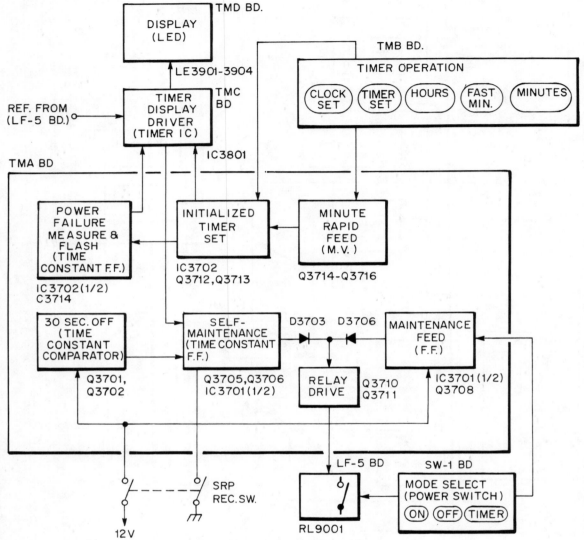

Fig. 7-36. *VCR timer block diagram.*

vation amendment. When a clock indication matches the reserved time (the second counter is "00"), a signal appears at wake 1 terminal (pin 20). This signal turns on the videocassette recorder only when the recorder is in the timer and the record modes. The start time of the VCR can be reserved on the timer only when its modes are the timer and record modes. The VCR turns off automatically about 30 seconds after the tape is

wound to its end and the record button is released automatically or manually.

This timer has a power suspension indicating function. The clock indication advances a little and the reserved time remains unchanged in case of a short power (below 2 seconds). In case of a longer power outage, the current time and the reserved time turn to 12:00 a.m. at the moment of the power recovery and the clock time advances from 12:00 a.m. When 12:00 a.m. is indicated, the power suspension indicator at the lower right corner of the timer flickers at 1-second intervals until the clock time and the reserved time are reset.

Clock Operation

The signal, obtained by frequency division of the OSC-1 produced in the circuit formed by R803, C803, and the internal circuit in the IC, and the 50/60 input, produced on the LF-5 board as a reference signal, are the inputs to the flip-flop shown in Fig. 7-37. The inputs become the F-Q output and the output enters the counter.

Reservation Function

Starting time can be reserved in the hour and the minute independently as alarm time with the help of the inc control. A real time is counted by the counter and advances. When the real time reaches the reserved hour and the reserved minute, a signal comes out from the comparator as wake 1 logic out. The timer functions as the 12-hour clock when ST = 1, SA = 1, and wake 1 = 1. ST and SA of IC801 open for the time constant (about 2 seconds) determined by R730 and C709 when the power is connected to the ac line.

The wake 1 terminal (pin 20) is plus when the ST and the SA open. The 50 Hz/60 Hz judgment

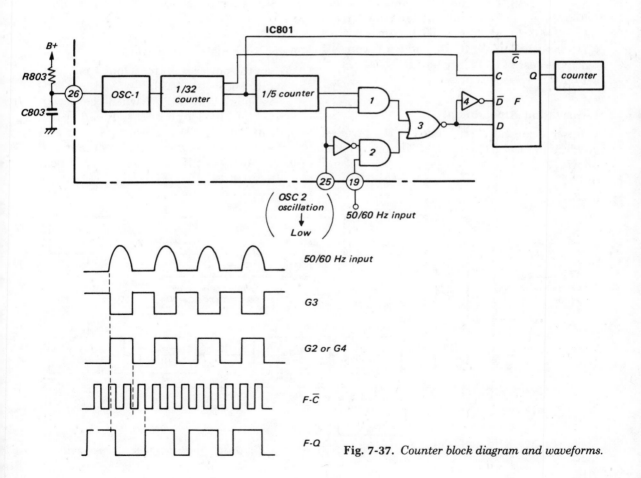

Fig. 7-37. *Counter block diagram and waveforms.*

183

circuit functions as follows. When ST = 1 and SA = 1, it judges for wake 2 = 1 that the line frequency is 60 Hz and for wake 2 = 0 that the frequency is 50 Hz.

The circuit judges that the frequency is 60 Hz when the wake 2 terminal (pin 19) is pulled by the B + through R804 in the ST = 1 and SA = 1 condition identical with that of the 12/24-hour judgment circuit. When the wake 2 terminal is connected to ground through R804, the frequency is judged as 50 Hz. The counter is set to count one second for 60 or 50 pulses for the reference signal counting in the IC.

Current Time Amendment

Current time is amended by the inc logic circuit in Fig. 7-38. The time can be amended every 1 step of the hour or minute in the condition that ST : SA = 0 : 1 and H or M. ST becomes 0 when Q713 is turned on by IC702 and SA becomes 1 when the base of Q712 is turned to 0 by the clock set button when it is pressed.

Reservation Amendment

Reservation can be amended in the reversed condition of the ST and SA condition as noted above.

Driving Indication

The supply of the LEDs is a ripple obtained by the half wave rectification of the ac 3 volts in D711 and D712. See waveforms in Fig. 7-39. The power supply is applied to combinations of LED1, LED3, LED2, and LED4. The cathode side appears as the waveform at the IC terminal. The segments of LEDs 2 and 4 are lighted for forming figures indicating a time.

Indicator for a.m./p.m.

Pin 3 of the IC is the p.m. indicator and also the number 4 digit as an IC terminal when it is discriminated from the IC terminal waveform. Because the com anode is number 4 digit, the LED is not driven directly. The signal inverted by Q801 is connected to the E-segment of LED 904 to indicate a.m.

Power Suspension Indication

When the power is off for a long time, OSC 1 and OSC 2 become 0 level. Pin 3 of the IC operates as shown in Fig. 7-40 when the power is restored. The IC side phase corresponding to the anode of Dp of LED 901 becomes open and 0 level alternately every second. Q, output of the flip-flop formed by IC702 ½ on the TMA board,

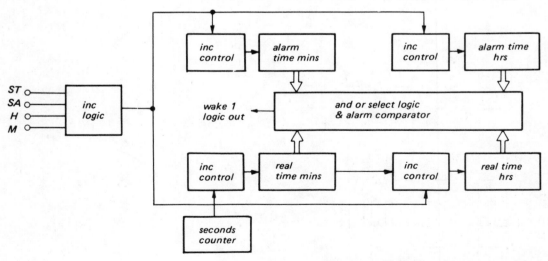

Fig. 7-38. *Block diagram of timer control.*

184

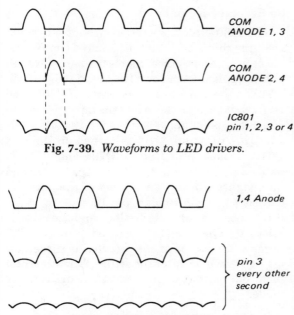

Fig. 7-39. *Waveforms to LED drivers.*

COM
ANODE 1, 3

COM
ANODE 2, 4

IC801
pin 1, 2, 3 or 4

1,4 Anode

pin 3
every other
second

Fig. 7-40. *Waveforms for IC clock operation.*

is tuned to plus by R742 and C713 at the moment when the flip-flop recovers from the power outage, and Q802 connected to the output turns on.

The Q802 emitter becomes 0 every other second by pin 3, and the phase of the emitter becomes the same with that of the anode of Dp, and Dp flickers every other second. The flicker of the power outage indicator is stopped by releasing pin 3 or by releasing the flip-flop with the hour and minute buttons, i.e., the indicator is released from flicker by pressing the clock set, time set, hours, or minutes button.

Current and Reserved Time Hold Function

This function holds the time indication and the reserved time occurrence of the power outage until the recovery of the power outage. A voltage stored in the C714 capacitor of $1000\,\mu F$ is supplied to Vp when the power goes off, which is contrary to the operation of the power suspension indication. The supply voltage for C9005 on the LF-5 board and the OSC circuit on the TMC board is stored in C801. The reserved time is maintained, and the time indication loses a little

when the power suspension recovers while the OSC 1 and OSC 2 are more than threshold.

Fast Minute Feeding

This function "feeds" the minute indication rapidly while the fast min button is pressed together with the clock set or the timer set button. This is for the purpose of amending the minute indication of the current time or the reversed time. It is because of the one-step method of amendment that it takes much time.

The power supply is fed to the MMV formed by Q714 and Q715 when the fast min button is pressed and the MMV starts oscillating (at about 20 Hz). The MMV output is connected to pin 23, M terminal of the IC. The M terminal (i.e., the fast min button) is tapped automatically instead of tapping the M terminal manually.

On-Off-Timer Indicator

The "on-off-timer" indicator is beside the power switch. The collectors of Q710 and Q711 are connected to ground, without regarding relay drive transistors Q710 and Q711 on the TMA board, in order to turn on the relay on the LF-5 board when the power switch is in the ON position. When the switch is in the OFF position, the bases of the transistors are connected to ground to turn off the relay. The hold circuit described next is reset by connecting pin 8 of IC701(1/2) to ground through D702.

A negative pulse is produced by C707 when the power switch is set to "timer" from "off." It is applied to the flip-flop of IC701(1/4) to make the Q output, pin 4, high. Then Q710 and Q711 turn on and the relay turns on. When the record button is depressed in the timer mode, 12 volts is supplied from the SRP board, Q708 turns on, and the flip-flop is cleared. The wake 1 output is expected in that state.

Recorder-On Hold Function

This function holds the VCR in the ON state when the time reaches the reserved starting time of the recorder and the recorder turns on in the timer mode. When the time coincides with the reserved time, the 0 level wake 1 described above appears at pin 20 of the IC. Receiving the signal, Q705 turns off. Then the collector of Q705 is pulled by +10 volts and a plus pulse is produced in Q703. Q706 turns on within the width of the pulse and a negative signal is applied to pin 13 of the hold circuit of IC701 (1/2).

Because C705 in the hold circuit turns low when the power is turned on and pin 8 turns low, the IC output at pin 10 becomes high and is inputted to pin 12. Pin 13 is high while Q706 is off because it is pulled by the power supply. The IC output at pin 11 is low. When the signal from the wake 1 comes in, pin 11 becomes high, pin 8 is high, pin 12 is low and the pin 11 output is kept at a high level, even if pin 13 returned to high. The pin 11 output is connected to the bases of Q710 and Q711.

30-Second Off Function

The hold circuit operates without fail when the time coincides with the reserved time in the timer mode and the recorder is turned on. Rec 12 volt is applied from the SRP board to Q701 via D715 on the TMA board and turns of Q701. The rec 12 volt becomes rec when the tape ends or the stop mode is set up. When the rec 12 volts is cut off, Q701 turns off. Q702 is turned on through D716, a pulse produced by Q704 and R713 is applied to the base of Q707, and Q707 is turned on instantaneously. Then pin 8 of the hold circuit goes low, pin 10 high, pin 13 high, the relay drive output at pin 10 goes low, and the relay is released. This system is named "30 sec. off," because the power of the recorder is turned off about 30 seconds after the setup of the stop (rec) mode of the recorder in the ON state in the timer mode.

8

Routine VCR Maintenance

This chapter covers VCR cleaning and other routine maintenance procedures. All of these tips can be used by the professional electronics technician or in many cases by the VCR owner. These maintenance points include how to properly clean the VCR heads and tape guides. The care and feeding of both the Beta-max and VHS video tape machines systems is covered. Other points of VCR service are degaussing the video heads and cleaning and checking the drive belt. The chapter concludes with what to do when video heads have been worn smooth and develop "stiction."

In many ways, the VCR is not too much more complicated for routine cleaning than some of the more sophisticated audio tape machines. However, some of the electronic and mechanical tape transport devices are considerably more sophisticated.

In an audio machine, the magnetic tape passes over stationary heads to record and reproduce sound. The VCR, by comparison, has *rotating* video heads along with fixed audio, control track, and erase heads. The rotating heads are needed because of the higher frequencies. The tape must travel across the heads at a much faster rate than for audio recording. This is accomplished by rotating the video heads at a very high speed past the slower moving tape.

The original video taping technique was perfected for the professional broadcast industry by Amprex. For home video recorders, the operation has been refined for use in a cassette format, eliminating the need to handle large rolls of tape.

DEMAGNETIZING THE HEADS

Every metal part of the VCR that comes into contact with the tape will gradually become slightly magnetized. This especially pertains to the video and audio heads, which should be demagnetized regularly. Residual magnetism affects all types of magnetic heads. Heads and other metal contact points in the tape path will become partially magnetized from such sources as the normal on-and-off surges from the recorder's electronics. Items such as a faulty bias oscillator, the use of an ohmmeter for VCR troubleshooting, or the use of some magnetized tools near the heads could easily cause unwanted magnetism.

A magnetized tape head (and other magnetized metal parts) will erase some of the high frequencies on prerecorded tapes or will cause a hiss or visible background noise. Also, magnetization of the video tape from these spurious sources can cause loss of color and partial erasure of the tape.

It is quite simple to perform VCR machine demagnetization. Just plug in the demagnetizing tool, turn it on, and then slowly bring it near the part to be demagnetized. Now move it slowly up and down a few times, and then slowly retract the tool to about 3 feet away. Bring the tip of the demagnetizer as close as possible to the head tip without actually contacting it. Should the tool

be turned off while close to the head, it will have the reverse effect and leave the head magnetized. If you are careful to avoid this, the head and other machine parts will be completely demagnetized. The Nortronics VCR-205 head demagnetizer is shown in action in Fig. 8-1.

Please note that you should only use an approved *VCR* head demagnetizer for this operation. The video head demagnetizers produce a weaker flux than most of those used for audio tape machines. Using a demagnetizer made for an audio tape player can actually *shatter* the video head chips.

WHY THE NEED FOR VCR CLEANING?

Even with normal operation, contaminants in the atomsphere can cause a VCR to perform poorly. Particles of dust, smoke, loose magnetic tape oxide and oxide binder combine over a period of time to form a hard buildup on the head surfaces. This buildup causes the tape to be physically spaced away from the proper firm contact with the face of the head, which reduces

Fig. 8-1. *Head demagnetizer in use.*

the amount of signal being recorded and played back. Normal spacing between the tape and head face is 0.000020 inch (for comparison, the thickness of a human hair is .004 inch). When this extremely close contact is lost due to dirt buildup on the heads, picture playback performance will be diminished. This signal loss can cause mushy, distorted sound with a noticeably lack of high frequencies. Visually, it appears as a lack of overall picture clarity or as a snowy picture. It is also necessary to clean tape oxide buildup from all the contact points in the tape path, like rollers and guides.

In extreme cases, where tape oxides are allowed to build up over a long period, the accumulation can become large and ragged enough to scratch or tear tapes or interfere with the precise tape speed required for VCR operation. Regular VCR cleaning is the only way to prevent oxide buildup and other major machine problems. Professional broadcasters and recording studios know this very well and clean all heads on a daily basis. It is recommended that the home model VCRs be cleaned and the heads demagnetized every 100 hours of operation, or 2 to 4 times a year.

The rotating video heads are the most delicate and expensive parts of the VCR. They must be treated with the utmost care and be kept very clean.

The video heads actually penetrate into the tape oxide when recorded and played back, thus there will always be a small amount of oxide shedding from the tape onto the heads and guides in the normal VCR running process. Thus, a good VCR cleaning procedure is required on a regular routine basis.

An oxide buildup on guides and heads will cause very poor record and playback of the video tapes. This shows up as streaks, noise (Fig. 8-2), or if built up enough, no picture at all. You might even suspect a faulty head drum. In fact, dirty heads will show up first in the playback mode because of the relative weak magnetic field from the tape that must be transferred to the head drum. Another streak condition you might think

is due to dirty heads is tape dropout. This is when a few horizontal lines across the picture are missing (Fig. 8-3) and appear as a line of interference. This dropout is caused by a streak of oxide missing from the tape or a faulty badly worn tape.

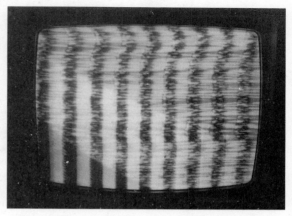

Fig. 8-2. *Poor tape playback due to oxide build-up.*

HOW TO CLEAN THE VCR

To clean the heads, use a special cleaning pad such as a chamois cloth or cellular foam swabs. If these are not available, a lintless cloth or muslin can be used. Cotton-tipped swabs should *not* be used for cleaning the video heads. The cotton strands can catch on the edges of the video heads and pull the small ferrite chip away from its mounting and ruin the head. However, cotton swabs soaked in cleaning fluid *can* be used to clean tape guides, control track, audio, and erase heads. To clean the heads and guides, you can use methanol or surgical isopropyl alcohol.

The cleaning pad should be liberally soaked in the cleaning fluid (alcohol) and then gently and firmly rubbed *sideways* across the heads. Never rub it up and down, as this may damage the heads. Clean the whole head in this same sideways motion. Make sure you then clean all places that the tape touches. Do not touch these parts with your fingers, as the oil will attract dust and dirt. But, it's a good idea to hold the head on top so it will not rotate as you rub it

Fig. 8-3. *Streaks due to drop-out.*

clean (in the direction of head rotation). This correct cleaning technique is illustrated in Fig. 8-4. The main thing to remember in head cleaning is to be very careful. The more often you perform this task the easier it will become.

Should you find a head that is very dirty and the head chip is plugged up with oxide, try the following tip for cleaning. The first thing to do is soak the area around the head chip with alcohol two or three times. Then, as shown in Fig. 8-5, use an old toothbrush also soaked in alcohol to clean out around the head chip. Do this very carefully, as the head can be damaged. (If it cannot be cleaned, replace the heads.) You might want to cut the bristles of the toothbrush shorter. A spray-can type video head cleaner is being used in Fig. 8-6.

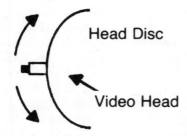

Fig. 8-4. *Correct rubbing technique for cleaning heads (horizontally).*

CLEANING BETA VCRs

Let's now go through the actual steps for cleaning a Beta machine. *Caution:* make sure the ac power cord is disconnected before cleaning the machine.

Step One. Remove the screws that hold the top cover of the VCR in place. These usually require a Phillips driver. Be careful not to damage the screw heads. If this is an older model Zenith or Sony Betamax, remove the tracking knob in the lower, left-hand corner by grasping it and pulling it up and off the shaft. Now, press the eject button to raise the cassette carrier platform, lift off the top cover, then push the cassette carrier down again.

Step Two. The two video heads are located within the rotating center ring shown in the lower left corner (arrow) in Fig. 8-7. Rotate this ring to bring each head (gap) into a convenient position by turning the black motor fan blower shroud shown by the arrow in the upper right corner. You might also want to use an inspection mirror to get a better look at the heads. *Do not* actually touch the head face with your fingers.

In order to clean these heads, first saturate one of the cellular foam cleaning swabs with a good tape head spray cleaner. Clean the heads using only a horizontal (side-to-side) motion. To

Fig. 8-5. *A toothbrush can be used for cleaning clogged heads.*

Fig. 8-6. *Spray type head cleaner being used.*

Fig. 8-7. *Location of the video heads. Courtesy of Nortronics.*

ensure that cleaning is done in a side-to-side fashion, hold the swab stationary against the head and use the fan motor shroud to rotate the head back and forth. *Caution:* do not clean heads with a vertical (up-and-down) motion. This can easily damage the very delicate VCR heads. The proper head cleaning technique is demonstrated in Fig. 8-8.

Step Three. The control track and audio heads are cleaned next as shown in Fig. 8-9. The control track and audio heads are located near the tip of the swab. Clean these heads in the same way you cleaned the video heads (using only a horizontal scrubbing motion.

Step Four. Now the erase head should be cleaned. The erase head is located by the tip of the swab shown in Fig. 8-10. Remember, use only a horizontal cleaning motion.

Step Five. After all of the heads have been

cleaned, perform the same cleaning function on all contact points (rollers, guides, etc.) that are in the tape path.

Step Six. Push the eject button to raise the cassette carrier and reassemble the unit by placing the top cover in position. Install and tighten all cover screws with the Phillips screwdriver. Replace the tracking control if there is one. The VCR should now be wiped clean with an anti-static dust cloth.

CLEANING VHS VCRs

Let's go through the actual steps for cleaning a VHS-format machine. *Caution:* make certain that the ac power cord is disconnected from the machine before removing the cover and cleaning.

Step One. First, remove the screws that hold

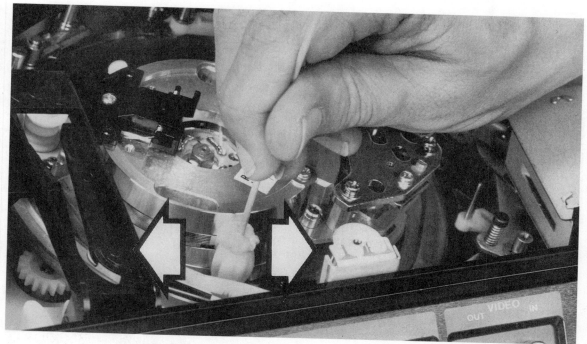

Fig. 8-8. *Proper head cleaning technique. Courtesy of Nortronics.*

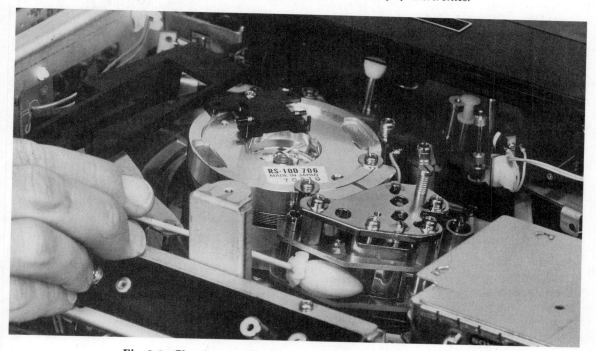

Fig. 8-9. *Cleaning the control track head. Courtesy of Nortronics.*

Fig. 8-10. *Erase head being cleaned. Courtesy of Nortronics.*

Fig. 8-11. *Location of head disc in VHS unit. Courtesy of Nortronics.*

Fig. 8-12. *Correct head cleaning technique. Courtesy of Nortronics.*

the top cover of the machine in place so that it can be removed. These are usually Phillips-type screws. Be careful not to damage the screw heads.

Step Two. The two very delicate video heads are located on the rotating head disc as shown by the arrow in Fig. 8-11. Gently rotate this wheel disc to bring each head into a convenient position for cleaning. You might want to use an inspection mirror to get a better look at the head faces. *Do not* actually touch the highly polished face of the disc wheel with your fingers. The control track, audio heads, and erase head are located on each side of the head cylinder.

To clean these heads, first saturate one of the cellular foam cleaning swabs with a good tape head cleaner fluid. Clean the heads using only a horizontal (side-to-side) motion. To ensure that cleaning is done in a side-to-side fashion, hold the swab stationary against the head and use the head wheel to rotate it back and forth. Refer to Fig. 8-12 for the proper head cleaning action. Caution: *do not* clean with a vertical (up-and-down) motion. This could damage the fragile heads.

Step Three. The control track and audio heads are located near the tip of the swab as can be seen in Fig. 8-13. Clean these in the same way you cleaned the video heads (using only a horizontal scrubbing motion).

Fig. 8-13. *Cleaning of control track and audio heads. Courtesy of Nortronics.*

Step Four. After all the heads have been cleaned, perform the same cleaning function on all contact points (rollers, guides, etc.) that are in the tape path.

Step Five. Now replace the top cover and install all screws. And if you wish to do a class-A job, wipe the machine clean with an anti-static cloth. *Caution:* do not use cotton swabs or other lint-producing materials, as the lint may be left during the cleaning process and could damage delicate VCR machine components.

Make sure the tape heads in the VCR unit are totally dry after the cleaning procedure before the machine is put into operation. Otherwise, the tape might not thread properly and jam up the machine. Nortronics VCR-103 spray tape head cleaner ensures completely dry heads, because this cleaner evaporates completely, leaving no oil or other residue.

NOTES ON TAPE TENSION

Remember that the tape tension is very important in the proper operation of a VCR. Too much tension causes excessive head and tape guide wear or tracking errors due to tape stretch that can permanently stretch the tape. Too much tension in the threading loop can actually stop the tape from traversing its path and even pull it out of its path. Too little tension allows the tape to fall out of its true path, preventing proper contact with the heads, which causes misalignment with the video tracks and results in picture dropouts. Many tape transport problems are due to the wrong tension on the tape, and checks should be performed on a regular basis.

BELTS AND DRIVE WHEELS

All of the drive belts and wheels should also be checked and cleaned when the VCR machine is apart for head cleaning. Check for loose drive belts and worn rubber drive wheels. If only the rewind and fast forward operate properly, then suspect worn drive wheels. The drive belts and wheels can be cleaned with isopropyl alcohol or any other cleaner made for this purpose. If new belts are installed, make sure they are put on properly and check them for proper tension. Always make a careful check when drive belts, wheels, or other mechanical parts are changed or adjusted.

Much of the VCR's mechanical alignment and parts replacement require using special jigs, gauges, and fixtures for correct operation. Make sure you are properly trained before attempting these procedures.

VIDEO HEAD "STICTION"

Video cassette recorders that have logged many hours of use might begin to exhibit a condition described as *stiction*. The word, a combination of the words *sticking* and *friction*, indicates a condition of the video head drum assembly that causes the tape to stop moving during record or playback. If this occurs and continues unchecked (as in recording with the timer mode), severe clogging of the video heads and tape damage could result.

The apparent cause of stiction is the loss of an air "cushion" between the tape and the record drum head. As the VCR is used, friction from tape travel polishes the drum surface smooth. This prevents the required air buildup, and the tape adheres to the drum. Obviously, replacement of the head drum assembly is the ultimate solution, but this is very costly.

The following alternative has been tested and recommended whenever the drum exhibits highly polished areas and is a last resort before the video head disc is changed. This procedure requires the use of a specially designed brush tool (Zenith part no. SD-21179) that can be obtained from your local Zenith distributor. *Caution:* do not use sandpaper, emery cloth or other similar abrasives to perform this repair.

If the head disc is not to be replaced after this procedure, use extreme caution when working around the video heads, as they can be easily damaged. Proceed as follows:

1. Remove the cassette lift assembly and threading arm guides (for Zenith and Sony machines).
2. Remove the arm lock bracket assembly and the tape retainer spring assembly.
3. Remove the rear panel assembly and perform the threading operation.
4. Fold the narrow side of a calling card (see Fig. 8-14), and insert this between the drum and head bracket assembly.
5. Position the video heads as for dihedral adjustment and lock the head in place with partially inserted dihedral screws, (if dihedral screws are not available, hold the rotating disc firmly with thumb and forefinger, taking care to not touch the heads).
6. Use the brushing tool (part no. SD-21179), with bristles set to approximately ⅛ inch. Brush horizontally across the upper (stationary) head disc, center (rotating) head disc, and lower (stationary) head disc. *Do not* come closer than approximately ⅜ inch to the head disc coil.
7. Continue in this way from the front of the head disc assembly to within ⅜ inch of the coil at the rear. Then rotate the video head disc clockwise until the rear coil is on line with the rewind sensing coil.

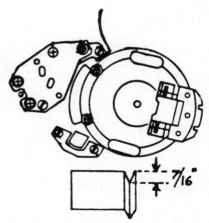

Fig. 8-14. *Protective shield made from a calling card.*

Fig. 8-15. *A typical VCR service bench.*

8. Hold the rotating disc in place and continue brushing toward the rear of the head disc assembly.

9. When the tape path around the drum has been brushed to a dull finish, blow any possible remaining aluminum particles out the rear of the VCR machine with a compressed air gun or use the canned air that can be found in some VCR head cleaning and maintenance kits. This brushing procedure should give many more taping hours without head replacement. A portion of my VCR service bench is shown in Fig. 8-15.

9

General Electric VCRs

Figure 9-1 shows a complete General Electric portable VCR system that consists of three units. The model 1VCD2021X is the VCR deck, the model 1CVT600 is the programmable tuner/timer unit, and the model 1CVA400 is the power supply. The specifications for all three units are given in Fig. 9-2. The information in this chapter is courtesy of GE.

DISASSEMBLY OF VCR

The first thing to do is take apart the VCR properly to obtain access to the circuits in question.

THE CABINET

The flowchart in Fig. 9-3 indicates disassembly steps of the cabinet parts and the PC boards in order to find the areas in need of servicing. When reassembling, perform the steps in the reverse order as those shown in the flowchart. *Note:* because this model is designed very compactly and uses locking tabs instead of mounting screws, work with extreme care when servicing these units.

THE VCR UNIT
Removal of the Bottom Case

1. Place the deck upside down so the bottom case faces upward.
2. Remove 4 screws (A) as shown in Fig. 9-4. Now remove the bottom case by lifting the rear portion of it. *Note:* when reinstalling, first insert the locking portion into the slot of the front panel. Final adjustments are required if the cassette guide and the cassette holder unit are replaced and/or removed.
3. Place a drop cloth or any soft materials under the PC boards or deck for preventing them from being damaged while servicing.

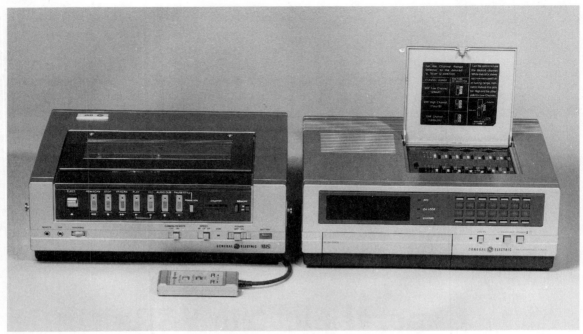

Fig. 9-1. *A General Electric portable VCR. Courtesy of GE.*

Removal of the Cassette Cover

1. Turn the deck over again so the cassette cover faces upward. Press the eject button to raise the cassette compartment Fig. 9-5.
2. Remove 2 screws (B). Then carefully lift and turn the rear portion of it to remove. Be careful not to damage the locking portion. *Note:* when reinstalling, first match the locking portion of the cassette cover to the tab on the cassette holder unit.

Removal of the Top Case

1. First confirm that the battery is inside the battery compartment or not. If it is, take the battery out.
2. Remove 2 screws (C). Then carefully lift the rear portion and pull toward the rear of the deck to remove (Fig. 9-6). While removing, keep the handle up and hold it with your hand. *Note:* when reinstalling, first insert the locking portions into the front panel.

Removal of the Front Panel

1. Stand the VCR deck so the control panel faces upward (Fig. 9-7).
2. Hold both the right and left ends of the front panel, and carefully lift and turn the top portion of it to remove. *Note:* do this step with extreme care so as not to damage the locking portions.

Removal of the Cassette Guide

Remove 2 screws (D) and the cassette guide. *Note:* when reinstalling, insert the cassette tape and ensure that the clearance between tape and projections on the cassette guide is more than 1 mm. Then tighten two screws (D) as shown in (Fig. 9-8).

Removal of the Cassette Holder Unit

Remove four red screws (E) and the cassette holder unit as shown in (Fig. 9-9). *Notes:* when this part is removed or replaced, the cassette

SPECIFICATIONS for 1CVD2021X

"SLP" described in this Service Manual means "EP".

Power Source:	12 VDC
	Battery 1CVA100
	Prog. Tuner Unit 1CVT600
	(Not available independently)
	AC Adaptor 1CVA400
Power Consumption:	Approx. 9.4 W at Play mode
Television System:	EIA Standard (525 lines, 60 fields)
	NTSC color signal
Video Recording System:	2 rotary heads helical scanning system
	Luminance: FM azimuth recording
	Chrominance: Converted subcarrier phase shift recording
Audio Track:	1 track
Tape Format:	Tape width 1/2″ (12.7 mm) high density tape
Tape Speed:	SP mode: 1-5/16 i.p.s (33.35 mm/s)
	LP mode: 21/32 i.p.s (16.67 mm/s)
	EP mode: 7/16 i.p.s (11.12 mm/s)
Record/Playback Time:	1 (SP), 2 (LP) or 3 (EP) hours with 1CAS060
	2 (SP), 4 (LP) or 6 (EP) hours with 1CAS120
FF/REW Time:	Less than 6 min with 1CAS120
Heads:	Video: 2 Rotary heads
	Audio/Control: 1 stationary head
	Erase: 1 full track
	1 audio track erase for audio dubbing
Input Level:	Video: VIDEO IN Jack (RCA type) 1.0 Vp-p, 75 Ω unbalanced
	Audio: MIC IN Jack −70 dB, 600 Ω unbalanced
	TV Tuner: VHF Input: VHF Ch2-Ch13, 75 Ω unbalanced
	(1CVT600) UHF Input: UHF Ch14-Ch83, 300 Ω balanced
Output Level:	Video: Video OUT Jack (RCA type) 1.0 Vp-p, 75 Ω unbalanced
	Audio: Audio OUT Jack (RCA type) −6 dB, 600 Ω unbalanced
	Earphone Jack: −20 dB, 200 Ω unbalanced
	RF Modulated: Ch3/Ch4 switchable, 72 dBμ (open voltage), 75 Ω unbalanced
Video Horizontal Resolution:	More than 230 lines
Audio Frequency Response:	SP mode: 100 Hz—8 kHz
	LP mode: 100 Hz—6 kHz
	EP mode: 150 Hz—5 kHz (10 dB down)
Signal-to-Noise Ratio:	Video: SP mode: better than 40 dB Audio: SP mode: better than 42 dB
	LP mode: better than 40 dB LP mode: better than 40 dB
	EP mode: better than 40 dB EP mode: better than 40 dB
	(Rohde & Schwarz noise meter)
Operating Temperature:	32°F—104°F (0°C—40°C)
Operating Humidity:	10%—75%
Weight:	13.5 lbs (6.1 kg) (with battery)
Dimensions:	12″(W) × 4-1/2″(H) × 9-11/16″(D)
	(304(W) × 114(H) × 245(D) mm)

SPECIFICATIONS for 1CVT600

Power Source:	120 VAC ± 10%, 60 Hz ± 0.5%
Power Consumption:	Approx. 52 W DC out 12 V 1.6 A
Television System:	EIA Standard (525 lines, 60 field)
Timer:	2 weeks/8 programs programmable timer
Input:	VHF Ch2-Ch13, 75 Ω unbalanced
	UHF Ch12-Ch83, 300 Ω balanced
	RF (Ch3 or Ch4)
Output:	Video: (10P connector)
	1.0 Vp-p, 75 Ω unbalanced
	Audio: (10P connector)
	−6 dB, 600 Ω unbalanced
	AC Outlet: 120 VAC Max 300 W unswitched
Operating	
Temperature:	32°F—104°F (0°C—40°C)
Operating Humidity:	10%—75%
Weight:	10 lbs. (4.6 kg)
Dimensions:	11-7/16″(W) × 4-3/8″(H) × 9-11/16″(D)
	(289(W) × 110(H) × 245(D) mm)

SPECIFICATIONS for 1CVA400

Power Source:	120 VAC ± 10%, 60 Hz ± 0.5%
Power Consumption:	Approx. 44 W
Output:	12 VDC Max. 1.6A for deck operation
	15 VDC Max. for battery charge
Operating	
Temperature:	32°F—104°F (0°C—40°C)
Operating Humidity:	10%—75%
Weight:	5.9 lbs (2.7 kg)
Dimensions:	4-3/16″(W) × 4-5/16″(H) × 9-11/16″(D)
	(105(W) × 109(H) × 245(D) mm)

Weight and dimensions shown are approximate.
Specifications are subject to change without notice.

Fig. 9-2. *Specifications of the model 1CVD2020X VCR. Courtesy of GE.*

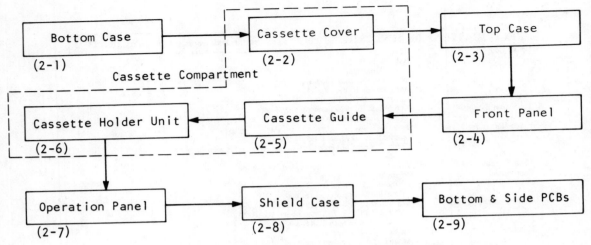

Fig. 9-3. *VCR disassembly flowchart. Courtesy of GE.*

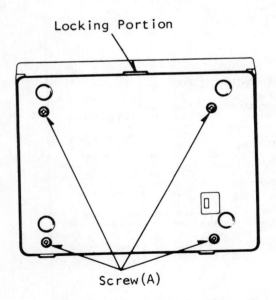

Fig. 9-4. *Bottom case removal.*

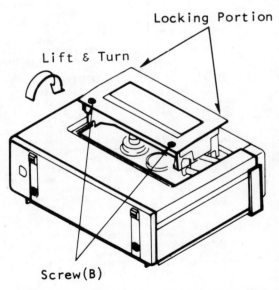

Fig. 9-5. *Removing the cassette cover. Courtesy of GE.*

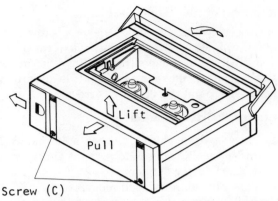

Screw (C)

Fig. 9-6. *Removal of the top case. Courtesy of GE.*

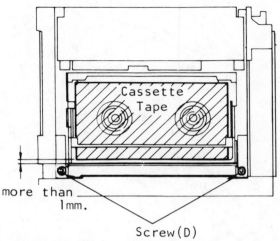

more than 1mm.

Screw(D)

Fig. 9-8. *Cassette guide removal. Courtesy of GE.*

holder adjustment should be performed. When reinstalling, ensure the pin located at lower left portion is engaged with the connecting rod. Refer to Fig. 9-10.

Removal of the Operation Panel

Note: as the space in this section is very compressed, work with care when doing each step. The operation panel need not be removed except when replacing the operation panel or servicing components mounted on it or on the system control boards.

1. Unlock two locking portions located on the lower part of each side. Refer to Fig. 9-11.
2. Unlock two locking portions located on top of each side.
3. Disconnect two connectors that are connected to the earphone jack and the battery meter.
4. Carefully unlock eight locking portions that lock the board to the operation panel. When reinstalling the operation panel, reconnect the two connectors. Make sure you reinstall the operation control buttons and knobs.

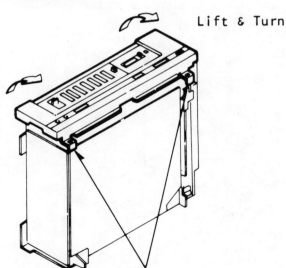

Lift & Turn

Locking Portion

Fig. 9-7. *VCR front panel removal.*

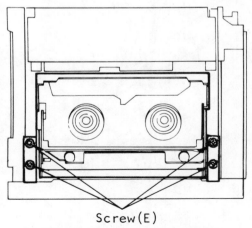

Screw(E)

Fig. 9-9. *Cassette holder removal.*

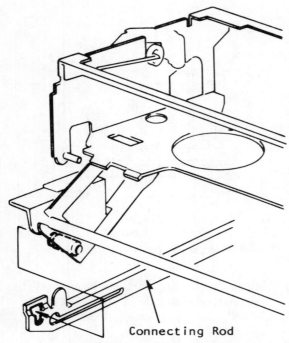

Fig. 9-10. *Installation of connecting rod.*

Removal of the Shield Case

Loosen two screws (F), remove a screw (G), pull the shield case toward the back of the deck, and lift it up to remove (Fig. 9-12). When reinstalling panel, make sure the red lead wire is restored around the dumper correctly.

Circuit Board Swing-out

1. Place a drop cloth or other soft material under the PC boards to prevent them from damage during servicing.
2. This procedure is required when the VCR deck is serviced with the operation panel in place. If this is not the case, first perform steps 1 and 2 of the "Removal of the Operation Panel" section.
3. When reinstalling, make sure the connectors are connected and any electrical components are not damaged.

Procedure for Opening PC Boards

1. Disconnect the connector (P21) that is connected to the tape counter.
2. Remove two screws (H), unlock the two locking portions, and carefully open the PC board. Support the PC boards with your hand to prevent them from being laid down. Refer to Fig. 9-13.

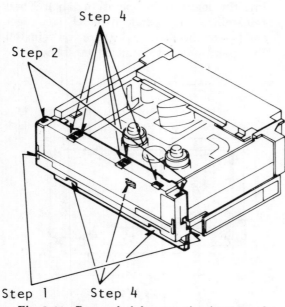

Fig. 9-11. *Removal of the operating front panel.*

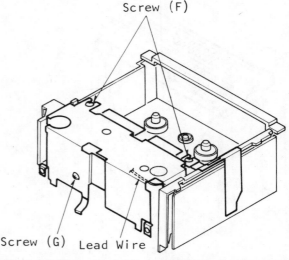

Fig. 9-12. *Removal of the case shield. Courtesy of GE.*

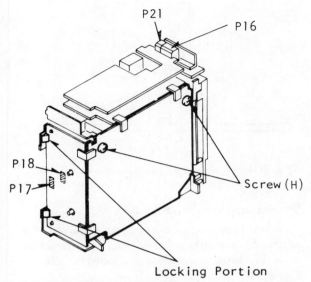

Fig. 9-13. *Opening the PC boards for servicing. Courtesy of GE.*

3. Disconnect the two connectors (P17 and P18) on the audio and chrominance board. Now release the leads connecting the jack panel board and P16 from the cable lead clamp. Carefully lay down the PC boards.

ELECTRICAL ADJUSTMENT PROCEDURES

The following test equipment is required to make these adjustments:

- DVM (digital voltmeter), voltage range 0.001 V to 50 V.
- Frequency counter, frequency range: 0 to 10 MHz

Refer to Fig. 9-14 for location of the power supply adjustments and components.

+12 Volt dc Adjustment

Use test point TP102 as shown in Fig. 9-15 (R122 +12 volt dc adjustment control).

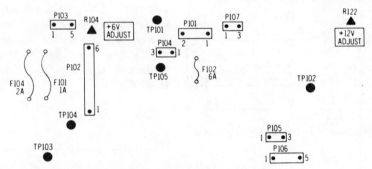

Fig. 9-14. *Power supply board. Courtesy of GE.*

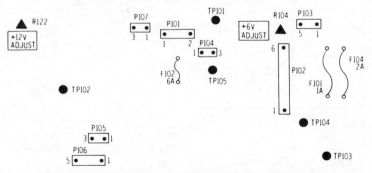

Fig. 9-15. *Power supply PC board component side view.*

205

1. Check the ac input voltage for 120 volt ac and then connect the electronic programmable timer/tuner unit to the deck as shown in Fig. 9-16.
2. Turn on the power switch on the VCR deck.
3. Connect the DVM between TP102 (+) and TP101 (GND) on the power supply board as shown in Fig. 9-17.
4. Adjust the +12 volt dc adjust (R12) for +12 volt +0.05 volt dc.

+6 Volt dc Adjustment

Use test point TP105 and adjustment control R104.

1. Connect the DVM between TP105 (+) and TP101 (GND) on the power supply board as shown in Fig. 9-18.
2. Adjust the +6 volt dc adjust (R104) for +5.65 ± 0.05 volts dc.

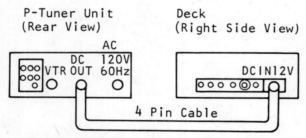

Fig. 9-16. *VCR deck and P-tuner unit connections.*

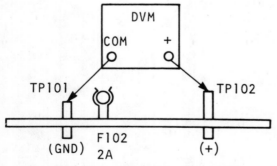

Fig. 9-17. *Power supply board connections.*

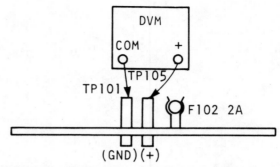

Fig. 9-18. *Power supply board voltage check points.*

Programmable Timer Section

When this section is adjusted, remove three screws (A) the channel select and potentiometer unit. The programmable timer board is shown in Fig. 9-19.

Clock Adjustment

Test point: pin 40 of IC6705 or IC6706. Adjustment: C6705 (clock adjust). Caution: because the trimmer C6705 (clock adjust) has already been critically adjusted at the factory, do not try to adjust the trimmer except after replacing the crystal (X6701) and trimmer (C6705).

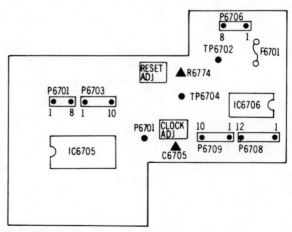

Fig. 9-19. *Programmable timer board component side view.*

1. Connect the frequency counter to pin 40 of IC6705 or IC6706.
2. Adjust the clock adjustment control (C6705) for 262,144 Hz ±5 Hz on the frequency counter. Note: because this frequency is used for the clock timer, it must be set as precisely as possible.

Reset Voltage Adjustment

Test points: TP6702 and TP105. Adjustment: R6774 (Reset voltage adjust).

1. Connect the DVM to TP6702 on the programmable timer board.
2. Adjust the +6 Vdc adj (R104) on the power supply board to 4.0 ± 0.05 volts at TP6702 on the programmable timer board.
3. Turn the reset vol adj (R6774) on the same board fully counterclockwise.
4. Slowly adjust the reset vol adj (R6774) counterclockwise and find the extinguishing point of the timer display tube.
5. Set the +6 volt dc adj (R104) to 5.65 volt dc at TP105 on the power supply board.

10

RCA VCRs

Information in this chapter is courtesy of RCA Consumer Electronics Manual VCR-1-S1, copyright 1978.

The RCA Model VBT200 VCR utilizes the VHS (Video Home System) recording format. This machine uses a direct-drive upper cylinder (head wheel) assembly that is powered by a three-phase, ac advanced motor. The combination of this motor and a low-mass headwheel unit is the key to the system's stability, required for color TV recordings of 4 hours. In addition, a very simple automatic tape threading system minimizes tape handling and ensures longer tape life. The mechanical operation of the VCR is controlled by the transport-control electronics system. Other circuits contain the power supply, signal processing, servo control and SP/LP switching.

The block diagram shown in Fig. 10-1 illustrates the various interrelationships among the several sections of the recorder. The video output from either a TV receiver-type "front end" (TV tuners and demodulator) or a direct video input are fed to the recording circuitry. This circuitry consists of separate luminance and chroma record sections whose outputs are combined in the record amplifier and fed to the video heads.

During playback, the recorded signal from the tape is fed to separate luminance and chroma playback sections after passing through preamplifier circuitry. The output of these circuits is combined in a video amplifier, whose output is fed to an RF converter. The converter remodulates the video information on either channel 3 or 4 for playback on a TV set. A section of the circuit (not shown) samples input video in the record circuit and couples it to the RF converter via part of the playback video amplifier. This set-up, called the E-E mode (electronics-to-electronics mode), lets the user monitor the video program being recorded.

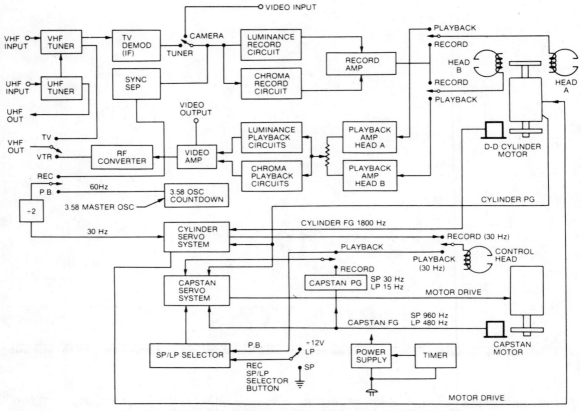

Fig. 10-1. *Overal block diagram of the RCA model VBT200 VCR.*

Cylinder head and tape transport speed control is provided by two separate servo systems called the cylinder head and capstan servo systems. The cylinder servo system receives reference signals from either vertical sync during recording or from a master 3.58 MHz oscillator (counted down to 60 Hz) during playback. These signals are compared against pulse-generated (PG) sample pulses (give head position) to maintain the exact cylinder head speed and position phasing required. Also, during recording, pulses that represent vertical sync are recorded by the control head along the bottom edge of the tape. These serve to synchronize the capstan servo system during playback.

The capstan (tape transport) servo system references instantaneous cylinder head position signals against a sample of capstan motor speed frequency generator (FG) signals to maintain constant tape transport speed during recording. During playback, the instantaneous cylinder head position information is referenced against the control track output signals to provide minute adjustments of capstan speed. In this way, the tape transport speed is dynamically adjusted to assure that the rotating heads properly align with the recorded video tracks.

The user selects 2-hour (SP) or 4-hour (LP) recording modes that determine transport speed during recording. The SP/LP selector circuit automatically determines the correct transport playback speed by sampling the control track pulse rate.

The power supply circuitry converts standard

the 120 volt ac power line voltage into three dc supplies for use by the recorder. Power consumption for this VCR unit is 45 watts.

POWER SUPPLY CIRCUITS

This VCR power supply uses a power transformer and two bridge rectifier circuits to develop unregulated 12 volts dc and 18 volts dc. A third unfiltered supply, using diodes D107 and D108, produces the "power-off-detector" supply. Refer to the power supply circuit in Fig. 10-2. This voltage is applied to the transport and control board logic system to sense a power failure and operate the stop solenoid so that the machine is not left in an operating mode in the event of ac power loss. A second power transformer supplies 16 volts ac and 3 volts ac to operate the digital timer.

The B+ distribution in Fig. 10-3 shows that the unregulated 12-volt supply output is applied to the transport and control board where it encounters some switching. This voltage is also supplied to the D-D motor board where it provides input power to operate the three-phase inverter that then drives the D-D cylinder motor.

The unregulated 18-volt supply is also fed to the transport and control board where it provides power to operate much of the logic circuitry contained on this board as well as the stop solenoid. Also derived from the unregulated 18-volt supply is regulated 12 volts dc that is generated by a series regulator circuit that utilizes low-level driver circuitry on the transport and control board and a power transistor that is chassis mounted. The regulated 12 volts appears at the emitter of the power transistor and can be measured at test point TP614. Note that the "power on" indicator (an LED) is powered from

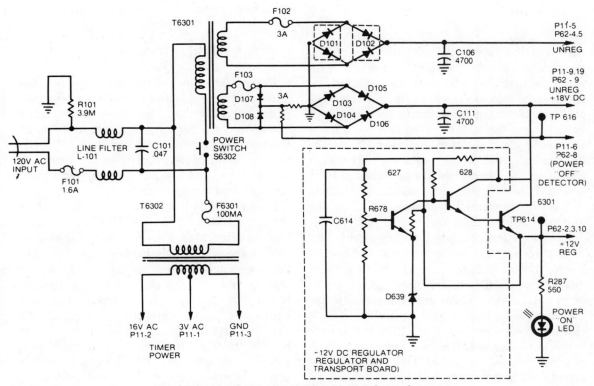

Fig. 10-2. *Power supply circuit and 12-volt regulator.*

211

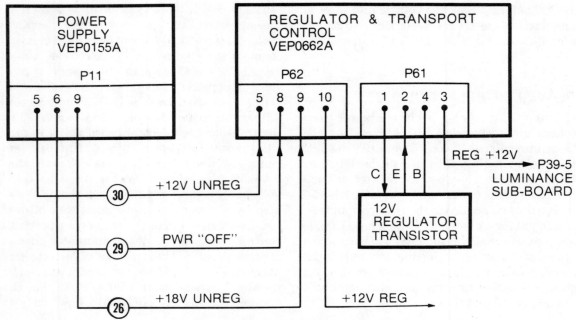

Fig. 10-3. *B+ distribution block diagram.*

the regulated 12 volts. Thus, from a servicing standpoint, if stop-solenoid action is heard when the machine is turned on and off but the power-on indicator does not come on, there is a problem in the regulator circuit because the 18 volts powers the stop solenoid and feeds input voltage to the regulator that then drives the "power-on" LED. Note that most of the VCR circuitry is powered from the 12 volts regulated source. However, the regulated 12 volts is then divided into several sub-sources via switching and various logic functions.

In summation, the power supply circuit provides three voltages to the transport and control board. These voltages are +12 volts unregulated, "power-off" indicator voltage, and +18 volts unregulated. These are all directed to the transport and control board. A driver circuit on this board supplies bias to the 12-volt regulator transistor. Also associated with plug P61, pin 3, is the regulated output of 12 volts from this board that is fed to the luminance subprocess board via the P39-5 plug.

B+ DISTRIBUTION

Refer to the block diagram in Fig. 10-4. Several additional 12-volt supplies are obtained from the regulator and transport control board. These voltages, for the most part, are routed to the servo board. An exception is the P63 plug that feeds unregulated 18 volts to operate the stop solenoid and a source of regulated 12 volts for the audio board of the VCR.

Several B+ sources come from Plug P65 via the servo board. At pin 1 of P65 is a +12 volts loading end source. This voltage becomes available at the end of tape loading in the machine because switch S6305 closes at the completion of loading and provides this voltage to the servo board. Pin 2 output is +12 volts regulated, which supplies the majority of the circuitry in the VCR that is common to the record and playback modes. Pin 3 feeds +12 unregulated, which supplies power to the D-D cylinder motor three-phase inverter circuit. A source of voltage known as "forward +12 volts" is available at pin

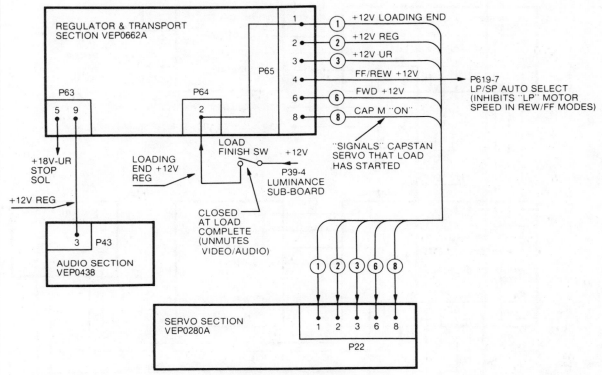

Fig. 10-4. *Voltage distribution.*

6. This voltage is present when the forward leaf switch on the transport board is actuated by the play button. This voltage is then processed into several different sources of "play" and "record" voltages. Pin 4 supplies a voltage — a "fast-forward/rewind +12-volt override voltage" — that is sent to the SP/LP auto select board. The purpose of this voltage is to inhibit LP motor speed whenever the machine is in rewind or fast-forward. Finally, emerging from plug P65 pin 8 is a voltage source known as "capstan-motor-on." This voltage becomes available at the instant the unload-finish switch opens to signal the capstan motor that the machine has started to load; thus, the capstan motor drive circuitry is enabled so the motor begins to run.

The B + Distribution, shown in Fig. 10-5, has a source voltage called "except record +12 volts" and is available at P21 pin 1. This voltage source

is present in the play or E-E modes of operation, but not present in record. Pin 8 of P21 supplies a voltage source called "delayed record +12 volts." This voltage is supplied a brief instant after loading is completed as signaled by the load-finish switch. The source of this voltage is logic and time delay circuitry located on the servo board that is triggered by the closure of the load finish switch. The record selection function is obtained by passing this voltage through a section of the record/play switch so that the voltage is only present during record.

Another voltage, "delayed forward +12 volts" (supplied to audio board via P24 pin 2), is fed from the same time-delay circuitry. The only difference is that this voltage does not pass through the record/play switch. Thus, the voltage is available any time the play button is depressed. Also, a source of record +12 volts is

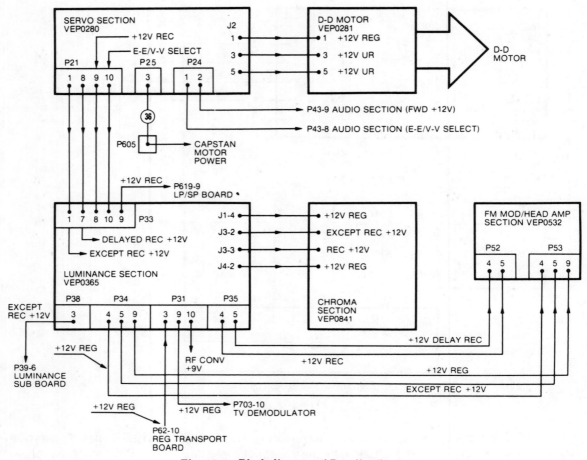

Fig. 10-5. *Block diagram of B+ distribution.*

tapped from the regulated B+ source via a section of the record/play switch on the servo board. This voltage is at pin 9 via plug P21.

At pin 10 of P21 is a voltage called the "E-E/V-V select voltage." This voltage is present when the machine is in the E-E mode of operation. Whenever record or play is started, this voltage source is defeated.

Other voltage sources emerging from the Servo board include the "delayed-forward +12 volts" that is fed to the audio section as well as "E-E/V-V select voltage" that is also supplied to the audio section. Emerging from plug 25 pin 3 is the dc voltage to operate the capstan motor.

The "P21" voltages are fed to the luminance board via plug P33. As seen in Fig. 10-5, the luminance board serves to distribute these various sources of B+ to other circuit boards in the machine. Power is applied to the chroma board via the interconnecting wire jumpers that link the chroma and luminance board. Additionally, "except-record B+ 12 volts" is fed to the luminance subprocess board via P38 pin 3. Plug P34 supplies power to the FM modulator head amplifier section of the machine (also, to P35 pins 4 and 5). The luminance board also supplies +12 volts regulated via P31 pin 9 to power the TV demodulator. Pin 10 of the same plug feeds +9 volts to the RF converter board.

DELAYED B+ SOURCE

This VCR uses two delayed voltages, provided to allow the system to stabilize before record or playback after tape loading. Also associated with the delayed B+ source is an interface with the pause switch. In addition to stopping the transport operation, it also interfaces with the mute circuitry.

E-E/V-V SELECT

The E-E/V-V select (see Fig. 10-6) shows an emitter-follower transistor (TR215) that is powered from the +12-volt regulated supply to provide a switched source of bias to operate electronic switches on the audio and luminance boards that either select the E-E (monitor) mode of operation or allow the machine to function in playback (V-V). When the VCR is in the monitor mode, transistor TR215 is conducting and thus supplies "E-E select" bias to the audio boards via plug P24 pin 1, and via diode D216 and P21-10, the voltage is supplied to the luminance board. When the machine goes into the

record or playback mode, the "E-E select voltage" is turned off, allowing the machine to function in play or record. Note that in the record mode, diode D215 supplies record +12 volts to the same circuitry on the luminance board that is active for E-E mode operation. This means that this particular circuitry must be deactivated when the machine is in the playback mode. Base bias (control voltage) for emitter-follower transister TR215 is obtained from another transistor (TR214) that implements an AND logic function between the load end +12 volts and the forward +12 volts. Note that base bias for this device is supplied via an RC time-delay network from the load-end +12-volt supply and the emitter of the device is grounded via the collector/emitter circuit of TR219 that conducts when forward +12 volts is available. Thus, in actual operation, depressing the play or the record/play buttons immediately causes conduction of transistor TR219 when forward +12 volts is applied from the leaf switch on the transport board. At the completion of loading, capacitor C248 charges through R267 until sufficient

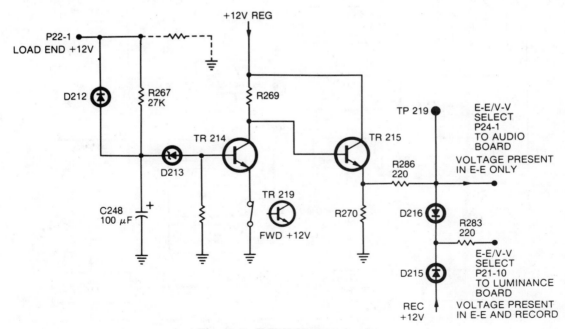

Fig. 10-6. *E-E/V-V select circuit.*

voltage is present on the capacitor. At this time, zener diode D213 conducts and drives the base of transistor TR214 into saturation to cause the collector of this device to go low, remove base bias from TR215, and turn it off.

The time-delay action provided by R267 and C248 delays the transition from E-E (monitor) mode into the record or play mode for a period long enough to allow tape transportation to stabilize.

PLAY/RECORD SIGNAL CIRCUIT OPERATION

In playback, the VCR's two video heads feed in individual preamps that are turned "on" when the particular head is in contact with the video tape. Preamplifier output is summed to-gether into a continuous RF signal prior to entering separate luminance and chroma circuits. Refer to the block diagram (Fig. 10-7).

SIGNAL CIRCUIT (PLAYBACK MODE)

The B&W signal is 3.4 to 4.4 MHz FM, which is fed to limiter stages that remove all amplitude variations including the chroma signal content. Limiter output drives an FM demodulator that recovers the original B&W video signal, complete with the overshoots introduced by the record emphasis. In order to correct frequency response of the video signal, the signal is passed through appropriate deemphasis networks that are switchable for 2- or 4-hour operation. After deemphasis, the video signal passes through some noise cancellation circuitry and emerges as 1 volt p-p negative sync video to drive the RF

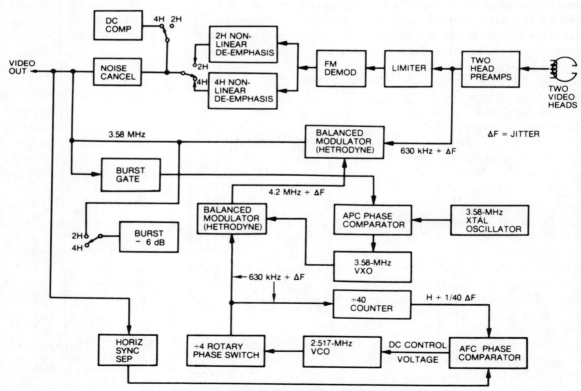

Fig. 10-7. *Signal circuit play block diagram.*

converter in the machine to provide channel 3 or 4 output to a TV receiver.

A dc compensation stage is switched in when in the 4-hour mode to remove the dc level shifts that were introduced in record to produce the half line offset. This dc compensation is a 30 Hz square wave that corrects the dc component of the video signal in the shifted fields to agree with the dc level of the video in the nonshifted fields.

The chroma signal consists of 629 kHz RF plus a "jitter-frequency" component. Jitter frequency is explained as follows: The video head speed and the tape-transportation speed should be held to a very tight tolerance, and in practice they are. However, an extremely small change in video head velocity, which would produce no noticeable change in a B&W picture, can produce chroma phase errors so severe that the picture is not viewable. Thus, this effect of minute speed variations, called the *jitter* component, must be removed in order to display proper color. If the 629 kHz down-converted chroma signal containing jitter was beat against a 4.2 MHz signal that contained the same jitter component, the difference would be a constant 3.58 MHz signal without the jitter component. This is accomplished by passing the 629 kHz signal into a balanced modulator where it is beat against 4.2 MHz with jitter to produce an output of 3.58 MHz chroma. In playback, this 4.2 MHz signal plus jitter is produced by the second balanced modulator.

Now, the 3.58 MHz VXO is locked to a locally generated constant 3.58 MHz signal derived from a crystal oscillator. The other input to the balanced modulator is 629 kHz plus the jitter factor. The 629 kHz signal is provided by the same 2.517 MHz VCO that is locked to the horizontal sync. Because in playback the horizontal sync pulse contains the small time-base errors caused by minute variations in headwheel speed, the playback horizontal sync pulse can be applied to the AFC phase comparator to produce dc control voltage with jitter modulation. Thus, this jitter is introduced into the 2.517 MHz signal whose output is divided to 629 kHz. In this way, the 629 kHz output contains the jitter component that cancels the jitter in the 629 kHz incoming chroma.

E-E OPERATION

Before the start of recording, the VCR is in a mode of operation known as E-E that couples the input signal from the TV demodulator via some of the record/play electronics and then to the video output circuitry where it modulates the RF converter for use of the TV receiver. The purpose of E-E operation is to allow the operator to view the picture that is to be recorded. For example, it allows the user to determine that the camera picture is adjusted properly or whether the tuner is set to the correct channel.

As shown in Fig. 10-8, the video signal from the TV camera switch is routed to the luminance board via P32-1. This 1-volt p-p negative-sync can be scoped at TP 301, which is the input to a buffer amplifier (TR301) that serves as an isolation and gain stage to prevent the following circuitry from loading the VCR video input.

The buffered video signal is fed to a low-pass filter comprised of LC components FL-301 with a filter cut-off frequency of about 5.8 MHz. The purpose of the filter is to remove any high-frequency components from the input composite video signal. Output signal from the filter passes into phase-compensator transformer T301, whose purpose is to correct the phase shifts introduced by the low-pass filter. Phase compensated video signal is then fed to pin 11 of IC 301, the video processing chip.

Blocks within IC 301 take the composite video signal (including color), feed it to an AGC amplifier whose purpose is to correct for any abnormal input level conditions and thus provide a constant 1-volt video signal to the actual video record circuitry. AGC regulated video is then coupled to a video amplifier in another part of this chip. The video amplifier output comes out of IC 301 at pin 7. This video signal is then fed to video buffer stage (transistor TR320) via "E-E" level control R3169. Buffer output is routed to a

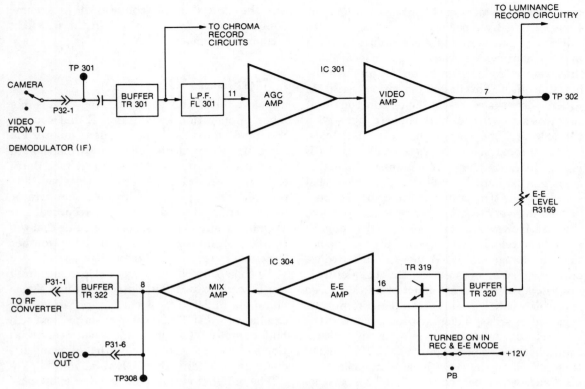

Fig. 10-8. *Simplified E-E block diagram.*

second signal-processing chip (IC 304) via electronic switch transistor TR319. This transistor is turned on during E-E and record operation. Switch output (video) is applied to IC 304 through pin 16 of the device.

Video processing IC 304 is primarily part of the playback circuitry. The composite video (including color) is fed to an E-E amplifier stage. Output of this stage is fed into a mixing amplifier whose output comes from the chip at pin 8. At this point, the video signal is 2 volts p-p and can be "scoped" at video test point TP 308. Signal from pin 8 is directed through video-phase-compensator transformer T307. A buffer stage following the phase the compensator makes the video signal available to the RF converter via plug P31-1.

LUMINANCE RECORD CIRCUITS

The input to the luminance record system closely resembles that of the E-E circuitry because the same electronics are used. Briefly, video signal from the input is fed to the buffer and filter circuitry via P32 pin 1 and TP 301. Video signal entering IC 301 pin 11 is fed to the same AGC and video amplifier stages. The output is pin 7 of the IC or TP 301. From this point, the signal takes a different path than it did in the E-E mode. Refer to **Fig. 10-9.**

Pin 7 video is at an amplitude near 1.5 volt p-p (as scoped at TP 302) and passed through a 3.58 MHz trap which removes the chroma signal components. This trap is switched "in" and "out" of the circuitry by the color-killer system.

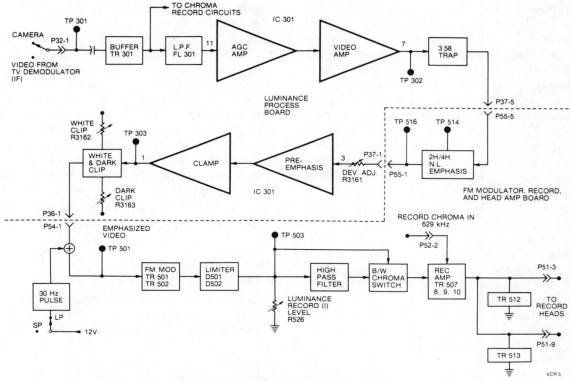

Fig. 10-9. *Luminance record block diagram.*

Basically, when the machine is in the "color" mode, additional low-pass filtering is introduced in the video circuitry to prevent chroma/luminance beats.

Luminance video signal is then routed to switchable 2-hour/4-hour nonlinear video emphasis circuitry contained on the head amplifier board via P37-5. This video is then returned to the luminance board where it is fed back into IC 301 via the FM deviation control. This control is used to regulate the peak-to-peak video level so that the proper signal amplitude is fed to the FM-modulator circuitry on the FM-modulator board.

The video, again processed by IC 301, enters the chip at pin 3, where it encounters additional amplification and preemphasis high-frequency boost. Output from the preemphasis circuitry leaves the chip at pin 1 and reenters the chip at

pin 15 where it encounters a sync-tip clamper circuit that provides dc restoration of the video signal.

Clamped video (TP 303) is a signal of somewhat over 1 volt p-p that has sizable overshoots (spikes). The purpose of preemphasizing (overshoots) the signal is to improve the high-frequency signal-to-noise performance of the machine as well as providing enhancement of picture definition. However, it is important that the overshoots be limited to proper levels to prevent overdeviation of the FM modulator. There are separate adjustable white-clip and dark-clip stages that are used to set the clipping level and thus regulate the amount of overshoot. Preemphasized and clipped video output is applied to the FM modulator board via P36-1.

Input to the FM modulator is emphasized clipped video from the luminance board that enters via P54-1. After passing through an adder stage, the video signal is fed to the FM-modulator circuit. When the unit is in the 4-hour mode, a 30 Hz square wave signal is added to the video to produce the alternate field level shifting necessary for generating the ½ frequency H offset from field to field. The FM modulator is just a multivibration (flip-flop) — transistors TR501 and TR502. When the input voltage corresponds to sync-tip level, the FM modulator runs at 3.4 MHz. When video is applied and the signal is driven towards white, the multivibrator frequency is driven upwards in response to the video signal. The FM modulator output is transformer coupled to a pair of limiter diodes that remove any amplitude variations from the FM signal. Limiter output can be viewed as an FM signal of about 0.6 volt p-p at test point TP 503. This signal, after being level regulated by record-current control R56, is fed to an amplifier stage consisting of TR503 that drives emitter-follower TR504.

Output from the emitter follower is fed through a high-pass filter that is selected by "color-mode" switch TR506 when the unit is operating with a color input signal. When the VCR is operating the B/W mode, switch TR505 routes the signal directly to the video-recorder amplifier. The purpose of the high-pass filter used in color "record" is to prevent FM modulation components from interacting with the 629 kHz chroma signal and producing beats. Thus, this filter has a cut-off frequency of approximately 700 kHz. The video record amplifier then boosts the level of the luminance record signal to provide 12 mA of luminance video record current at the input to the D-D assembly.

Note that the 629 kHz down-converted chroma enters the record amplifier via P52-2 where it is mixed with the luminance component. Thus, the output of the record amplifier is luminance FM with about a 3 mA chroma record current mixed. This signal is then routed to the individual video heads that are driven in parallel.

Note that in the record FM modulator block diagram, the play/record head switching is done electronically. When the unit is in record, the record signal is applied to the heads (via individual transformers) through PS1-3 and -9. The ground ends of the video heads are returned to the circuit board via P51-1 and -10. Note that the heads are grounded through switching transistors TR514 and TR515, which are on when the unit is in record. Notice too, that these switches also ground the input of the playback preamplifiers and peaking capacitors C530 and C545 to prevent them from interacting with the record current. Also evident is that the playback switches TR512 and TR513 are off when the VCR is operating in the record mode.

PLAYBACK HEAD AMPLIFIER CIRCUITS

In playback, the video heads must recover the information impressed upon the tape and present it to the luminance and chroma systems to replicate the original video signal. Due to various electrical and physical phenomenon, it is necessary to shape the response of and compensate for the response of the record amplifier and the inherent limitations of the video tape when the signal is played back. Thus, preamplifiers with special response characteristics are used to restore proper frequency relationship to all of the sideband components when the signal is played back. And due to slight manufacturing tolerances between video heads, not only is it necessary to compensate for the overall video signal, but it is also necessary to match, or equalize, the individual heads so that the field-to-field variation of video signal (RF head signal) is minimal. Referring to the head amplifier play block diagram (Fig. 10-10), observe that the two video heads drive individual IC preamplifiers consisting of IC 501 and IC 502. During playback, one side of each head is grounded through electronic switches consisting of transistors TR512 and TR513, which are on. The head signals are applied to terminal 2 or IC 501 and IC 502 respectively.

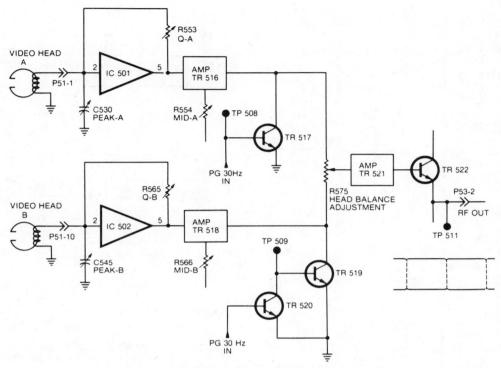

Fig. 10-10. *Head amplifier playback block diagram.*

Associated with the input to these ICs are two trimmer capacitors called "Peak-A" (C530) and "Peak-B" (C545) that set the resonance of the video heads for a peak response at 4.5 MHz. This response peak, and the frequency-response adjustments of the individual preamplifiers, create playback preamplifier frequency characteristics that properly restore the sideband levels to their proper frequency relationships.

Associated with each of the IC preamplifiers is a feedback frequency response control network consisting of RC components and adjustable resistors R553 ("Q-A") and R565 ("Q-B") that are adjusted for proper playback frequency response characteristics. The outputs of both preamplifiers are fed to additional individual stages of amplification consisting of TR501 and TR518 and then to a summing point consisting of mix control R575.

It is most desirable in video playback to turn "on" only the amplifier that is in contact with the video tape in order to prevent the inactive video head from contributing noise to the input signal. The signal levels at the video heads are incredibly small, and any stray noise after amplification will degrade performance. Thus, switching transistors TR517 and TR519 are gated on and off to allow video signal to pass whenever the appropriate head is on the tape. The "summed" output of the heads receives additional amplification by transistors TR521 and emitter follower stage TR522. The summed RF head signal (luminance FM and chroma) is routed off the FM modulator/head amplifier board to the luminance and chrominance circuitry via P53-2. This signal can be scoped at test point TP 511.

As shown in the luminance play block diagram (Fig. 10-11), the RF head signal is fed to the luminance processor board at P34-2 for input to the luminance and chroma processing circuitry.

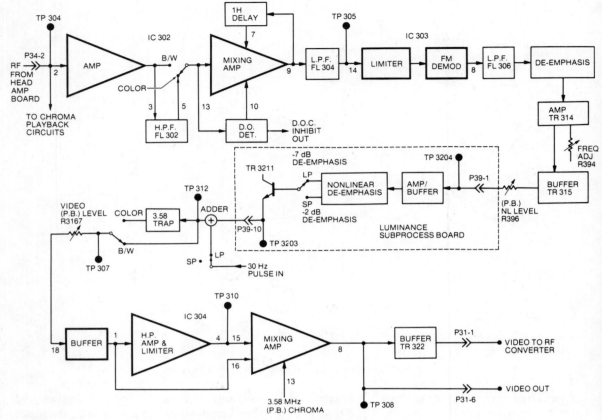

Fig. 10-11. *Luminance playback block diagram.*

THE COLOR OSCILLATORS

The VBT200 VCR uses three separate oscillators to process the chrominance signal. These are the 3.58 MHz XTAL oscillator (crystal), 3.58 MHz VXO (variable crystal oscillator), and the 2.517 MHz VCO (voltage-controlled oscillator). The 3.58 MHz chroma information is down-converted (hetrodyned) to 629 kHz in the record mode by beating the signal against a 4.2 MHz CW signal in a balanced modulator (mixer) circuit. Refer to color oscillators block diagram (Fig. 10-12).

Not only is the record chroma signal down-converted to 629 kHz, but it is also recorded in a configuration where on the first field (Head-1 pass), the phase of the chroma signal is ad-

vanced 90 degrees per line. When the second field is recorded (Head-2 pass), the chroma signal is *retarded* in phase by 90 degrees for each horizontal line. This system of chroma signal recording is called "chroma rotary-phase" recording. This system of chroma recording allows a comb-filter circuit in the playback electronics to effectively cancel chroma crosstalk signal that is present in LP recordings because there is a negative guard band in the LP mode.

In order to accomplish the chroma rotary-phase recording technique, the 4.2 MHz signal used to down-convert the chroma information must change phase by 90 degrees for each horizontal line. Thus, this signal must somehow be keyed by the horizontal sync pulse. The 4.2 MHz cw signal is actually produced by beating to-

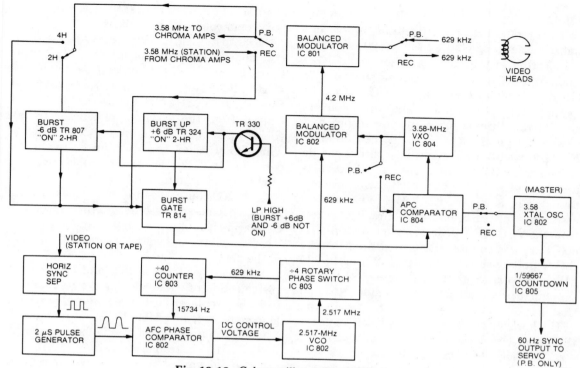

Fig. 10-12. *Color oscillators block diagram.*

gether a 3.58 MHz signal obtained from the variable crystal oscillator (VXO) and a phase-rotary 629 kHz signal. These signals are hetrodyned in a second balanced modulator (mixer), and the sum of the two frequencies is the 4.2 MHz cw signal used to down-convert the chroma to 629 kHz. During record, the 3.58 MHz VXO is locked to the input signal burst by an APC system as found in some color TV sets. The 629 kHz rotary-phase CW signal is extracted from a divide-by-4 counter clocked by a 2.517 MHz signal produced by the voltage-controlled oscillator (VCO) that is phased-locked to the record signal horizontal sync pulse.

In playback, the 629 kHz down-converted chroma signal is up-converted back to the 3.58 MHz frequency by using the same oscillators and balanced modulators in reverse. The incoming 629 kHz signal is fed to the balanced modulator where it is beat against 4.2 MHz phase-rotary CW signal. The difference frequency (3.58 MHz) is the regenerated chroma signal. During playback, the 3.58 VXO (beats against 629 kHz phase-rotary CW to produce 4.2 MHz for up-conversion) is referenced to the 3.58 MHz XTAL (crystal oscillator).

The APC circuit associated with the 3.58 MHz VXO receives output from the 3.58 MHz XTAL oscillator as a reference and the burst component of the up-converted 3.58 MHz color signal is compared against this locally generated 3.58 MHz reference signal to generate an error voltage that is used to correct 3.58 MHz VXO phase and frequency so that the output 3.58 MHz chroma signal has a high degree of phase stability. Also, during playback, the 2.517 MHz VCO is locked to playback horizontal sync. Thus, any jitter component present in the playback signal that could affect chroma phase (color hue) can be cancelled by introducing this jitter component "out of phase" into the 629 kHz phase chroma. At the same time, the hori-

223

zontal sync pulse triggers the gated counter circuit to produce the proper phase of chroma signal required to up-convert the chroma signal.

Referring to the block diagram (Fig. 10-12) again, you will note that four integrated circuits (IC 801 to 804) provide most of the circuitry required for up-conversion and down-conversion of the chroma signal during record and playback. Contained in IC 802 is the 3.58 MHz XTAL (crystal) oscillator whose output can be scoped at TP 807. During "record" this oscillator is switched "off." The other oscillator contained in IC 802 is the 2.517 MHz VCO and its associated phase-comparison circuit. Output from the VCO (IC 802 pin 17) is fed to pin 6 of IC 803, the gated counter circuit that produces the rotary-phase 629 kHz CW signal. One of the 629 kHz signals is fed to a divide-by-40 counter in the chip to produce a horizontal frequency signal that leaves the chip at pin 2 to provide an "FH"(horizontal frequency) feedback signal to the phase-comparator circuitry located in IC 802. This signal enters the chip at pin 18. The other input to the AFC phase comparator is the 2-microsecond pulse that is timed by horizontal sync. This signal is produced by circuitry in IC 803 and comes out of the chip at pin 14. This pulse (TP 811 signal) is sent to the phase-comparator circuitry in IC 802 and enters the chip at pin 7. Thus, this feedback loop locks the 2.517 MHz VCO to a multiple of 160 times the horizontal scan frequency (2.517440 MHz).

This signal is fed to pin 6 of IC803 where it is counted down to a nominal 629.360 kHz in the chip. Then, gating circuitry selects the proper output from the counter to provide a signal that represents phase zero with respect to the chroma signal phase, depending on which head is on the tape at a particular time.

The direction of phase rotation, related to which head is on the tape at a given time, can be determined and preset by sensing whether the PG pulse is positive-going or negative-going. The cylinder head PG (pulse generator) pulses from J1 pin 1 is fed to pin 7 of IC 803 where it selects the direction of rotation for the four-bit

counter. The outputs of the four gates that select the proper counter signal phase are summed together into a single 629 kHz rotary-phase signal that emerges from the chip at pin 4. This output signal is sent back to the balanced modulator stage in IC 802 where it beats with 3.58 MHz CW from the 3.58 MHz VXO contained in IC 804. Output from the balanced modulator IC 802 comes out at pin 12 as a nominal 4.2 MHz signal. This signal is fed to the balanced modulator contained in IC 801 where the actual conversion of chroma information is performed.

The VXO contained in IC 804 is locked to the input chroma-signal burst during record by comparing its output to burst in an APC circuit. The 3.58 MHz burst signal enters IC 804 at pin 3 of the chip.

Transistor switches TR815 and TR816 make the selection of which signals are compared to lock the oscillator during record and playback. During record, 3.58 MHz VOX output emerges from the chip at pin 14 and is fed back to the APC circuit via transistor switch TR815. The other signal applied to the APC system is burst, which is fed via burst-gate transistor TR814 and some limiter circuitry into chip pin 6. This burst signal can be scoped at TP813. During playback, switch TR816 is turned on to supply reference signal from 3.58 XTAL oscillator located in the IC 802 chip. This signal (from TR816) is fed into pin 3 of IC 804. The other signal (up-converted playback signal burst) enters the comparator via IC 804 pin 6. Thus in playback, the regenerated chroma signal is phase-locked to the locally generated 3.58 MHz signal provided by the 3.58 MHz crystal oscillator contained in IC 802. The crystal oscillator in IC 802 also drives a frequency counter (IC 805), which counts the 3.58 MHz signal down to 60 Hz to provide a reference signal to lock the head cylinder servo system during playback.

CYLINDER HEAD SERVO SYSTEM

The cylinder head servo system has two basic feedback control loops. The first loop (shown at the bottom of Fig. 10-13) is the speed control loop that maintains the head cylinder rotation at very nearly the nominal 1800 rpm. Cylinder head speed is sensed by sampling output from the D-D motor FG assembly. This is a 1.8 kHz sine wave signal that is fed to a frequency amplifier whose output is shaped into a 900 Hz square wave by a divide-by-2 counter.

The square wave signal is fed to a logic AND gate along with the output of a standard-time generator. This generator is a one-shot multivibrator whose pulse width is preset by the cylinder free-running speed control. This constant width pulse is compared with the pulse produced by the cylinder FG in an AND gate . The output of the gate is the difference between these two

pulses. If the motor tends to run fast, the width of the gate input pulse (motor-speed sample) becomes narrower; thus, the output pulse that represents the difference between the sample pulse and the standard-time generator pulse also becomes narrower. The gate output pulse is integrated (filtered) to provide a dc signal to a motor drive circuit. In case the motor slows down, the pulse width becomes wider, more motor drive is produced, and the motor speeds up. In this way, the motor speed is held close to 1800 rpm.

The top section of the block diagram shows the phasing (position) control part of the servo system. During record, the video signal is fed to a sync separator that separates out the 60 Hz ver-

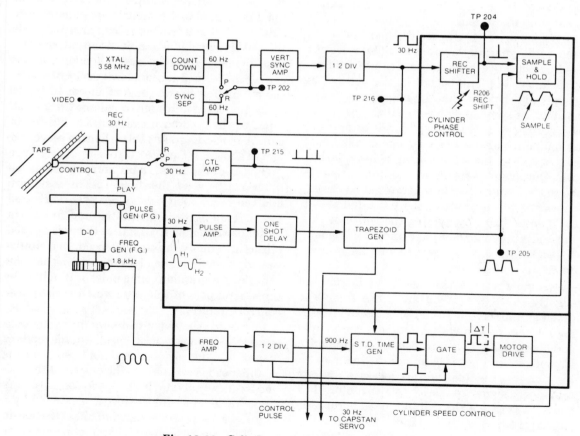

Fig. 10-13. *Cylinder servo system block diagram.*

tical sync pulse. Vertical sync is fed to an amplifier and through a divide-by-2 counter to produce a 30 Hz square wave signal. This signal follows two paths. One path is an additional amplifier that generates the 30 Hz square wave control-track signal which is recorded along the bottom edge of the tape. The other path is through a time delay circuit that allows the video heads to be phased with respect to vertical sync. The output of this stage, known as the "record shifter," is a narrow sample pulse that is fed to a sample-and-hold circuit as the reference signal.

The feedback signal (represents speed and position of the cylinder head) is taken from the cylinder motor pulse generator coils (PG). This 30 Hz signal tells two things: which head is on the tape and where the head is on the tape at any particular instant. After passing through a pulse-amplifier circuit and a "one-shot" delay, the signal drives a trapezoid generator. (The trapezoid signal is a waveform that has a definite leading-edge rise time and a definite trailing-edge fall time.) This signal is also fed to the sample-and-hold circuit. When the cylinder head motor speed and phasing are exactly right, the trailing edge of the trapezoid is sampled at its midpoint. Each time a sample is taken, it is held at a dc voltage in the sample-and-hold circuit. Output of the sample-and-hold circuit is fed to the standard-time generator and it determines the pulse width of the pulse fed to the AND gate. Thus, by varying the width of this pulse, it is possible to provide a small increment of speed control necessary to accurately position the heads on the tape.

Assume that the cylinder head is running slightly slow. In this instance, the sample is taken at a higher point on the trapezoid resulting in more dc voltage output from the sample and hold circuit. This is translated into a change in conduction time of the standard-time generator such that the comparison process in the gate provides a signal to speed up the motor. The converse is true if the cylinder head is running slightly fast. In this instance, the sample is taken lower on the ramp and the motor is slowed down.

CAPSTAN SERVO SYSTEM

The capstan servo circuitry is much like that used for the head cylinder servo. Again, two loops are used — a speed control loop and a position or phasing control loop. Capstan motor speed is sensed by the capstan FG assembly, which produces a sine wave of 480 Hz in the LP mode and 960 Hz in the SP mode. After amplification and processing by a divide-by-2 counter, the 240 Hz (LP mode) or 480 Hz (SP mode) square wave is available as input signal to a standard-time generator/AND gate circuit similar to that of the head cylinder servo. When the capstan motor is running at the correct speed, the input signal to the standard-time generator/gate circuit is 240 Hz. Note that in the LP mode, the 240 Hz signal is available at the output of the first divider. When the machine is operated in the SP mode, an additional frequency divider in the circuit forces the capstan motor to run at twice the speed in order to supply the 240 Hz input to the standard-time generator/gate circuit. A block diagram of the capstan servo system is shown in Fig. 10-14.

Output of the gate is a pulse representing the difference between the standard-time generator and the input square wave representing motor speed. This pulse is integrated and fed to the motor drive system so that the entire circuit stabilizes at roughly the correct speed. In a similar way to the cylinder head servo system, the pulse width of the standard time pulse is modified by the second part of the loop, which is the phase control circuitry.

In record, the capstan motor provides constant tape transportation speed. In this mode of operation, the feedback signal representing motor speed is taken from the output of the second frequency divider in the speed control chain. Thus, after passing through a divide-by-8 counter, a square wave signal of 30 Hz (SP) or 15

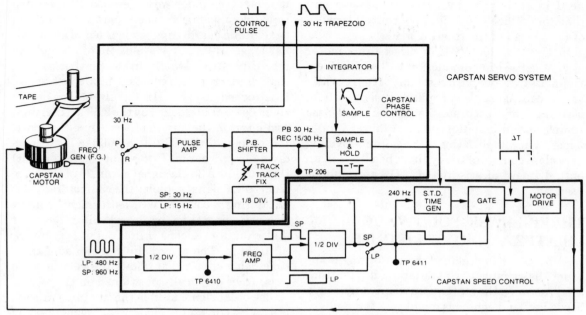

Fig. 10-14. *Capstan servo system block diagram.*

Hz (LP) is available to represent motor speed. This signal, after processing by a pulse amplifier and some delay circuitry, is a sample pulse that is applied as one input to the sample-and-hold circuit. The reference signal input to the sample-and-hold circuit is developed from the 30 Hz trapezoid produced by the head cylinder PG. As with the head cylinder servo, the output of the sample-and-hold circuit is a dc voltage that represents small increments in speed (phase) changes in the capstan motor operation. This dc voltage is fed to the standard-time generator to modify the pulse width and thus produce a vernier-change in motor speed.

During playback, the 30 Hz control-track pulse on the video tape is fed to the pulse amplifier in place of the output of the divide-by-8 counter. This signal (after processing) becomes the sample pulse. Thus, any changes in transportation speed are sensed by a change in the sample point so that the changes in output of the sample-and-hold circuit provide appropriate dc

control to the capstan motor to correct the speed/position errors.

VCR CYLINDER HEAD DRIVE CONSIDERATIONS

One of the problems in designing a good video tape recorder for home use is to accurately control the operation of the video scanning process. Most industrial VCRs, as well as many home VCRs, use a belt drive system to operate the head cylinder. The belt drive system drives the head cylinder at a speed slightly faster than the nominal 1800 rpm. Locking in the cylinder head thus depends on using a servo-controlled braking system that drags the cylinder head down to the required exact 1800 rpm, depending on belt slippage to provide the degree of isolation between the main drive and the servo-controlled cylinder head. This sytem works well with most TV sets, but on some you might see some horizontal instability (jitter). Also, the cylinder head in some machines can be massive and therefore difficult to accurately control because of the inertia of the rotating mass.

In the VBT200 VCR, a direct-drive motor system is used in conjunction with a relatively low-mass cylinder head assembly. This provides very precise control over cylinder head rotation, assuring very accurate horizontal time-base timing and elimination of most "jitter" problems. To accomplish this technique of providing very close control of the cylinder head motor, an entirely new direct-drive (D-D) assembly is used.

THREE-PHASE FULL-WAVE BILATERAL DRIVE CIRCUIT

Although at first glance the motor might appear to be a simple 12 volt dc motor, it is far from that. Actually, the motor used to drive the video heads in this machine is a multiple three-phase motor that is driven by a very precisely controlled three-phase ac inverter. The windings, called the *main coils* of the motor, provide the motive power. Feedback necessary to sustain oscillation of the three-phase inverter is sampled by position indicator coils that are designed so that the feedback signal always tells the torque instruction circuit which transistor combination should be turned "on" next to sustain rotation of the motor. Refer to the drive circuit block diagram (Fig. 10-15).

The motor speed is modified by control input from the cylinder-servo system that is compared against a standard voltage in the drive instruction logic circuit. The output of the drive instruction logic circuit is then fed to the position indicator, torque direction circuit, and torque instruction circuitry so that the rotation of the motor becomes locked to the cylinder servo system instructions.

DETAILED CYLINDER SERVO SYSTEM ANALYSIS

The cylinder servo system operates differently in play and record modes. In the record mode, the cylinder servo is locked to vertical sync taken from the sync separator located on the chroma board. Input sync is fed to an amplifier stage in IC 201 via plug 21 pin 3 and the record/playback switch. In the playback mode, the cylinder runs at a constant 1800 rpm because it is locked to a 60 Hz signal obtained via a countdown chip clocked by the 3.58 MHz crystal oscillator on the chroma board. In either case, the output of the IC 201 amplifier is fed to a divide-by-2 counter (one-shot multivibrator) that outputs a 30 Hz signal at pin 26 of the chip. This 30 Hz signal supplies the control track signal during record, as well as furnishing input (via chip pin 21) to the record shifter one-shot multivibrator. Record shifter ouput is at pin 18 of the chip. The record shifter is an adjustable time-delay circuit that allows the vertical sync pulse to be physically positioned on the tape.

The other signal used in the sampling process is a 30 Hz trapezoid. This signal is generated by the cylinder PG (pulse generator) circuit. The cylinder PG signal is an alternating positive and negative pulse taken from the PG magnetic pick-up located in the D-D motor assembly. The PG signal enters the servo system through the motor drive board that contains a differentiator circuit and amplifier that receives input via P27-1. Output from transistor TR2115 is a positive pulse signal when head-one contacts the tape and a negative pulse when head two contacts the tape. These pulses are fed to the servo board through wire jumper J2. The input signal is viewable at TP 211. Following TP 211, the positive and the negative pulses are separately processed.

The positive pulse if fed to IC 201 pin 2 (via diode D206) where it triggers one of the two PG shifter one-shot multivibrators. Output of this one-shot multivibrator triggers (sets) a flip-flop (FF) in IC 201. The other PG shifter multivibrator located in IC 202 is triggered by the negative PG pulse that is inverted by transistor TR211. Transistor TR211 signal resets flip-flop FF in IC 201.

Output from flip-flop FF is a 30 Hz square

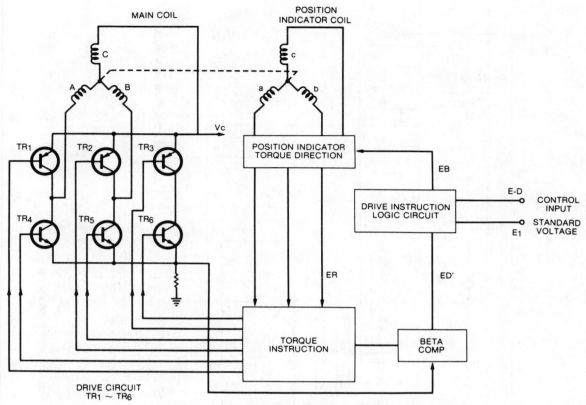

Fig. 10-15. *D-D cylinder motor bilateral drive circuit.*

wave signal in which the duty-cycle of the positive and negative half-cycles are variable via the individual PG shifter controls. This signal (scoped at TP 201) drives a trapezoid generator that is an integral part of IC 201. The 30 Hz PG signal also provides head-switching signals to turn on the individual head preamplifiers on the FM modulator/head amplifier board.

The PG flip-flop output is fed to the trapezoid generator via IC 201 pin 15. The trapezoid generator generates a 30 Hz trapezoid signal that is timed so that the sample pulse (IC 201 pin 14) samples the center of the trailing edge of the trapezoid signal when the cylinder head (upper cylinder) is running at the correct speed and locked to the system signals. Sample gate action can be viewed with a dual-trace scope via test points TP 204 (sample gate-pulse signal) and TP 205 (trapezoid signal). See Fig. 10-16.

The sample gate output (IC 201 pin 12) is a dc voltage of about +6.5 volts that swings up or down, depending on where the sample is taken on the trapezoid, as determined by the physical position of the heads on the video tape at the instant of sampling. This voltage, limited by diodes D209 and D210, is fed to pin 10 of IC 202, where it modifies the pulse width of a one-shot multivibrator that serves as the standard-time generator. Associated with this one-shot multivibrator is the cylinder free-run control whose adjustment sets the pulse width of the standard-time generator so that the cylinder head free-running speed is very close to 1800 rpm. Standard-time generator output (IC 202 pin 2) is sent to a logic AND gate located on the motor control board via jumper J1.

Contained on the motor drive board is the motor speed control circuitry associated with

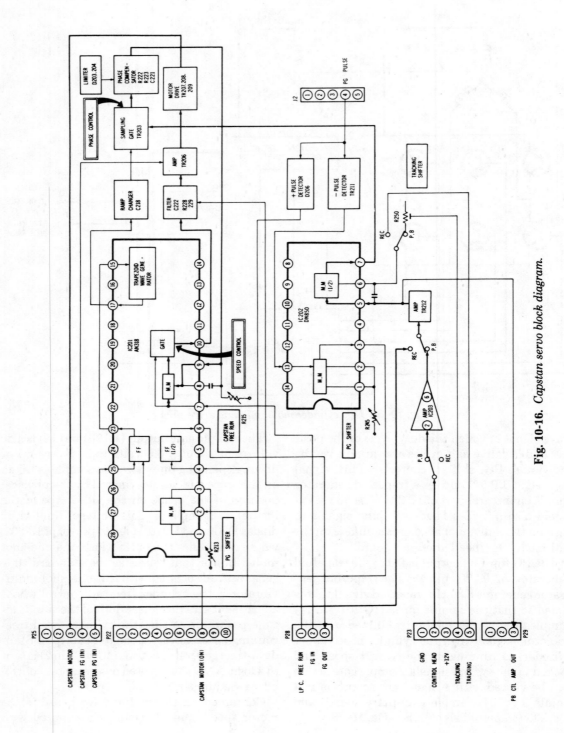

Fig. 10-16. *Capstan servo block diagram.*

the cylinder head motor FG pick-up. The cylinder head FG signal is a 1.8 kHz tone that is fed to the input of a three-transistor "FG amplifier" (transistors TR2101, -02, and -03). The input signal can be scope checked at TP 2101. FG amplifier output is counted down to 900 Hz by counter IC 2101. The 900 Hz output is fed to gate transistor TR2104 and also fed back to pin 12 of IC 202 on the servo board where it triggers the standard-time generator (one shot) contained in the IC. Also entering the standard-time generator is the output of the phase-control circuitry of IC 201 (pin 10 of IC 202) along with the cylinder free-run control voltage. The combination of the bias introduced by cylinder free-run control (R245) and the dc phase-control signal sets the output pulse width of the standard-time generator (output is via pin 9 of IC 202). This signal is sent back to the gate transistor (TR2104) on the D-D motor drive board via jumper J1.

The speed control gate (TR2104) compares the counted-down (900 Hz) FG pulse and the standard-time generator pulse, producing an output that represents the time difference between the two. Circuit parameters are such that the output of the filter amplifier that drives D-D motor-control IC 2102 pin 5 is a dc voltage of about 4.8 volts when the D-D motor speed is correct and the servo system is locked. The feedback loop conditions are such that if the system is not locked or the motor tends to run slow, the dc input to pin 5 of IC 2102 will be somewhat greater than the nominal 4.8 volts. If the motor speed tends to be fast, the IC 2102 input will be somewhat lower than the nominal 4.8 volts.

CAPSTAN SERVO OPERATIONS

The capstan servo system is similar in concept to the cylinder servo. As with the cylinder head servo, the capstan servo has two modes of operation. In record mode, the capstan servo system maintains constant tape speed for either 2-hour (SP) or 4-hour (LP) operation. In playback, the system maintains constant speed as well as maintaining position control of the tape

so that the video heads on the cylinder unit will properly track the recorded information on the video tape. Refer to the capstan servo block diagram (Fig. 10-17).

To maintain constant tape speed in record, the capstan servo is referenced to the cylinder PG derived trapezoid signal. The trapezoid signal, after some integration by "ramp-changer" capacitor C218, is fed to a sample-gate transistor (TR203) along with the sample-gate signal derived from the capstan PG signal. The sample-gate signal is derived from the counter system on the SP/LP select board (30 Hz in "SP" and 15 Hz in "LP" mode). Output from the sample gate is an amplitude-limited dc voltage (limiter diodes D203 and D204) which modifies the pulse width of the capstan standard-time generator (MM) via IC 201 pin 9. The basic pulse width is set with capstan free-run controls. Output from the standard-time generator drives a logic AND gate, along with the capstan motor FG signal. The output of the speed-control AND gate leaves IC 201 via pin 10. This pulse, after integration in a filter network consisting of capacitor C222 and associated components, is base bias for amplifier transistor TR206. Also associated with amplifier TR206 is a logic input from the transport/control board that indicates the logical sequences necessary to initiate various modes of operation have been completed before the capstan motor is allowed to operate and move tape. Output from the amplifier TR206 is fed to a three-transistor motor drive circuit that supplies dc voltage to the capstan motor. This voltage leaves the servo board via P25-3.

The rough speed control of the capstan motor is accomplished in basically the same manner as was described for the cylinder servo. In this case, the capstan FG signal (960 Hz or 480 Hz, depending upon the mode of operation) is fed to the servo board via P25-4 where it is applied as input to a flip-flop (FF) contained on IC 201. This input signal can be scoped at TP Y203. Output of this flip-flop (half input frequency) comes out at pin 6 as a 480 Hz or 240 Hz signal (SP or LP), which is directed to the SP/LP auto

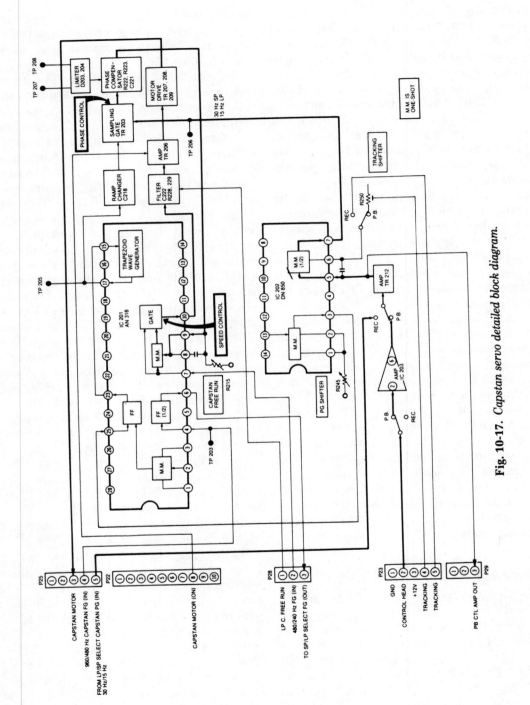

Fig. 10-17. *Capstan servo detailed block diagram.*

select board via P28-3. An amplifier/inverter stage on the SP/LP board processes the signal to a level sufficient to drive a multivibrator back on the servo board that is part of IC 201. However, more importantly, this serves as a pick-off point where the additional stage of frequency division can be switched "in" to feed a 240 Hz signal to the standard-time generator when in the SP mode.

Also on the SP/LP board is the divide-by-8 counter that produces the capstan pulse generator (PG) signal for input to the servo phase-control circuitry with the VCR is in the record mode. A pick-off from this counter supplies the SP mode 240 Hz FG signal.

The 30 Hz or 15 Hz counter output PG signal is processed into the sample pulse that samples the trapezoid signal to produce the control voltage necessary to provide correct motor speed. The signal frequency is 30 Hz in the SP mode and 15 Hz when the unit is in the LP mode.

The capstan PG signal enters the servo board on P25-5 where it is routed to the record/play switch. When the VCR is in the record mode, the signal is amplified by transistor TR212 and then serves to trigger the tracking shifter one-shot multivibrator part of IC 202. Output of this stage, which interfaces with the tracking control when the machine is in playback, comes out of IC 202 via pin 7. After some additional processing (differentiation), the signal is fed to the sample gate as the sample pulse, which is indicative of the motor speed error.

SP/LP AUTO SELECT CIRCUITS

This VCR machine uses special logic and control circuitry to allow the operator to manually select 2-hour or 4-hour operation when recording a tape. During playback, logic and timing circuitry samples the control-track signal and directs electronic switching, causing the machine to play back the tape at the correct speed. Note the auto select timing diagram shown in Fig. 10-18.

In addition to setting the capstan motor speed

for SP (Standard Play) or LP (Long Play) operation, the SP/LP auto select circuit controls four additional functions that are necessary for proper SP and LP performance during record and playback. Through use of a control voltage known as "LP high" (+12 volt source in LP mode), the audio record and playback equalization are changed to optimize frequency response in both modes of operation. Also, during the LP record mode, additional video pre-emphasis is introduced and the "½-FH" frequency interleave circuitry is activated. In playback mode, the LP-high voltage switches in complementary video-frequency de-emphasis and an ½-FH interleave cancellation square wave signal.

The SP/LP auto select board also provides a muting voltage that is present when the machine senses erroneous control-track frequency (indicative of wrong speed operation). For the interval of time necessary for the machine to change speeds and stabilize its operation, the muting voltage is present. Finally, and most important, the SP/LP auto select circuitry electronically switches the capstan motor speed from the SP (2-hour) to the LP (4-hour) mode or vice versa.

CAPSTAN MOTOR SPEED SELECTION

The capstan motor SP/LP speed selection is accomplished by switching in or switching out an extra stage of frequency division in the capstan FG (frequency generator) circuit. To satisfy the requirements of the capstan servo system, the FG signal input to the standard-time generator and logic AND gate must be a constant 240 Hz. Refer to the block diagram (Fig. 10-19). During SP record, base voltage is applied to transistor TR6405, causing it to conduct and load down the signal output of amplifier TR6411. Thus, it can be seen that the signal input to the standard-time generator (signal can be scoped at TP 6411 on the SP/LP select board) must come from the output of IC 6403 (pin 3), which is an additional divide-by-2

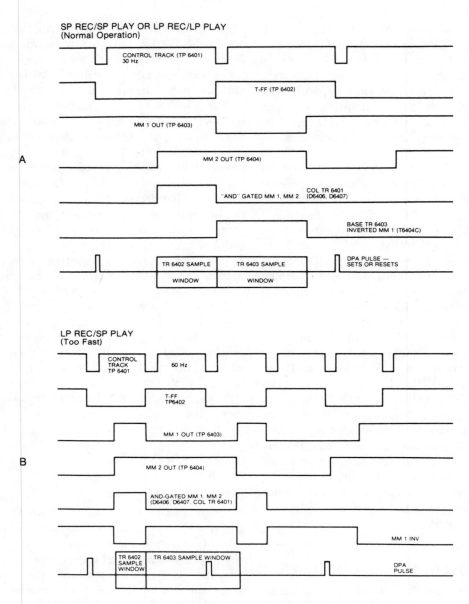

SP REC/SP PLAY OR LP REC/LP PLAY
(Normal Operation)

CONTROL TRACK (TP 6401)
30 Hz

T-FF (TP 6402)

MM 1 OUT (TP 6403)

A

MM 2 OUT (TP 6404)

"AND" GATED MM 1, MM 2   COL TR 6401
(D6406, D6407)

BASE TR 6403
INVERTED MM 1 (T6404C)

TR 6402 SAMPLE   TR 6403 SAMPLE   DPA PULSE —
SETS OR RESETS

WINDOW   WINDOW

LP REC/SP PLAY
(Too Fast)

CONTROL
TRACK
TP 6401   60 Hz

T-FF
TP6402

MM 1 OUT (TP 6403)

B

MM 2 OUT (TP 6404)

AND-GATED MM 1, MM 2
(D6406, D6407, COL TR 6401)

MM 1 INV

TR 6402
SAMPLE   TR 6403 SAMPLE WINDOW   DPA
WINDOW   PULSE

Fig. 10-18. *A, B & C. SP/LP auto select timing diagram waveforms.*

counter. Under these conditions, the capstan motor is forced to run at the SP speed. When LP operation is desired, transistor TR6406 is turned on and thus shorts the signal from the pin 3 output of IC 6403. At the same time, the signal from the collector of TR6405 is fed to the standard-time generator via OR gate diode D6418. When the unit is in play, the appropriate choice of FG input signal from the OR gate is automatically selected by the logic circuit contained on the same board. The action of the circuit in this area is such that either transistor

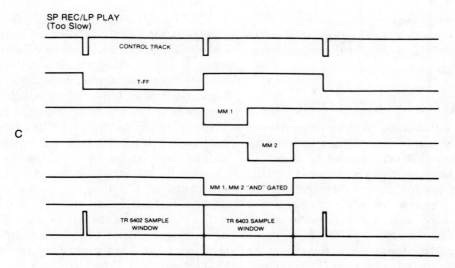

SP REC/LP PLAY
(Too Slow)

CONTROL TRACK

T-FF

MM 1

MM 2

MM 1, MM 2 "AND" GATED

TR 6402 SAMPLE WINDOW

TR 6403 SAMPLE WINDOW

C

Fig. 10-18. *(continued)*

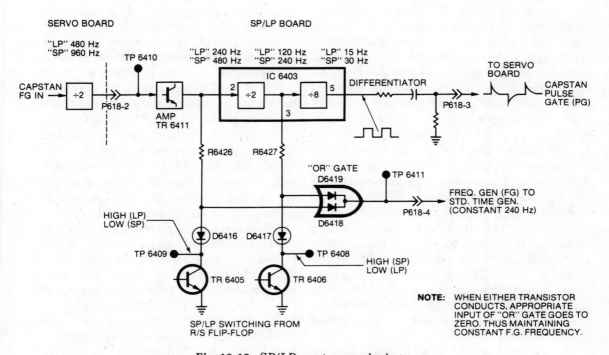

Fig. 10-19. *SP/LP capstan speed selector.*

TR6405 is driven into conduction or transistor TR6406 is conducting, making the proper choice of FG signal available to the standard-time generator and logic gate.

SP/LP AUTO SELECT LOGIC

The logic circuit used to sample the control track and control the SP/LP operational modes is somewhat complicated as shown in Fig. 10-20. Basically, two transistors (TR6402 and TR6403) are connected in a set/reset (SR) flip-flop. This means that when the flip-flop is triggered into one mode, it will remain so until it is triggered into the opposite state. Note in the circuit diagram that the output of this circuit is a line that is designated as LP high (TP 6406 voltage). This line is applied to a pair of OR gates consisting of diodes D6408 and D6409, D6410, and D6411 respectively. Outputs from these gates ultimately drive two switching transistors designated TR6405 and TR6409.

During LP record mode, + 12 volts applied to the base of transistor TR6406 via logic diode D6409 causes this device to saturate. This action shorts out the FG signal from amplifier transistor TR6411 and forces the FG input to the standard time generator to be that obtained from the extra stage of frequency division in IC 6403. This action also cuts off switch transistor TR6405 causing its collector voltage (TP 6409) to go high (+ 12 volts), providing sources of voltages designated *video LP high* and *audio LP high* to become available at P619-5 and P619-6. These voltages are routed to the luminance process board and the audio board.

When in the SP record mode, switch transistor TR6406 is turned off, and switch transistor TR6405 is thus saturated, driving the TP 6409 LP-high line to the low (OV) logic state. At the same time, diode D6416 shorts the signal that comes directly from transistor TR6411 making the FG signal available to the standard-time generator that came from the extra stage of frequency division in IC 6403.

In play, the operating mode is determined by the "PL high" output of the SR flip-flop (dc voltage at TP 6406). As shown in Fig. 10-20, this voltage is high in the LP mode and low in the SP mode. Whenever the machine is in the LP mode of operation, the output of the SR flip-flop is high at TP 6406. Under these conditions, bias voltage is fed to the base of switch transistor TR6406 via OR gate diode D6408. Now switch transistor TR6405 is cut off and the proper conditions required for LP operation are set up. Conversely, when the machine is in the SP mode of operation, the flip-flop output is low at TP6406 and switch transistor TR6406 is cut off.

The remainder of the circuitry is dedicated to controlling the state of the SR flip-flop. Note that two input lines control the flip-flop — *set* (S) and *reset* (R). When the machine is operating in the correct mode (playing back tape correctly), signals are not present on either of the control lines. In the event of an error condition, a pulse voltage appears on either the (S) or the (R) line, which causes the flip-flop to switch to the other state and set up the necessary conditions for the proper playback mode.

For example, if the machine is running too slow (such as if the machine was running in the LP mode but the tape to be played back was an SP recording), a pulse will appear on the FF (S) line (TP 6407), which causes the flip-flop to be driven to the opposite state to establish the SP playback mode. The opposite error condition (LP recording with machine in SP speed mode) Q causes a pulse to appear on the (R) line (TP 6405), which causes the machine to enter the LP mode. The actual pulse that is used to set or reset the flip-flop is produced by transistor TR6401 (differential pulse amplifier, or DPA). The base of this device is timed by the control-track signal after it is processed by a pulse clipper stage contained in IC 6401 and a "T-type" flip-flop. The T type flip-flop is an IC device (IC 6402) that changes stage every time it is triggered. Thus, the device acts as a divide-by-2 counter.

As can be seen in the timing diagram (normal operation), the 30 Hz control track signal at TP 6401 triggers the T flip-flop to change state on

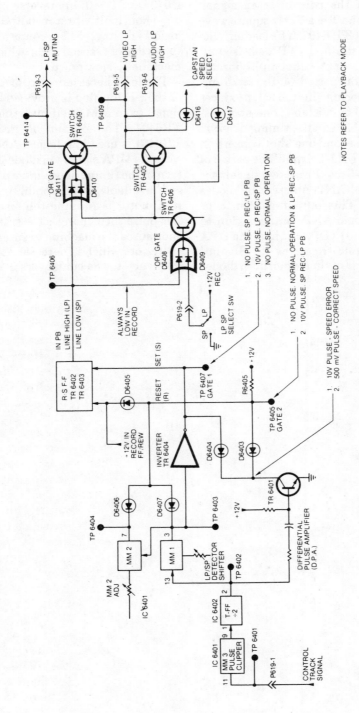

Fig. 10-20. *SP/LP auto select logic circuit.*

237

the leading edge of the control track signal. Thus, the output of the T is a 15 Hz square wave that can be scoped at TP 6402. The leading (rising) edge of the T flip-flop (TP 6402 signal) triggers a monostable, or one-shot multivibrator that has a conduction period of about 25 ms. This multivibrator, located in IC 6401, produces a trigger for a second monostable, one-shot designated as "MM 2." Note in the timing diagram (Fig. 10-18) that the second one-shot triggers on the rising edge of the MM-1 signal. The period of this one-shot is also 25 ms. These two signals are combined in a logical AND function by diodes D6406 and D6407. This output is then connected in another logic AND configuration to the collector of TR6401 (DPA). Thus, the DPA pulse is prevented from triggering the SR flip-flop except in intervals of time when both diodes

D6406 and D6407 are reverse-biased due to both one-shot multivibrator outputs being in the logic-high stages. This condition only occurs in the event of a speed error which is causing the machine to slow down.

The logic necessary to control the set (S) line of the SR flip-flop is provided by taking output signal from MM1 and inverting it in transistor TR6404. The output of inverter transistor TR6404 is used in a logical AND configuration with the DPA signal via diode D6404. As can be seen in the timing diagrams, a pulse at TP6407 is only available under conditions when the capstan motor is running too fast. Application of this pulse then changes the state of the flip-flop and causes the machine to enter the LP mode of operation, which is necessary to restore correct frequency to the control track signal.

11

JVC VCRs

Information in this chapter courtesy of the JVC Corporation (Victor Company of Japan Limited), copyright 1980.

The JVC HR 7300U uses the VHS format and achieves a very low tape consumption. Recording time in the standard mode is 2 hours. Increased recording time results from the narrow gap video heads, high sensitivity video tape, and the slant azimuth recording head configuration that eliminates the need for a guard band between recorded tracks. Also, the VHS format takes into consideration special operating modes such as still picture, slow motion, and speed playback.

The VHS format presents several technical challenges. First among these were obtaining high picture quality and high resolution despite the slow relative speed between the tape and video heads, improving signal-to-noise ratio (S/N) and preventing black to white reversal phenomena to the short recording wavelength of 1.2 μm. Also, the ± 6 degree azimuth angle of the video heads alone is not sufficient to eliminate crosstalk from the low-band converted color signal.

Steps for solving these difficulties included adoption of a nonlinear emphasis circuit and selecting the emphasis amount for optimum S/N. The reversal problem was overcome by using a double limiter circuit, while a phase shift system has been designed for eliminating color crosstalk. By using a four-head upper drum, the HR-7300U can record and play back in both 2-hour standard VHS and 6-hour extended modes. The design also permits playback of pre-recorded tapes in the 4-hour mode.

This machine features "feather touch" operation. Just a light touch of the machine buttons or remote control unit keys supplies mode command signals to the various circuits, motors, switches, and solenoids to set up the selected mode. In order to protect the set and the tape, various internal sensors are provided. By continuously monitoring these, the decision is made to continue or stop the mode in progress or shift to another mode.

A built-in microprocessor assists in detection and control of the operating modes. The microcomputer is preprogrammed, and the program can be altered. While a basic understanding of the principles of a microcomputer would be helpful, an understanding of which input signals result in which specific output signals is more important for practical purposes in servicing. In the following circuit description, "mechanism control" is shortened to "mechacon."

BLOCK DIAGRAM DESCRIPTION

In this circuit description, refer to Fig. 11-1. The mechanism control, hereafter referred to as "mechacon," receives mode *command* signals from the operation keys and mode *detect* signals from the various sensors. It then produces signals for driving the motors and solenoids to set up the required modes. Mode control signals are also sent to the appropriate circuit boards. These control functions are performed by the central processing unit (CPU) of IC2, which is a one-chip, four-bit microcomputer. In the block diagram, the input signal generators from the function keys and sensors are shown at the left of the CPU, while the motors and solenoids controlled by the CPU output signals are located at the right, together with the circuit board connectors supplied by CPU signals.

Note that the CPU has only two sets of input ports, (A and B) totalling eight bits, while a total of 14 inputs are obtained from the operation keys and sensors. For this reason, IC1 multiplexer is provided as an input expander. Using the three-bit bus select signal from CPU port H (strobe data irrelevant), one output is selected from among four inputs and sent to the CPU input ports.

Thus, the four 4-to-1 multiplexer circuits of IC1 supply four-bit outputs to the CPU input A ports from 16 inputs, and the remaining operation and sensor signals go directly to the input B ports. At the same time, the four seven-bit latches of IC4 function as output expanders. The four-bit outputs of the CPU D ports are expanded to 28 latched outputs. Latch positions are determined by the three-bit bus select from the CPU H ports and the strobe data.

IC4 outputs are supplied through open collector inverters as either low or open (high when connected to other boards) outputs to the other circuit boards. Drive signals for motors and solenoids are obtained directly through inverters from CPU output ports E, F, and G.

OVERALL SIGNAL FLOW

During a selected mode, high from IC3 comparator goes to CPU input port B to signify that the mode is being held. At this time, the four-bit operation key scan data are obtained from CPU port C in cycle of 7.5 msec (133 Hz) for the binary range of 0000 to 1111. The digital sequence goes through IC11 buffer and RA6 digital to analog converter (DAC) to become a sequential voltage in 16 steps from 0 volts to about 9.5 volts, which is fed to the comparator at 133 Hz.

When an operation key is pressed, a fixed dc voltage, determined by the resistance combination of function board RA1 (DAC), goes to the non-invert input of the comparator. At the same time, the 16-step scan data is applied to the invert terminal. The comparator produces either a high or a low output.

If the scan data is higher than the fixed input from the operation key, low output from the comparator goes to CPU port B. The CPU interprets this as pressing of an operation key, resets port C and sends a new four-bit output sequence of 0000, 0001, etc. As a result, the comparator output again becomes high. When one of the 16 steps from the operation key to the comparator becomes low, a low comparator output goes to CPU port B. By detecting this low together with the output status of port C, the CPU can detect which specific operation key has been pressed. This process will be covered in more detail later.

The CPU also detects other input port data. These include timer, cassette switch, cassette lamp, and the various sensors. Data pertaining

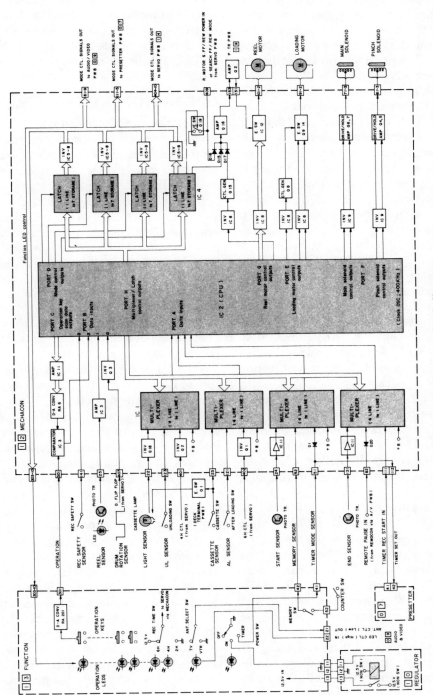

Fig. 11-1. *Mechacon block diagram.*

241

to the depressed key are checked, such as whether it is the same as the mode in progress or if a shift to a newly selected mode is possible from the present mode. Outputs corresponding to the operation key are then sent to the motors, solenoids and circuit boards.

From port D, the four-bit output goes to IC4 latch, resulting in 28 latched outputs (in practice, five of these are not used) that are sent through open collector inverters to the circuit boards. Control signals also go to the function board for lighting the LEDs corresponding to the depressed operation key control.

Three bits of the four-bit port E output are for loading motor control. These are supplied via inverters to the control generator for determining motor torque and to electronic switches Q9 to Q14 that select motor direction. For example, in the play mode, rotation is in the loading direction. When the loading mechanism begins operation and the unloading (UL) switch is off, the CPU detects the start of loading. At the end of loading, the "after loading" (AL) comes on, at which time the CPU detects the end of loading and stops the loading motor. The unloading process is the reverse of this.

The four-bit port F output is divided into two bits each. These go via inverters to the solenoid drive and hold amplifiers for switching the main and pinch solenoids on and off according to mode. The pinch solenoid is driven after completion of loading (AL switch on) during play and recording.

Four bits from port G are supplied through an inverter to electronic switch IC12 for controlling forward and reverse rotation of the reel motor. Rotational torque is controlled by drive voltage from the servo board to D16 during search, both fast forward and rewind (S-FF and S-REW). During ordinary fast-forward and rewind, supply is from port D with control via an inverter and control generator (Q15) at D17. Latch IC3 provides control via D15 during unloading, at short rewind, at the start of FF/REW, and in the idler mode (during which the reel idler shifts toward the supply or take-up reel disk).

During play/rec, take-up is driven mechanically by the capstan motor. The select bus form port C selects the sensor state data required for the particular mode and supplies them to the CPU as auto-stop input data.

OPERATION KEY-IN CIRCUIT

Note: the mechacon circuit is completely controlled by the CPU, eliminating the need for a complex drive and other circuits. Thus, only the operation key-in, DAC, and comparator functions outlined.

Referring to Fig. 11-2, when the sub-power switch is set from off to on, a high pulse of approximately 60 ms duration from IC3 (pin 2) goes to the reset terminals of the CPU, resetting the CPU and producing the stop mode. All port outputs are at stop mode at this time.

The four-bit operation key scan data output from port C covers the binary digits from 0000 to 1111 in 0.16 msec increments, then produces 1111 for about 5 ms. Each cycle is about 7.5 ms (133 Hz) in duration.

Via IC11 buffer converter, the four-bit output goes to RA6. The RA6 DAC converts the data to a 16-step output from 0.15625 V to 9.63125 V, which goes to the invert pin 6 of IC3 comparator. Ten volts goes to the non-inverting input via R45 and R46, so long as an operation key is not pressed. The normally high comparator output goes to CPU port B, instructing the CPU that an operation key is not pressed. Pressing an operation key sends a fixed voltage corresponding to the particular key to the noninvert input for the comparator, where it is compared with the key scan data.

The voltage obtained by pressing each operation key is limited to one of the 16 steps according to the four-bit key scan data. As an example, assume the PLAY key is pressed, as is indicated in Fig. 11-3. This applies 2.99 volts to the comparator non-inverting input. At this time, when the voltage at the inverting input from the key scan data rises beyond 2.99 V (as when the four-bit data exceed five), the comparator low output

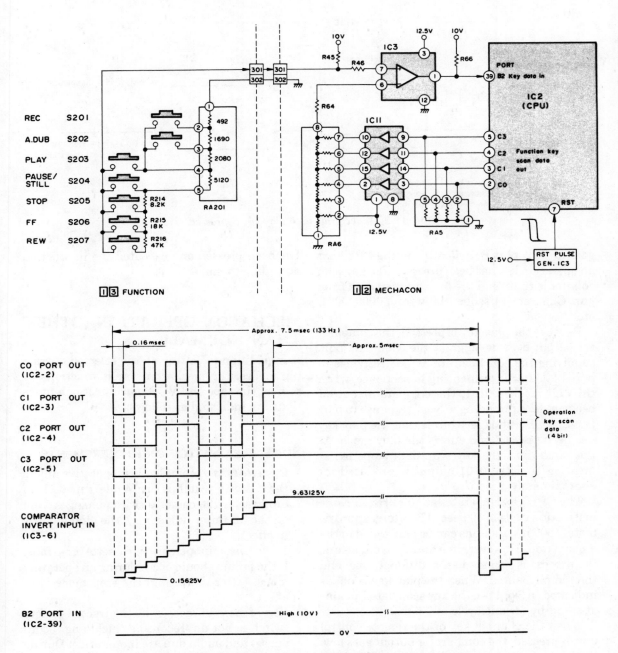

Fig. 11-2. *A & B. Mechacon functions and DAC comparator outputs.*

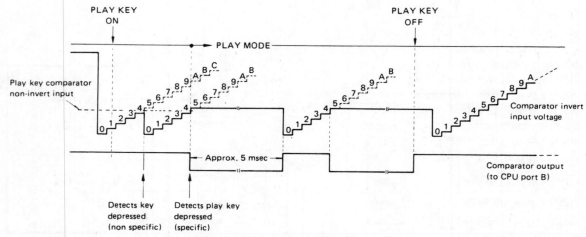

Fig. 11-3. *Waveforms with play key depressed.*

goes to CPU port B, indicating to the CPU that an operation key has been pressed. The key scan counter is reset, and the four-bit scan data from port C increment sequentially from 0000, 0001, etc.

Because the play key is pressed, the comparator output becomes high at the 0000 data poll, supplying the equivalent of a "no" response to port B2. After incrementing in sequence, when the poll reaches 0101, the comparator output becomes low, providing a "yes" response to port B2. The CPU thus recognizes that the play key has been pressed and enters the play mode. At the same time, the key scan data stops incrementing, and the 0101 output is produced for approximately 5 ms.

The CPU clock operates at 400 kHz, and one instruction cycle is 10 μsec. Therefore, approximately 500 instructions can be performed in the 5 ms period. However, in order to avoid errors, the process is repeated an additional time, and then implemented. When the play key is off, as indicated in Fig. 11-3, the key scan data remains at stand-by for one cycle of 7.5 ms.

When the channel key of the remote control unit is pressed, the comparator output goes low. With port B low for two polling cycles, the channel up command is then recognized. In the above manner, the CPU is able to determine which key

has been pressed and executed the instruction according to mode.

MECHACON OPERATION (OTHER THAN NORMAL)

In addition to setting up the selected operating modes, the mechacon circuit functions to protect the tape and VCR machine. Such functions are indicated as follows:

SUB POWER SWITCH-ON REQUIREMENTS

- *Cassette sensor:* completely inoperative unless the cassette switch is on. While a mode is in progress, if the cassette switch changes from on the off (after unloading), the stop mode is in effect.

- *Light sensor:* inoperative if cassette lamp fails. If the failure should occur during an operating mode, (after unloading) the stop mode is in effect.

- *Record safety sensor:* if the record safety switch is not on, recording (including record pause) and audio dub are inoperative. During these modes, if the switch state changes from on to off, unloading is performed and the stop mode is in effect.

- If the unloading switch is off when the sub power is set to on, unloading will not occur.
- The channel key of the remote control unit is operative only during the stop, unloading REC pause, FF, and REW modes.
- *Start sensor:* if on during the REW mode, the REW mode becomes inhibited. When the start sensor switches on during rewind or search rewind (after unloading), the stop mode is in effect.
- *End sensor:* when the end sensor is on, play, recording, and FF are inoperative. If these modes are in progress and the end sensor switches on (after unloading), the stop mode is in effect.
- *Reel sensor:* with the after-loading switch on during play, recording, audio dub, search FF and search REW, and if the take-up reel and disk rotation stops, the tape will unload and the stop mode will be in effect after about 3 seconds. This limitation does not apply to pause/still. During FF/REW while the unloading switch is on, if take-up reel disk rotation stops, after about 4.2 seconds the stop mode is in effect.
- The drum motor rotates when the unloading switch is off.
- *Drum rotation sensor:* with the unloading switch off, if the drum motor (video heads) rotation stops, after about three seconds, unloading is performed and the stop mode in effect.
- Memory sensor: during rewind with the memory switch of the function board on, when the tape counter indication decrements from 0000 to 9999, the stop mode is in effect.
- The unloading motor rotates in the forward direction when the play LED is lighted and the after loading switch off. When the stop LED is lighted and the unloading switch off, the motor rotates in the reverse direction.
- *AL sensor:* During unloading, if the period between unloading switch off and after loading (AL) switch exceeds 10 seconds, unloading is performed and the stop mode is in effect.
- *UL sensor:* during unloading, if the period between after loading switch off and unloading (UL) switch on exceeds 10 seconds, the mechanism stops and the emergency mode is entered.
- *Pause/still overtime sensor:* if the pause or still mode continued for more than 5½ to 6 minutes, unloading is performed and the stop mode is in effect.
- *Auto rewind:* during play/rec, audio dub, FF, and search FF, when the end sensor functions and auto stop is produced, (after unloading) the rewind mode is in effect. However, if the stop key is pressed during unloading, the auto rewind mode is not entered.
- *Short rewind:* when the stop mode is produced from FF or REW, short rewind is performed for 240 ms then the stop mode is in effect.
- The capstan motor rotates when the unloading switch is off. However, this limitation does not apply to pause/still.
- When the remote control unit is connected, control can be performed by either the local or remote operation keys. If both are used, the most recently pressed key instructs the mode. However, when the stop is pressed, the stop mode has priority regardless of the pressing sequence.
- During recording, the remote pause mode has priority over the local and remote operation keys. However, this does not apply to the stop mode.

SUB POWER SWITCH TO TIMER

- If the record safety switch is off, record start is inhibited and at pre-start, the stop mode is in effect.
- While a mode is in progress with the sub power switch on, if the switch is set to timer, the mechanisms stops. However, at the timer start time, record or record pause becomes the record mode.
- During timer recording, the pre-start (high) signal initiates loading 10 seconds prior to the timer start time, then at record start, the pinch roller engages to yield the recording mode.

12

Special VCR Circuits and Stereo Audio

This chapter delves into the concept, operation, and circuitry of the Sony Super Beta VCR recording system and Sony's Audio-Beta Hi-Fi record/playback system. Then information is presented on the HQ (high quality) and Super VHS circuits found in some VHS machines. This information is followed by the details and principles of the MTS (multi-channel television sound) techniques that are now being used in some TV and VCR recorders. I also talk about how the TV stereo sound standard was established and the workings of the multichannel TV sound (MTS) system.

SONY BETA HI-FI SYSTEM

The original Betamax was comprised of a composite video that consisted of luminance and chroma. The chroma occupies a band of frequencies from 0 Hz to 4.5 MHz. Note recorded spectrum shown in Fig. 12-1. The chroma occupies an area within the luminance band centered around 3.58 MHz. In order to record this composite video information, the luminance and chroma must be separated and dealt with individually. The luminance signal is FM modulated. This was done to reduce the number of octaves required to record the luminance signal. The luminance signal is recorded by causing an FM modulator to deviate from 3.6 MHz (for sync tips) to a high frequency of 4.8 MHz (for peak white levels). This 1.2 MHz deviation of the FM carrier produces upper and lower sidebands. The width of these sidebands determines the resolution of the signal recorded.

The chroma part of the composite video occupies an area within the frequency spectrum that overlaps the record luminance. To avoid these frequencies from interacting and causing chroma beats, the chroma signal was down-converted to 688 kHz. This combination of down-converted chroma and FM luminance comprises the RF signal recorded by the video heads.

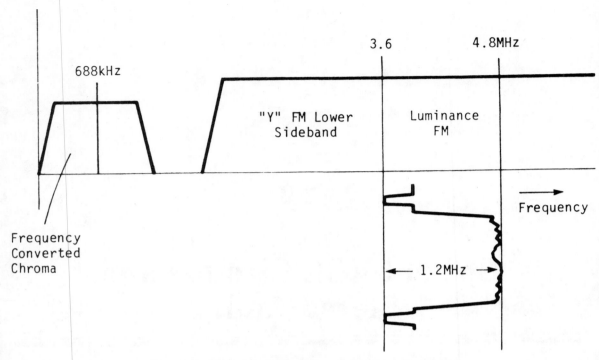

Fig. 12-1. *Recorded spectrum for original Sony Betamax.*

RECORDED SPECTRUM BETA HI-FI

Beta hi-fi is a method that enables the video tape recorder to record left and right channel audio, separately using the video heads with a much higher fidelity than previously possible using the conventional longitudinal tracks. To do this, the left and right channel audio signals were FM modulated and divided into four pilot audio carriers. Four carriers are required to maintain the separation between the left and right channel audio, as well as to reduce the crosstalk between agjacent tracks of the video information. These four audio carriers are centered about 1.5 MHz and are mixed with the luminance and chroma information to be recorded by the video heads on the tape. See waveform drawing of Fig. 12-2.

The addition of the four audio FM pilot carriers required the FM luminance signal to be shifted upward by 0.4 MHz to make room be-tween the chroma and luminance information for these carriers. The high-frequency limitations of the video heads resulted in a loss of some of the FM sidebands due to this 0.42 MHz shift in the luminance frequency. This caused a slight reduction of the amount of resolution in the picture produced by Beta hi-fi units.

RECORDED SPECTRUM SUPER BETA AND BETA HI-FI

The Super Beta system overcomes the limited resolution of not only Beta Hi-Fi but also of conventional Betamax units. This increased resolution is achieved by using narrower gap heads that results in an improved high frequency response, and a 0.8 MHz upward shift of the FM Luminance carrier results in a larger lower sideband. The increased bandwidth of the total luminance signal results in resolution

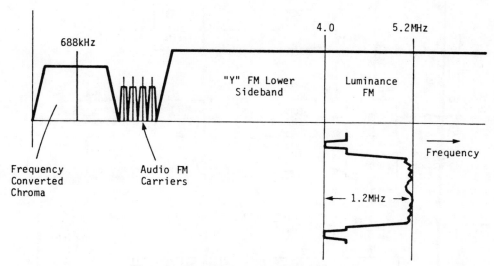

Fig. 12-2. *Recorded spectrum for Beta Hi-Fi.*

greater than achieved by both Beta Hi-Fi and the conventional Betamax system. Refer to waveform drawing of Fig. 12-3.

SUPER BETA RECORD BLOCK

There are few changes to the conventional video processing chain needed to record Super-Beta (highband) or Beta Hi-Fi. They are high-lighted as the overall record system is explained. Use the block diagram in Fig. 12-4.

Composite video from the VCR tuner or the rear panel line input is selected for the AGC stage. The AGC stage either increases or decreases its gain to produce a video signal with a constant amplitude horizontal sync pulse, and therefore a constant output level.

The composite video leaves the AGC stage

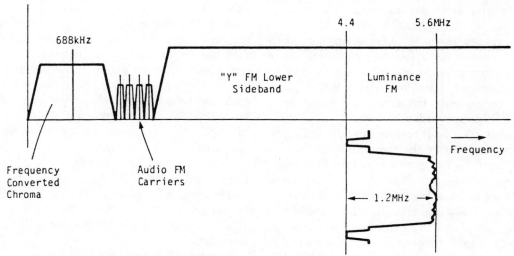

Fig. 12-3. *Recorded spectrum for Super Beta and Beta Hi-Fi.*

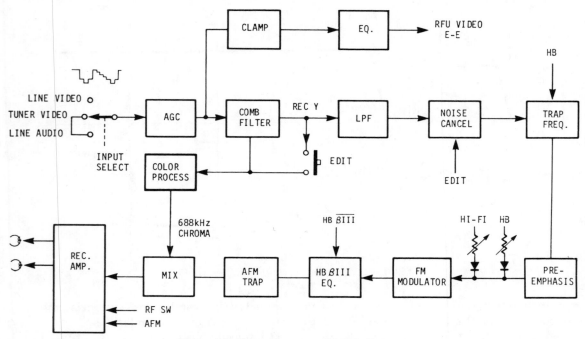

Fig. 12-4. *Super Beta record block diagram.*

and takes two paths. One path is through the E-E RFU video output line for an external monitor to display the recorded picture. The other path is into a comb filter that separates the luminance and chroma components of the composite video signal and outputs them. The chroma down-converting processing is conventional and is not explained here. When receiving composite video from another VCR, the EDIT switch can be engaged to negate the effects of the comb filter, permitting individual band-pass filters afterwards to separate the composite video for a clearer, edited recording.

The low-pass filter restores an overall flat frequency response of the luminance signal that was altered by the previous comb filter stage. The luminance is acted upon by the noise cancel stage that follows. When the EDIT switch is engaged, less noise cancellation is achieved in this stage. This permits the full band width of the playback signal to pass during the recording EDIT mode.

The luminance signal's upper frequency limit is fixed by a trap in the next processing stage. During high-band or Super Beta recording, a wider bandwidth is desired. Therefore, a higher frequency trap to raise the upper frequency limit is selected with the front panel Super Beta switch.

With the amplitude and frequency limits of the luminance signal now controlled and therefore known, an increase in gain at higher frequencies is brought about in the pre-emphasis stages. Upon playback, the high frequencies are returned to normal levels in the de-emphasis stages. This boost in record and reduction in playback reduces the loss of detail during the record/playback process.

The FM modulator changes the AM luminance signal to an FM signal containing upper and lower sidebands. Beta Hi-Fi and Super Beta modes add a voltage to the incoming luminance signal to shift the carrier frequency of the AM modulator.

The equalization stage that follows not only rolls off the upper sideband that would contribute noise to the signal but also balances the level of lower side band (lower frequencies) compared to the luminance FM carrier signal (higher frequencies). This prevents black streaks in the playback video commonly called over modulation noise. The Super Beta switch and the BIII mode connect to this stage. The extended bandwidth in Super Beta requires that the equalization emphasis be changed to maintain frequency spectrum balance. However, this is only necessary in the Super Beta BII speed. The normal high-frequency losses at the slower Super Beta BIII speed maintain proper balance without the Super Beta equalization emphasis. Therefore, this Super Beta emphasis is not used in Super Beta BIII.

The AFM trap stage removes any noise in the frequency band between the chroma and the luminance's lower side band, so the four pilot carriers used to record Beta Hi-Fi can be inserted here later on by the record amplifiers.

Luminance signal that has been controlled in amplitude and bandwidth, boosted, modulated, and cleaned is mixed with processed chroma in the next stage. The new luminance and chroma signal is mixed with the AFM audio if present, amplified, and stored on magnetic tape using video heads.

SUPER BETA CARRIER SHIFT

Of the three high-band changes to the record luminance stages, the most important is the FM modulation frequency shift. Super Beta gains a larger side band and therefore detail by shifting the record luminance signal $+0.8$ MHz. The record carrier shift diagram (Fig. 12-5), shows this is done by adding a small voltage to the input of the modulator at IC1 (pin 30) from the front panel Super Beta switch through D212 and series resistors.

Beta Hi-Fi also shifts the modulator frequency, but only $+0.4$ MHz. This is also done by adding a small voltage to the input of the modulator from the front panel Beta Hi-Fi switch. In record, when the Beta Hi-Fi switch is turned on, 0 Vdc appears on the base of Q801, turning it off.

The resultant HIGH at Q801/collector turns ON Q504 and Q505, which luminates the front panel Beta Hi-Fi light. When the Beta Hi-Fi lamp is lit, voltage is also applied to Q610 and Q611 turning them both on. The voltage from Q611's conduction is controlled by RV204 to set the maximum luminance carrier frequency in Beta Hi-Fi. This voltage forward biases a diode in the D212 package and is then applied to the modulator at IC1 (pin 30) to shift the modulator frequency to 0.4 MHz for Beta Hi-Fi.

SUPER BETA PLAYBACK BLOCK

The Super Beta and Beta Hi-Fi modes make few changes to the playback signal path. They will be highlighted as the overall record system is explained. Refer to the block diagram in Fig. 12-6.

The signal from the tape is picked up by the video heads, amplified by the head amplifier, and mixed together. The luminance and chroma signals are separated using bandpass filters before taking different paths. (The chroma process path is not affected by Super Beta nor Beta Hi-Fi modes and therefore is not discussed.)

The luminance path is through an equalization amplifier stage that provides a flat response from the higher frequency FM luminance carrier signal to the lower frequency FM luminance lower sideband signal. This stage is necessary because the video heads cannot play back with a linear (equal) output over the entire luminance spectrum.

During Super Beta (high band), a wider luminance bandwidth is played back. The balance between the lower frequencies and extended higher frequencies must be maintained with a Super Beta attenuation stage.

The RF luminance signal path divides after the equalization amplifier. One path is into the dropout compensator stage. This section produces an output pulse if the input RF luminance signal falls below threshold level. During Super Beta mode, the greater amount of short-time-constant, high-frequency signals might not reach the DOC's threshold level and incorrectly

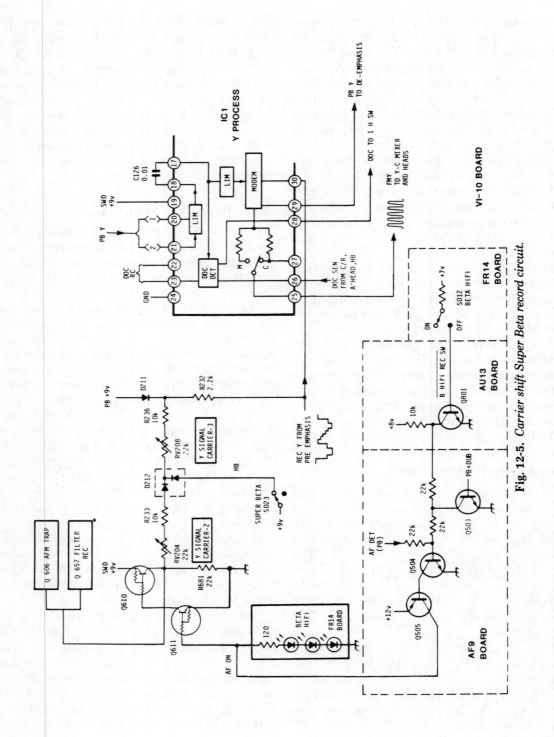

Fig. 12-5. *Carrier shift Super Beta record circuit.*

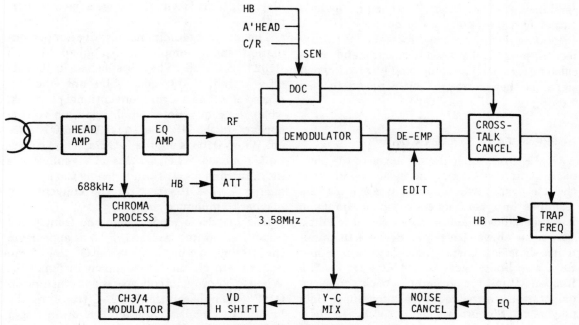

Fig. 12-6. *Super Beta playback block diagram tie-in.*

signal a dropout. Therefore, the DOC sensitivity is reduced in Super Beta. During pause, cue, or review, the DOC is disabled by further reducing DOC sensitivity.

The main path of the RF luminance signal is through the demodulator stage. During the recording process, the low-level, high-frequency components of the luminance signal were boosted, so they can be attenuated in the playback de-emphasis stage. Thus, the low-level high-frequency losses that occur when recording and playing back on magnetic tape are reduced.

The crosstalk cancellation stage compares the past and present horizontal lines of signal, derives a difference of the two signals, and subtracts it from the present active horizontal line. High-frequency crosstalk cancellation can be done using this comparision method, because one horizontal line is similar to the next one.

The crosstalk cancel stage also receives signals from the dropout compensator stage. When the DOC signal is HIGH, a loss of RF from the tape has been detected. This DOC signal toggles

a switch in the crosstalk cancel stage that selects the last active line of horizontal luminance from a 1H delay line and continues to insert it into the luminance chain until the DOC signal toggle returns LOW.

After the crosstalk cancel stage, a trap frequency stage is used to set a maximum frequency limit to the playback signal. This is necessary to eliminate high-frequency noise outside the luminance bandpass. In the Super Beta mode, a higher frequency trap is chosen because of the wider bandwidth of the playback signal. An equalization circuit follows the trap to maintain a flat frequency response across the luminance band.

The noise-cancelling stage removes high-frequency noise within the luminance spectrum. This completes the processing of the luminance signal from an RF signal played back from the video heads through demodulation, frequency response equalization, and noise reduction stages.

The processed luminance and chroma are

combined in a Y-C mixer. The composite video output is acted upon by the next VD/H shift stage only during special effects modes to correct for possible distorted horizontal and vertical sync signals. The completed playback video signal is then converted by a modulator for viewing on the TV receiver.

SUPER BETA PLAYBACK CIRCUIT

Of the three Super Beta changes to the playback luminance stages, the most important is the selection of trap frequencies. Each trap determines the upper frequency response limit and is set just above the luminance signal to eliminate noise above that point. Because the luminance signal is higher in frequency during Super Beta, two limits must be set. One trap sets a frequency limit for normal playback and a second trap sets a frequency limit for Super Beta playback. If the correct selection of these traps is not made, Super Beta frequency response could be limited or noise could be added to regular

playback. This is why this stage is most important.

The incoming luminance signal after demodulation and de-emphasis is applied through R102 to L-C traps. They establish an upper cutoff or termination frequency, but only one trap is used at a time. During conventional playback, when the Super Beta switch is off, no voltage appears on the HB line, so Q108 and Q722 are off. When Q108 is off, the L-C trap at its collector is not used. When Q722 is off, its collector is HIGH, turning on Q107 and placing the low-frequency L102-C103 trap in the conventional playback luminance circuit.

In Super Beta mode, when the front panel Super Beta switch is closed, +9 Vdc appears on the HB line, turning on Q108 and Q722. Refer to Super Beta circuit tie-in shown in Fig. 12-7. When Q722 is on, its collector is LOW, turning off Q107, which disconnects its L-C trap from the luminance circuit. When Q108 is on in Super Beta, the high-frequency L104-C161 trap is used. After the upper frequency limit has been

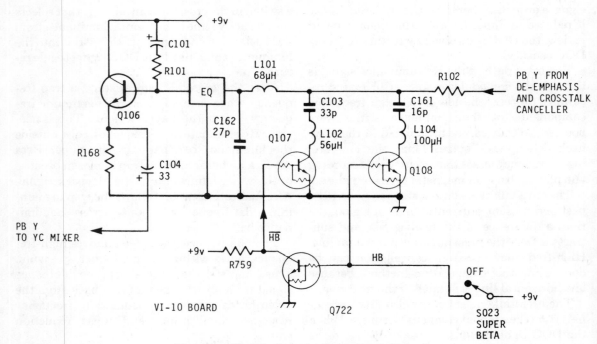

Fig. 12-7. *Circuit tie-in for Super Beta playback.*

set, noise above this limit has been eliminated, and signal below this limit is allowed to pass using a low-pass filter consisting of L101 and C162. The luminance signal is then passed through an equalization stage, a buffer transistor (Q106) and on to a mixer to combine with chroma to complete the return to a composite video signal.

CAPSTAN CONTROL FOR SLOW SPEED EFFECT

Refer to Fig. 12-8. Speed and direction signals leave System Control 3 and are made available to the main Systems Control 1 (IC401) for capstan control. The front panel pause, play, cue, review, and other push buttons are also made available to the same System Control (IC401) and are key-scanned to identify which pushbutton in the matrix was pushed. When the pause button is pressed, the jog dial and shuttle ring are enabled. Information is clocked through System Control 1 (IC401) using an RF switching pulse coming in at pin 69 (there is a 6 MHz clock at pins 59 and 60).

Pause, step advance, 1/5 speed, and X1 speeds are controlled by outputs from IC401 and are labeled SLOW, CAPS FWD and STEP. They are fed into IC801 where they come out with an important change in timing. During Pause, step, and 1/5 speeds, the SLOW input at IC 801 (pin 34) goes HIGH, causing the output of IC801 (pin 9, TPS) to go HIGH. TPS is delayed by IC513, depending on the tape speed, and emerges at IC513 (pin 3) as the CAP CUT signal. CAP CUT turns ON Q509, which grounds out the capstan servo drive, breaking its servo loop so step advance or slow motion can be accomplished. When the jog dial is rotated, a step pulse entering IC801 (pin 41) and the RF switching pulse at pin 35 are used to manufacture a step pulse that leaves pin 22 to advance the capstan motor to the next field.

The jog control IC801 not only starts the capstan motor moving into the next field, but also must stop it on the A field to ensure a noiseless still-picture. Braking the capstan motor is done by changing the capstan direction signal CAP RVS from HIGH to LOW and applying step drive pulses (step pulse) from drive pulses (step pulse) of a controlled duration. In order for IC801 to develop step pulses to stop the tape exactly on the A field, it must first know how far it is to the new A field, and second, how fast the capstan is moving.

These two questions are best answered by referring to the slow servo/head timing chart, (Fig. 12-9), which shows the inputs and outputs of IC801 during a step advance. From the jog dial and through IC401, IC801 receives a step advance signal at IC801/pin 41 to advance the tape. A short time later when the RF switching pulse again makes a LOW-to-HIGH transition, a step pulse at IC801 (pin 22) is output to drive the capstan motor. When the motor turns, the tape also moves. The tape provides feedback to IC801 (pin 30) and 31 called CTL. The CTL signal tells IC801 how close the tape is to the A track. This answers the first question (how far is it to the A field?). The motor provides feedback to the IC801 (pin 7), called 8FG. IC801 uses 8FG to tell how fast the motor's going by counting the time between the 8FG pulses. This answers the question about how fast the capstan motor is turning. IC801 also knows the number of 8FG pulses from one A field to the next and can therefore guage how close it is to the next A track, even if CLT is missing. Now that both questions have been answered, pulses determined by 8FG and CTL can be manufactured to stop the tape on the A track. This signal is called the step pulse from IC801 (pin 22).

As the jog dial is rotated faster, normal PB (X1) speed is reached. At this time, the SLOW input to jog control IC801/pin 34 then goes LOW causing the TPS output to go LOW, which causes TPS IC801 (pin 9) to turn off Q509. This restores the capstan servo loop, and the tape is locked to X1 speed.

AUDIO-BETA HI-FI

In the Beta Hi-Fi system, the audio signals are converted to modulated FM carriers and re-

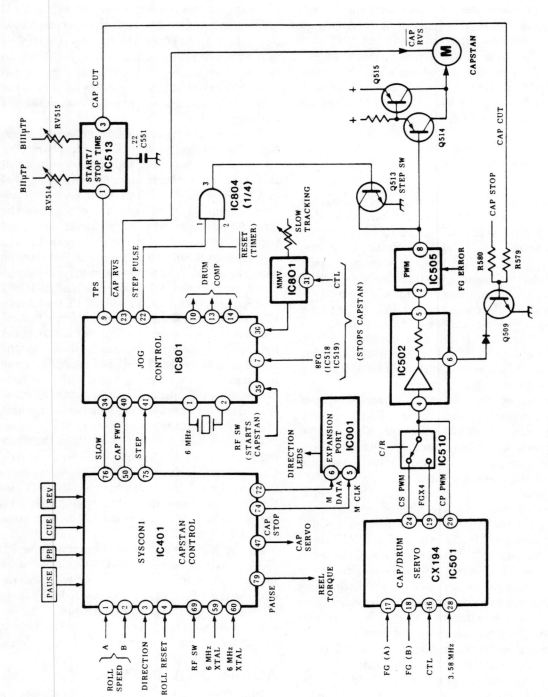

Fig. 12-8. *Capstan control (slow) diagram.*

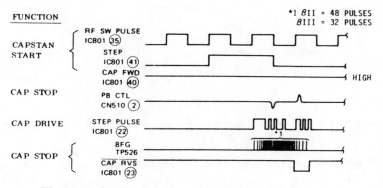

FUNCTION

CAPSTAN START
- RF SW PULSE IC801 ㉟
- STEP IC801 ㊶
- CAP FWD IC801 ㊵ — HIGH

CAP STOP
- PB CTL CN510 ②

CAP DRIVE
- STEP PULSE IC801 ㉒

CAP STOP
- 8FG TP526
- CAP RVS IC801 ㉓

Fig. 12-9. *Slow servo/head control timing chart waveforms.*

corded on the video track of the video tape using the two rotary video heads. The advantages to using the FM recording process are as follows:

- Excellent frequency response
- Low noise and distortion
- Wide dynamic range

Using the rotary video heads results in a relative tape-to-head speed that is about 350 times that of a conventional audio recording system.

Wow and flutter specifications are below the measurable limit.

The Y signals have been shifted up slightly from 0.4 MHz to make room for the AFM signals to be inserted between the chroma and Y signals on the video track. Refer to recorded spectrum signal drawing in Fig. 12-10.

The AFM signals consist of four different carriers—two for each channel—separated by 150 kHz. Four carriers are used to reduce cross-

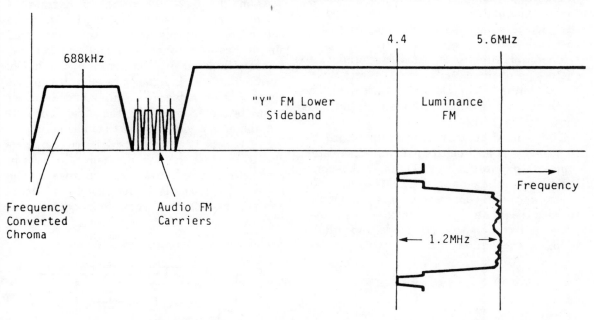

Fig. 12-10. *Recorded spectrum—Super Beta and Beta Hi-Fi.*

Carrier	Frequency	Video Track	Audio Channel
f1	1.380682 MHz (87.75fH)	A	L
f2	1.530157 MHz (97.25fH)	B	L
f3	1.679633 MHz (106.75fH)	A	R
f4	1.829108 MHz (116.25fH)	B	R

Fig. 12-11. *Beta Hi-Fi pilot frequency chart.*

talk between the left and right channels and between "A" and "B" video tracks.

The frequency assignments of the four FM carriers are shown in the chart of Fig. 12-11 for the pilot frequencies. The F1 and F3 signals represent left and right channel audio and are recorded on the A video track.

In the playback process, the four AFM carriers are reconstructed into left and right channel audio by alternately switching the A and B tracks in synchronization with the video heads.

VHS HQ (HIGH QUALITY) VIDEO

The High Quality system is a new technology that enables high picture quality in every part of the picture while maintaining superior VHS compatibility.

The two HQ improvements incorporated into the design of these VHS recorders are; WC — White Clip and DE — Detail Enhancement. The block diagram (without detail enhancement) is shown in Fig. 12-12.

The bandwidth of magnetically recordable frequencies is an important factor that determines the picture quality of VCRs. With technological advancements in the fields of material, device, and circuit engineering, it is now possible to raise the white clip level by 20 percent over previous recorders to accommodate recordings of higher frequencies.

By raising the white clip level, the rising flank of a waveform from dark to white, where high-frequency signals are concentrated, is made clearer than before. This results in sharper edges on vertical objects in the picture, improving the overall quality of the picture.

DETAIL ENHANCEMENT

The detail enhancement circuit is a low-level high-frequency, non-linear emphasis circuit that reinforces the luminance signal. This allows recording of more detail of a scene such as strands of hair or pebbles on a beach.

As noted, Fig. 12-12 shows a block diagram of a video signal processor circuitry without detail enhancement. In this circuitry, the video signal simply passes through an equalizer. The circuitry with detail enhancement is shown in Fig. 12-13. The video signal passes through the equalizer plus a high-pass filter and a limiter before being recombined in the adder. The output signal is a video signal that has enhanced high-frequency components. The circuit operates to increase the enhancement of objects in the picture in proportion to "faintness." The fainter the object is in relation to its background, the more the detail enchancement in-

Circuitry without detail enhancement

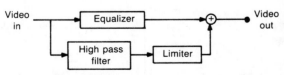

Fig. 12-13. *Circuitry with detail enhancement.*

Circuitry without detail enhancement

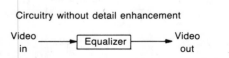

Fig. 12-12. *Circuitry without detail enhancement.*

creases. The chart in Fig. 12-14 illustrates this relationship.

HQ QUESTIONS AND ANSWERS
What Does HQ Do?

1. Increases white clip level (Fig. 12-15).

   - HQ increases white clip level by 20 percent without overmodulating the recording.
   - This procedure results in sharper edges, particularly vertical edges and profiles.

2. Reduces color noise (Fig. 12-16).

   - A noise reduction circuit doubles color output while increasing noise only 1.4 times.
   - By improving color signal-to-noise ratio, color streaking and patches are reduced, even in minute details.

3. Reduces luminance noise (Fig. 12-17).

   - A noise reduction circuit doubles the luminance output while increasing noise only 1.4 times.
   - The circuitry improves the signal-to-noise ratio by 3 dB, markedly reducing noise throughout the picture.

SUPER VHS VIDEO RECORDERS

The Super VHS VCRs contain new systems and circuitry that, when connected to a digital TV receiver, provide in excess of 400 lines of horizontal resolution in the RBG mode. When connected in the normal manner to a TV receiver, a significant improvement in the quality (resolution) of the picture is noticable.

These new recorders look just like any other VHS recorder. The mechanical part of the deck is almost identical to those used in "standard" VCRs. Inside, most of the circuitry is the same for the mechacon, audio, servo, and tuner. The Super VHS feature is accomplished by the following:

- New narrower gap heads
- New oxide formula cassette
- Expanded 1.6 MHz Y FM deviation
- New non-linear emphasis signal-processing circuitry

The recorder has been designed for what might be considered a dual system. That is, it will record and play standard VHS tapes in addition to SUPER VHS tapes. This is accomplished by a special video cassette with a new oxide formula. The cassette cartridge has an identification hole at the bottom (see Fig. 12-18) that functions to place the recorder in the Super VHS mode. In this manner, it will record and play Super VHS tapes and will also accept standard VHS tapes for record or playback.

In the video signal processing circuitry, changes have been made to increase the deviation of the luminance signal. In Fig. 12-19, the frequency range of the Y FM signal in a normal

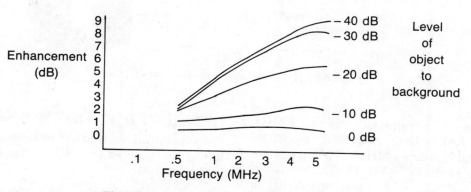

Fig. 12-14. *Detail enhancement (dB) curve chart.*

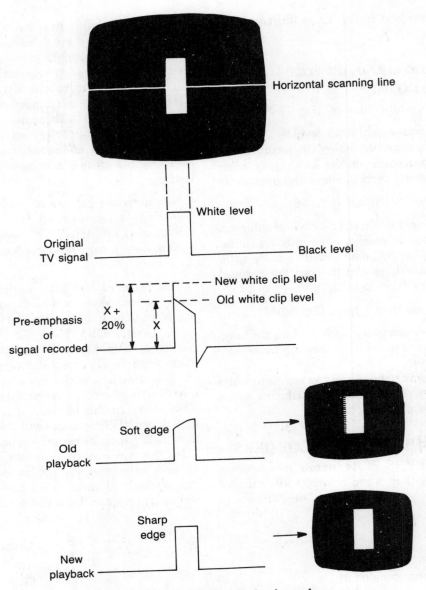

Horizontal scanning line

Original
TV signal

White level

Black level

New white clip level

Old white clip level

Pre-emphasis
of
signal recorded

X +
20%

X

Old
playback

Soft edge

New
playback

Sharp
edge

Fig. 12-15. *Increased white clip level waveforms.*

VHS recorder is from 3.4 MHz to 4.4 MHz, a range of 1 MHz. In the Super VHS mode, the Y FM signal is from 5.4 MHz to 7 MHz, a range of 1.6 MHz as shown in Fig. 12-20. These are the major differences between standard and SUPER VHS.

DETAILED HQ SYSTEM ANALYSIS

Such newly developed technologies as the following are adopted in VHS VCRs of the HQ picture system.

• Level up for white clip by 20 percent

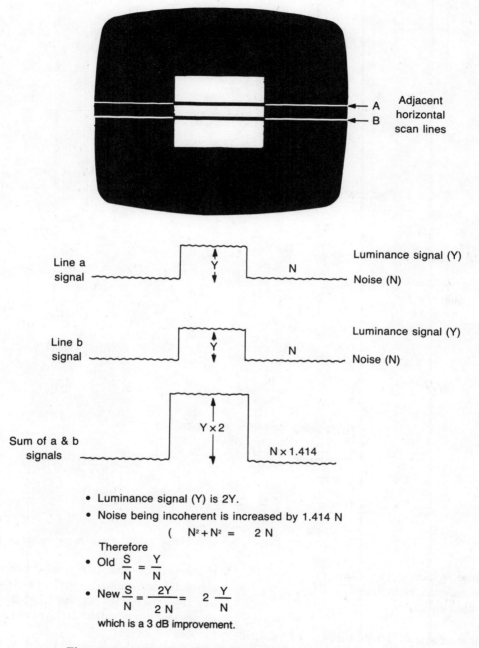

Fig. 12-16. *Luminance noise reduction waveform illustration.*

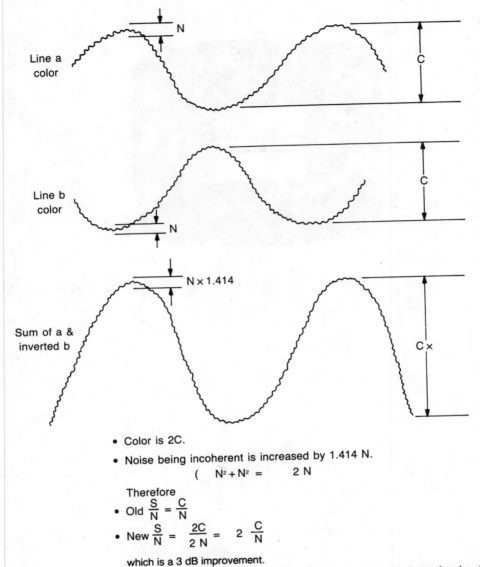

- Color is 2C.
- Noise being incoherent is increased by 1.414 N.

$$(\quad N^2 + N^2 = \quad 2\,N$$

Therefore
- Old $\dfrac{S}{N} = \dfrac{C}{N}$
- New $\dfrac{S}{N} = \dfrac{2C}{2\,N} = \quad 2\,\dfrac{C}{N}$

which is a 3 dB improvement.

Fig. 12-17. *Color noise reduction illustration.*

- Y noise reduction (Y/NR)
- Chroma noise reduction (C/NR)
- Detailed enhancer

Characteristic of TV pictures can be considered and analyzed from various points of view, however main factors to influence quality of VCR pictures in playback are pulse response characteristic and S/N ratio. Pulse response characteristic shows how much recorded signals are reproduced in playback, while S/N ratio shows amount of noise on the plane and edges of playback pictures.

Among the above mentioned four items, A and D improve pulse response characteristic, while B and C improves S/N ratio.

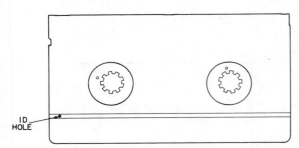

Fig. 12-18. *Drawing of Super VHS cassette.*

LUMINANCE NOISE REDUCTION (Y/NR)

VHS VCRs are equipped with an emphasis circuit as in the past. The emphasis circuit is composed of a pre-emphasis circuit and de-emphasis circuit.

Incoming video signals first go to the pre-emphasis circuit, which emphasizes high-frequency components of the signal. It then goes to the clipping circuit for frequency modulation as the input signal is recorded in FM waveform.

In playback, the demodulated video signals contain high-frequency noise components that are mainly generated in FM recording and playback. The de-emphasis circuit suppresses the high frequency to reduce noise, and the video signal returns to its original level because its high frequency was emphasized in recording.

The emphasis circuit improves S/N ratio in the manner as stated above. Conventional emphasis circuits are composed of C and R components that can be replaced with transversal filters of delay elements. The drawings in Fig. 12-21 and Fig. 12-22 illustrate pre-emphasis and de-emphasis circuits and relationships between them. As shown by the drawings of transversal filter circuits, previously output horizontal signals are composed of outputs of conventional circuits. If a delay line of transversal filters is a delay element to delay the horizontal scanning time at a unit, emphasis that vertical signals are composed is effected. According to the above

VHS

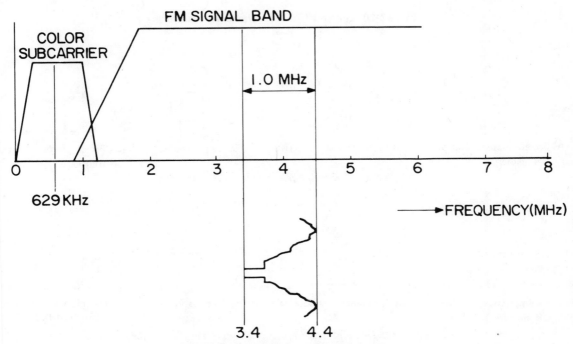

Fig. 12-19. *VHS FM signal band waveforms.*

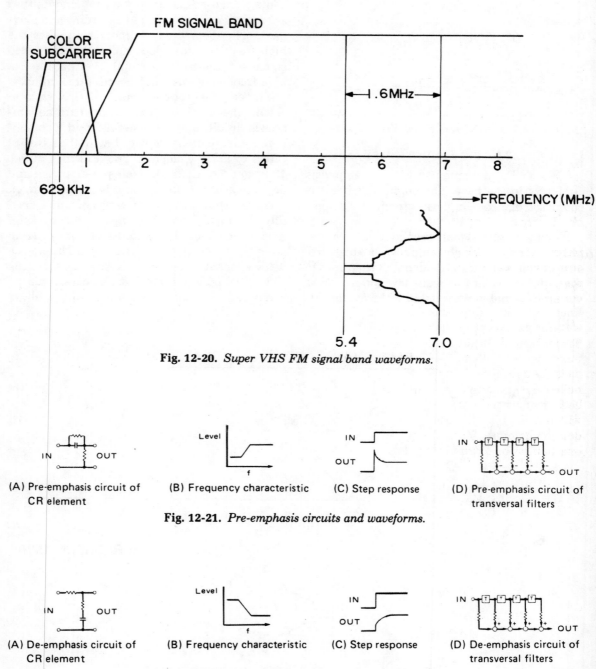

Fig. 12-20. *Super VHS FM signal band waveforms.*

(A) Pre-emphasis circuit of CR element

(B) Frequency characteristic

(C) Step response

(D) Pre-emphasis circuit of transversal filters

Fig. 12-21. *Pre-emphasis circuits and waveforms.*

(A) De-emphasis circuit of CR element

(B) Frequency characteristic

(C) Step response

(D) De-emphasis circuit of transversal filters

Fig. 12-22. *De-emphasis circuits and waveform illustrations.*

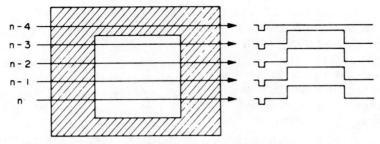

Fig. 12-23. *A drawing of how S/N ratio improvement is obtained.*

principle then, a Y/NR circuit has been developed.

S/N RATIO IMPROVEMENT

The next explanation is the idea of S/N ratio improvement. A TV picture is comprised of 525 scanning lines. In Fig. 12-23, for example, every scanning line is a signal waveform, shown by the corresponding arrows. In the figure, neighboring lines show little difference between correlation waveforms, except in the case of n-4, which is a bordering portion of the picture. If a noise is mixed with the signals in recording and playback (as shown later in Fig. 12-24), and the signals of line n and line n-1 are added, the *signal* becomes double, but the *noise* will be $\sqrt{2}$ times (it will not double, because noise occurs at random). Therefore, when the signal is reduced to one-half of the original waveform, the noise becomes $\sqrt{2}/2$, and the S/N ratio is improved to $1/\sqrt{2}$ (3 dB in amount).

As the correlation between lines exists not only between two neighboring lines but also between the other two lines next to the neighboring ones, S/N ratio can be improved by adding these lines using a circuit like that in Fig. 12-25.

In the above case, the adding is not done on the same rule, but the rate increases proportionally for two closer lines. The rate can be set uniformly such as shown in the example of Fig. 12-24. In the case of the addition on the same rule, this idea is realized by using a 1H delay circuit. As an example, Fig. 12-26 shows an example.

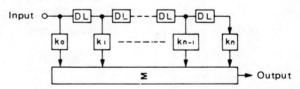

Note: DL is a 1H delay circuit

Fig. 12-25. *S/N ratio is improved by adding lines using this circuit.*

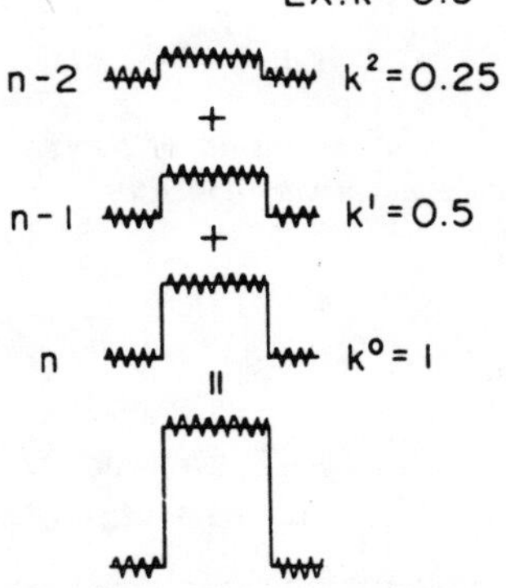

EX. k = 0.5

n-2 $k^2 = 0.25$

+

n-1 $k^1 = 0.5$

+

n $k^0 = 1$

‖

Fig. 12-24. *How S/N ratio can be set uniformly.*

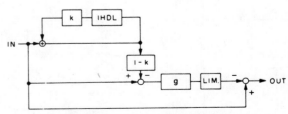

Fig. 12-26. *Block diagram of S/N limiter circuit.*

If a circuit such as in Fig. 12-27 is applied, vertical signals are added on the picture as shown in the waveforms of (Fig. 12-28).

However, if an edge portion of a picture is vertically displayed, the waveform is moderate at its rise portion as in Fig. 12-29a, because the neighborning scanning line is added. This causes deterioration in its vertical frequency characteristic.

To solve this problem, apply the principles of pre-emphasis and de-emphasis explained previously for the vertical lines. Treat the signal as the waveform shown in Fig. 12-29b to correct the rise portion and increase it in vertical frequency characteristic. By this treatment, the waveform similar to the original, shown in Fig. 12-29c, is obtained in playback.

If this treatment is performed slightly just for low-level signals, S/N ratio is improved without deterioration in changeability. This treatment is done by the limiter see block diagram of Fig. 12-26.

PRINCIPLES OF Y/NR

Refer to (Fig. 12-27) for a circuit of the recording system. In recording, this circuit functions for precompensation of decrease in vertical resolution in playback. Waveforms observed at vertical rate are shown in Figs. 12-30 A through E.

Refer to Fig. 12-31. When a square waveform is input to a signal source, the output from ADD(1) is moderate in its rise portion due to the cyclic low-pass filter that consists of ADD(1), 1H DELAY, and ATT(1). As this output is larger in amplitude than the original one, ATT(2) corrects it to have the same level as that of the original signal. The output of SUB(1) is a high-pass signal that is the difference compo-

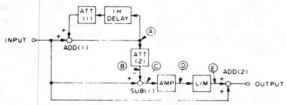

Fig. 12-27. *Y/NR circuit in recording system.*

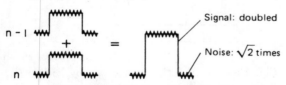

Fig. 12-28. *Added vertical signals.*

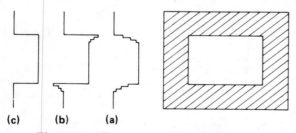

Fig. 12-29. *Treatment of signal waveform.*

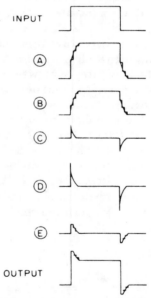

Fig. 12-30. *Precompensation of decrease in vertical resolution observed at vertical rate.*

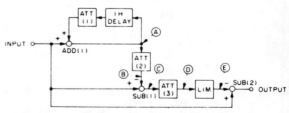

Fig. 12-31. *Block diagram of NR playback circuit.*

nent between the original and low-pass signals. This high-pass signal is amplified to have a level that compensates the playback signal, and after its amplitude is controlled by the limiter to secure the changeability, it is mixed with the original signal by ADD(2), to be sent as the REC signal.

Figure 12-31, the playback circuit, functions to remove noise generated in recording and playback. When P.B. signals containing noise are fed to the input terminal, the cyclic low-pass filter composed of ADD(1), 1H DELAY, and ATT(1) removes the noise, and ADD(1) outputs a noiseless signal. If this signal is sent to ADD(2) in the same manner as in recording, the output of SUB(1) is high-pass signal containing noise.

Therefore, the circuit is designed so that the signal passes ATT(3) (which decreases noise effectively), the limiter (to secure changeability), and SUB(2) (which adds noise in reversed polarity). Through the above process, the same output signal as the original can be obtained, except it's without noise. The waveforms found throughout the playback circuits are illustrated in Fig. 12-32 A through E.

PRINCIPLES OF COLOR NR CIRCUIT

The principle of the C/NR circuit is the same as that of the Y/NR circuit. As the color signal is a 3.58 MHz signal whose phase is inverted every 1H, the addition of lines is performed by inverting the phase every 1H. For the edging portions, it is treated by applying the correlation with the luminance signal. The C/NR circuit is provided only for the playback system.

NR CIRCUIT CHIP

Refer to chip diagram shown in Fig. 12-33. The IC407 chip contains a dropout compensation circuit and a non-linear de-emphasis circuit for playback in EL/LP mode, besides the Y. NR circuit.

The Y signal is input through pin 22 of IC407 and is fed ADD(2) via the non-linear de-emphasis circuit. The ADD(2) is a cyclic comb filter

composed of ADD(1) and 1H delay circuit, which is used for dropout compensation also. Signal output from ADD(2) is supplied to pin 13 through pin 15. This signal is sent to the noise reduction block and then output from pin 10. The signal output from pin 12 in REC mode is used for regular pre-emphasis.

MTS TELEVISION SOUND SYSTEM

The MTS (Multi-channel Television Sound) broadcast system features the transmission of stereo, bilingual, and voice/data signals. The spectral diagram of the MTS signal is shown in Fig. 12-34.

The MTS system is compatible with current transmissions and mono receivers. The L + R signal has 75 ms pre-emphasis and deviates the aural carrier $25 \pm$ kHz on 100 percent peaks. The next component of the MTS signal is the 15.734 kHz stereo pilot. The pilot deviates the aural carrier 5 kHz and is used in the receiver to detect the L − R stereo sub-channel. The L − R

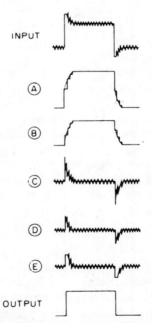

Fig. 12-32. *Waveforms found throughout the NR playback circuit.*

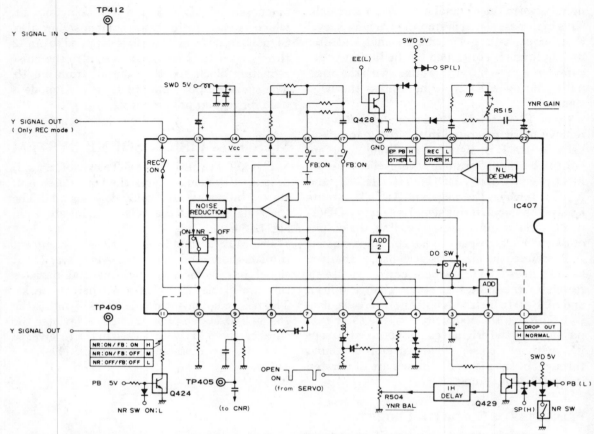

Fig. 12-33. *Details of Y/NR circuit and chip. Courtesy Zenith Electronics Corporation.*

channel is an AM double sideband suppressed carrier signal and deviates the aural carrier 50 kHz. The SAP sub-carrier is an FM, 10 kHz deviated signal centered at 78.670 kHz (5FH). The SAP subcarrier deviates the aural carrier 15 kHz. Both the L − R and SAP audio signals are dbx encoded to reduce buzz and noise.

CIRCUIT COMPOSITION

The MTS demodulator is shown in Fig. 12-35. The audio demodulator is comprised of eight ICs. IC301 demodulates the IF signal to yield the baseband signal. Sub audio SAP demodulators and the L/R matrix are contained in IC302. IC303 and IC304 are for dbx noise reduction.

The peripheral circuits required by IC303 and IC304 are contained in IC305, IC306, IC307, and IC308.

IF TO MTS SIGNAL FLOW

The IF signal from the tuner/IF circuit is amplified by Q301 and Q302. The SAW (surface acoustic wave) filter (SAW 301) removes adjacent channel signal components, after which the signal goes to pins 8 and 9 of IC301. In IC301, the IF signal is amplified and applied to the video demodulator (AM detector). The PLL (phase-locked loop) circuit provides fully synchronous detection and features less audio noise than ordinary detector systems.

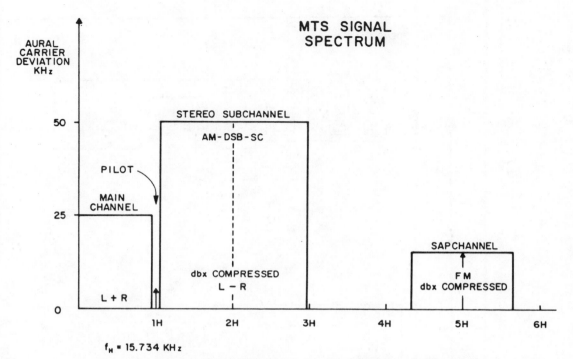

Fig. 12-34. *MTS signal spectrum. Courtesy Zenith Electronics Corporation.*

The signal from IC301 pin 28 goes to a band-pass filter (CF301), yielding the 4.5 MHz ±250 kHz FM audio signal. This is applied to IC301 pin 13. A limiter removes the AM component and the FM signal is demodulated to form the MTS signal.

The MTS signal from IC301 pin 19 is sent to IC302 pin 18, where it is amplified and distributed in two lines from pin 19. One of these becomes the main and sub signals, and the other becomes the SAP signal.

MAIN (L+R) AND SUB (L−R) SIGNALS

The MTS signal from IC302 pin 19 goes through LPF 301 (50 kHz) to remove the SAP component and then to pin 20 of IC302 to remove the pilot signal. It is then distributed in two lines. One goes as the main (L+R) signal to the pin 16 output and the other as the sub (L−R) signal to the stereo demodulator circuit. The main signal is sent via LPF 302 (15 kHz) to

pin 12 of IC301, a voltage follower amplifier. The signal exits pin 15 and is routed to pin 12 of IC302, where it applied to the matrix circuit. At the stereo demodulator in IC302, the sub signal is applied to the AM detector and mode switch and exits at pin 8. It passes through low-pass filter LPF303 (15 kHz) and is routed to pin 3 of IC303. The signal is inverted and distributed in three lines from pin 4. One of these goes to IC305 pin 4, the high-pass and low-pass filter, then from IC305 pin 7 to RMS DET-1, to IC303 pin 7. RMS DET-1 controls the spectral expander VCA at pin 24.

The second route is via the wide-band filter between IC305 pins 4 and 9 to pin 9 of IC303, then to RMS DET-2 at IC303, which controls the wide-band expander VCA at pin 18.

In the third line, the signal from pin 4 of IC303 is sent to pin 26 and through a buffer stage, exiting pin 25 of IC303. IC306 contains a fixed de-emphasis network between pins 10 and 5. A variable de-emphasis circuit, which is composed

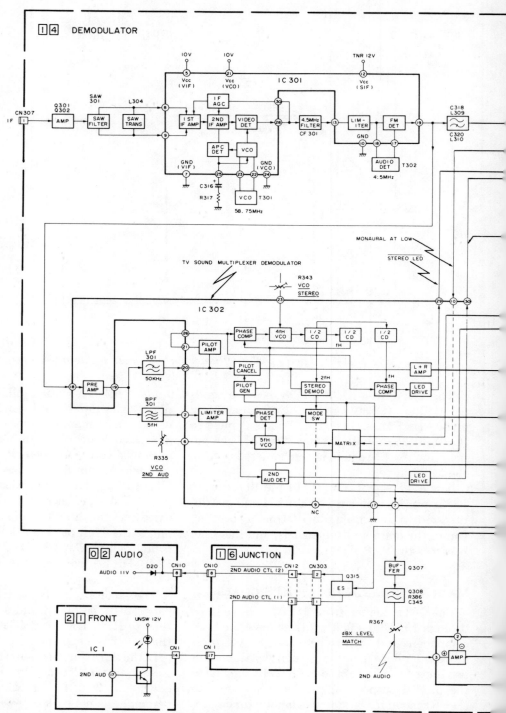

Fig. 12-35. *MTS decoder block diagram. Courtesy Zenith Electronics Corporation.*

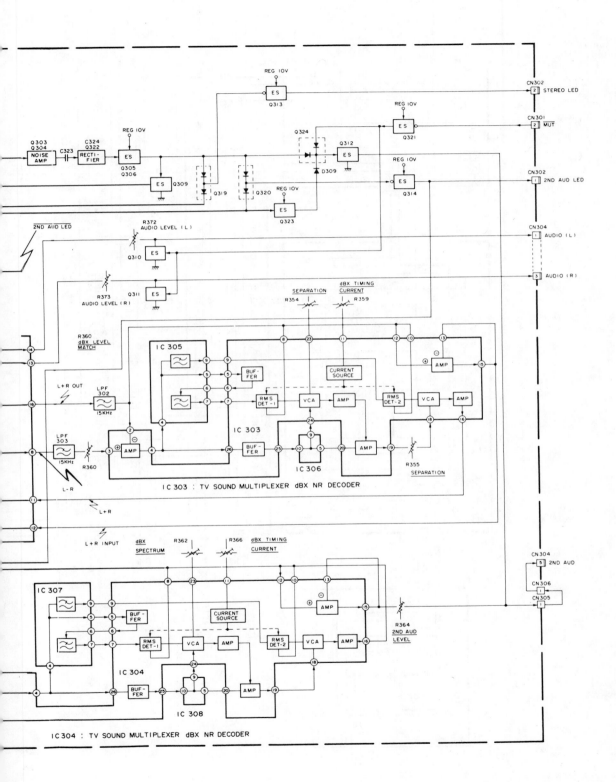

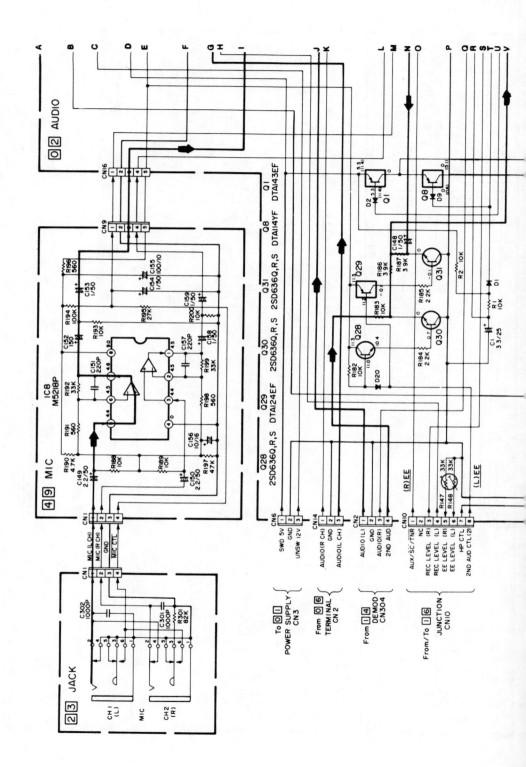

272

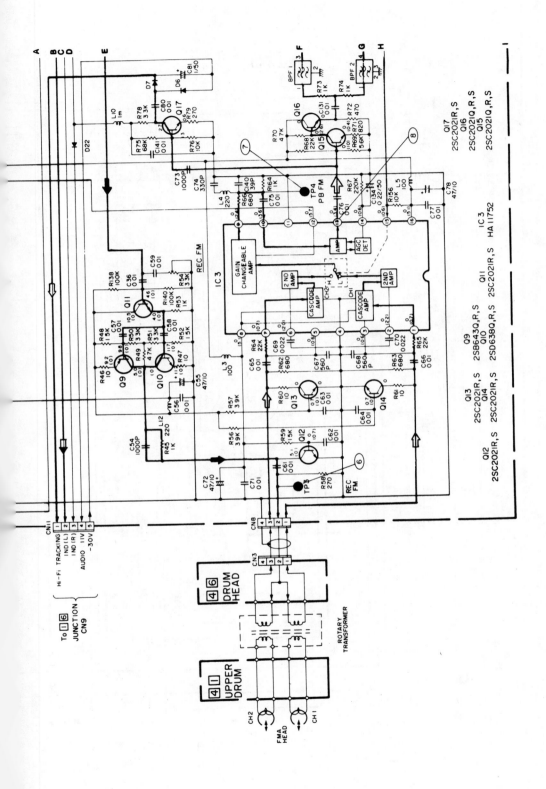

Fig. 12-36. *Complete VR3300 VCR audio schematic diagram. Courtesy Zenith Electronics Corporation.*

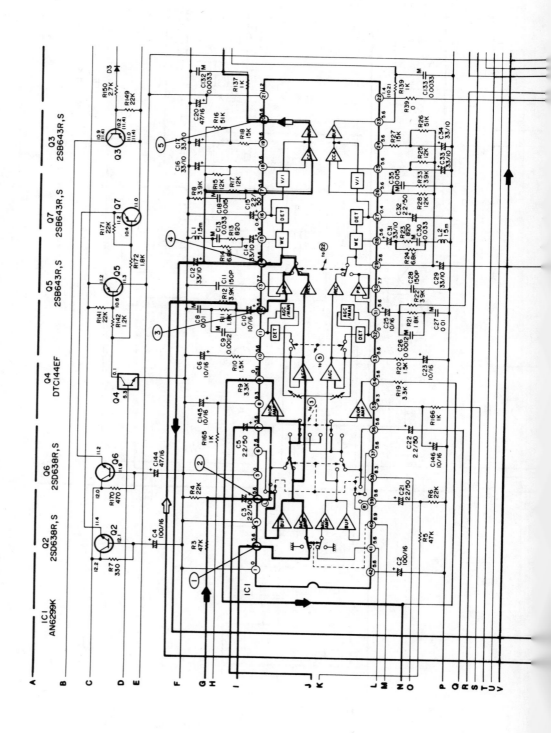

274

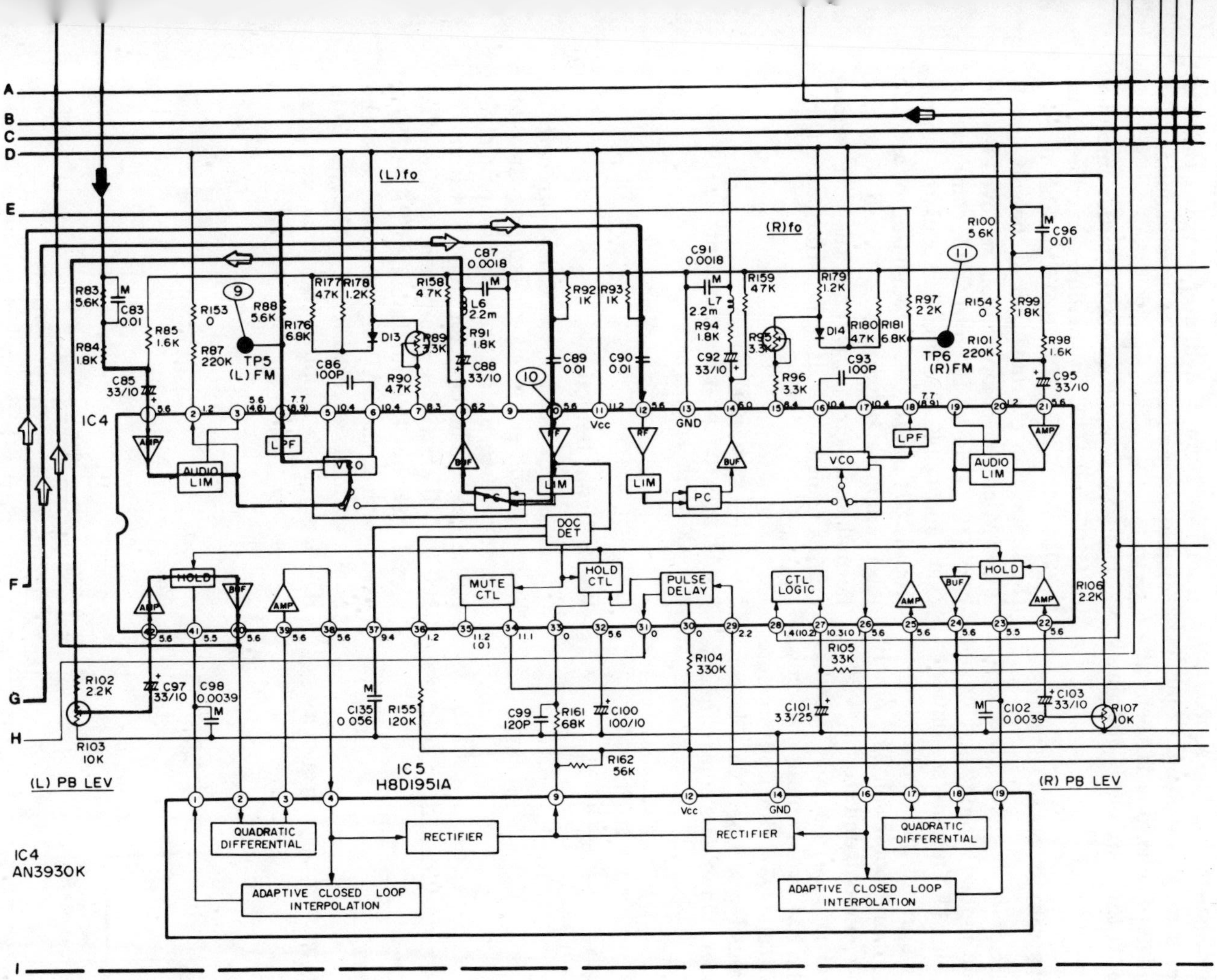

Fig. 12-36. *(continued)*

of a frequency divider, includes VCA at pin 24 of IC303. Frequency response is precisely opposite to the variable pre-emphasis of the compressor circuit at the broadcast station. The signal from IC303 pin 19 goes through wideband expander VCA at pin 18 and fixed de-emphasis, then exits pin 16 to be routed to the matrix at IC302 pin 11.

MATRIX CIRCUIT AND SIGNAL OUTPUTS

The matrix circuit of IC302 processes the main (L + R) signal applied to pin 12 and the sub (L − R) signal at pin 11. Outputs are obtained from pins 13 and 14. In this model, pin 9 is open and pin 17 is at ground potential. The pin 10 input selects between mono and stereo.

The audio signal outputs from pins 13 and 14 are supplied to the FM audio amplifier circuit. The circuit of Q310 and Q311 functions to mute noise during channel selection. This is controlled by the Tuner/Timer data control microprocessor.

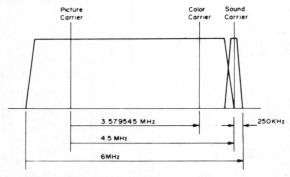

Fig. 12-37. *TV signal frequency spectrum.*

THE SAP SIGNAL

The bandpass filter BPF301 (5fh) extracts the SAP component from the MTS signal output from IC302 pin 2. A limiter removes the AM component, and after FM demodulation, the SAP signal appears at pin 7.

The SAP signal goes via the buffer amplifier of Q307 and Q308 to the dbx demodulator of IC304, IC307, and IC308. Within this circuit, the signal flow overlaps that of the sub signal. The FM carrier component is removed by the LPF (19.5 kHz) of R386 to C345, contained in Q307 and Q308. The dbx demodulator output from IC304 pin 16 goes through the voltage-follower amplifier between pins 13 and 15 to the normal audio circuit. Q312 is the SAP signal muting transistor. Muting is performed in three situations:

- channel selection
- SAP program absence from broadcast
- weak signal strength

Weak Signal Detection

This circuit detects weak field strength in order to avoid erroneous STEREO/SAP LED lighting, and stereo/SAP signals with impaired signal-to-noise. The detector functions from the 100 to 150 kHz noise component of the MTS signal. At about 15 dBu, the stereo/SAP LED is extinguished, the SAP signal muted, and the monaural mode selected. The MTS signal from IC301 pin 19 is supplied to a bandpass filter

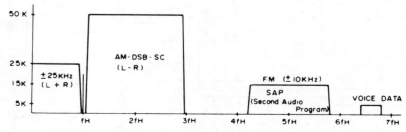

Fig. 12-38. *Stereo sound baseband spectrum (United States).*

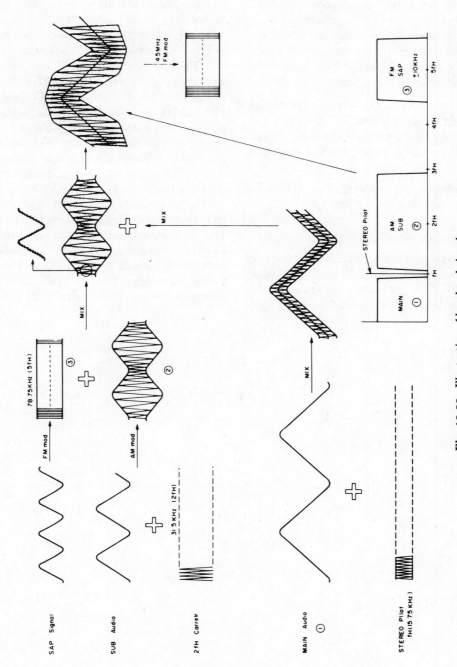

Fig. 12-39. *Illustration of baseband signals.*

composed of C318, C320, L309, and L310. The noise component output is amplified by Q303 and Q304, rectified by Q322 and smoothed by C324. The hysteresis amplifier of Q305 and Q306 shapes the dc signal waveform. Under weak signal conditions, the Q306 output is HIGH. This is distributed in three lines:

- To the stereo/SAP LED drive circuit for extinguishing the indication.
- To switch on Q312, to mute the SAP signal.
- To swtich on Q309, to engage the monaural mode.

The complete stereo audio SAP signal circuit is shown in the schematic of Fig. 12-36.

MTS TV Sound System Reviewed

As previously noted, the MTS system features the transmission of stereo, bilingual, and voice/data signals. The waveform diagram in Fig. 12-37 illustrates the TV signal frequency spectrum.

The MTS System is compatible with current transmissions and mono receivers. Figure 12-38 shows the spectral diagram of the MTS signal.

The L+R portion or main channel of the MTS signal is identical to the current mono signal. The L+R signal has 75 ms pre-emphasis and deviates the aural carrier 25 kHz.

The next component of the MTS signal is the 15.734 kHz stereo pilot carrier. The pilot carrier deviates the aural carrier 5 kHz and is used in the receiver to detect the L−R stereo sub-channel. The L−R sub-channel is an AM modulated double side band suppressed carrier signal and deviates the aural carrier 50 kHz.

The SAP subcarrier is an FM modulated, 10 kHz deviated signal centered at 78.670 kHz. The SAP subcarrier deviates the aural carrier 15 kHz. Both the L−R and SAP audio signals are dbx encoded to reduce buzz and noise.

The waveforms for the MTS Baseband signals are illustrated in the drawings of Fig. 12-39.

13

Color Video Camcorders

This chapter covers three different brands of camcorders. For practicality, I only highlight some of the special circuits that are not found in basic VCR machines. The chapter begins with the SONY model CCD-V8 camcorder that contains special circuits such as the autofocus and power control. Then the Zenith model VM6200 camcorder system and troubleshooting chart is covered. The chapter ends with coverage of the RCA model CC011 camcorder circuit operations.

SONY MODEL CCD-V8 CAMCORDER

The following information and illustration are courtesy of Sony Corporation of America.

Charge-coupled device (CCD) solid-state pickup elements are now used in most camcorders. The CCD camera can be divided into eight main circuits as follows:

- the CCD
- timer

- sync generator
- CCD driver circuits
- process
- white balance and iris control
- matrix
- encorder

All of these circuits are integrated circuits with support components for adjustments, effectively simplifying the overall complex circuitry of the CCD camera.

TIMING

The block diagram in Fig. 13-1 shows the timing generator (IC707) and its relationship to the other main circuits of the camera. Its function is to generate all the necessary timing pulses for the operation of the camera circuits. The output signals on the left of the timing IC block (CX23047A) are used to develop all the pulses needed to operate the CCD imager. XSG1 and XSG2 are used to develop a 60 Hz pulse. The combination of this 60 Hz pulse and the 180-de-

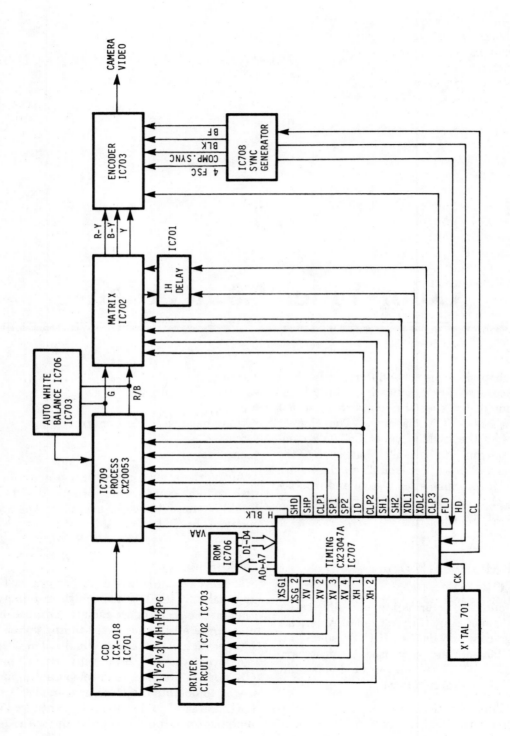

Fig. 13-1. *Overall block diagram of CCD camera. Courtesy of Sony Corporation of America.*

gree out-of-phase 15.75 kHz pulses from XV1 to XV3 produce the V1 and V3 pulses that shift charges out of the sensors to, and down, the vertical registers.

Also shown on this side of the block are the XV2 and XV4 outputs. These two signals are 180 degrees out of phase at a frequency of 15.75 kHz. They are processed to produce the V2 and V4 pulses, whose only purpose is to shift charges down the vertical registers.

The final two outputs shown on the left side of the block are XH1 and XH2. The pulses originating at these points are 180 degrees out of phase and have a frequency of 9.55 MHz. They are processed to develop the H1 and H2 pulses for shifting, or clocking, the charges from the H1 and H2 registers at the bottom of the CCD imager to the output.

To operate the output circuit of the CCD, which functions similar to a sample-and-hold circuit, a switching pulse is needed. This switching pulse, called the PG pulse, is developed from the XH1 pulse.

On the right side of the timing IC block, the outputs to all the other operational blocks of the camera are shown. The first seven outputs are used primarily by the process IC. Their purposes are as follows:

- H.BLK (horizontal blanking) introduces the horizontal blanking pulse in the signal.
- SHD (sample/hold data) extracts the CCD imager pixel data from the output signal.
- SHP (sample/hold precharge) extracts the CCD imager precharge level from the output signal.
- CLP1 (clamp 1) is used to clamp the green, red, and blue signals prior to multiplexing and insertion of horizontal blanking.
- SP1 and SP2 (separation) separate green picture information from the red and blue information.
- ID separates the red and blue information. It is also used in the color multiplexing circuits of the matrix IC.
- VAA Pulse (vertical area available) is used together with CLP1 in the clamping process.

The last six outputs of the timing generator are used mainly in the matrix circuits.

- CLP2 and CLP3 (clamp) clamp G and Rg signals.
- SH1 and SH2 (sample and hold) operate the sample and hold circuits in the matrix IC that develops the Y signal.
- XDL1 and XDL2 (delay) drive a clock in the 1H delay IC that is instrumental in matrixing the G and RB signals.

At the bottom of the timing IC block are two input pulses: HD (horizontal drive) and FLD (field drive). These two pulses from the sync generator inform the timing generator of line and field timing.

The main support circuits of the timing generator are the 28.6 MHz clock and IC706. IC706 is a ROM, programmed to correct flaws created in the CCD during manufacture. Each CCD has its own ROM specially programmed, because the flaws are not at the same spot on every CCD.

POWER CONTROL CIRCUITS

The power control circuits (see Fig. 13-2) are located on the DS-10 board, which uses the inputs from the CAMERA ON and VTR ON switches together with the output from the mode control CPU to switch power to the appropriate circuitry in response to the input commands.

The unregulated 6 Vdc is applied to CN12 (pin 8), providing B+ to the DS-10 board power control circuit is present. When the CAMERA POWER ON switch is activated, the CAM POWER SW at CN012 (pin 1) goes low. This input is applied to Q602/B, turning Q602 on and producing a high at Q602/C. This high is applied to an inverter within IC601 (pins 10 and 11) and a second inverter, IC601 (pins 12 and 13), and the resultant high output at pin 12 is applied to the clock input IC602 (pin 11). Q602 and the two inverters within IC601 form a chattering prevention circuit that prevents chattering of the switch contacts from affecting power control and prduces a fast rise time to clock the flip-flop in IC602. A similar chatter prevention circuit consisting of

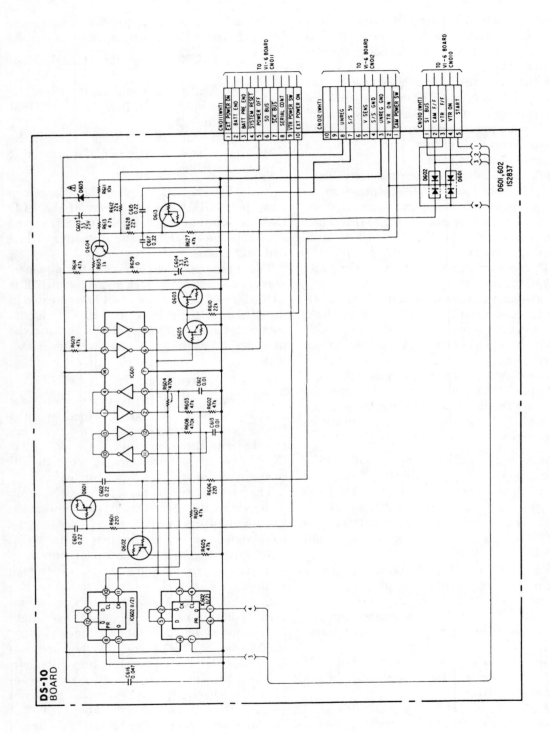

Fig. 13-2. *Power control circuit. Courtesy of Sony Corporation of America.*

Q601 and two inverters in IC601 is used for the VTR POWER SWITCH input. The output of the second circuit is applied to the clock input of the second flip-flop at IC602 (pin 3).

When power is applied to the CCD-V8, the two D-type flip-flops in IC602 must be reset, so the Q outputs at pin 1 and pin 13 are both low. This is done by the reset circuit, consisting of Q604 and an inverter in IC601 (pins 8 and 9). When the UNREG voltage is applied to the DS-10 board, power is applied to the power control circuit, and the increasing UNREG voltage is coupled across C603 to the base of Q604, turning it on. This produces a low at Q603/C while C603 is charging. This low is inverted by IC601 (pins 8 and 9), and the high at IC601 (pin 8) is applied to the CLEAR inputs of the D-type flip-flop (pins 4 and 10). When the CLEAR line is high, the Q outputs (pins 1 and 13) are both made low. When the capacitor C603 fully charges, Q604 is no longer held on, and the collector goes high, removing the high reset from the D-type flip-flops.

When the CAMERA POWER switch is operated on the CCD-V8, the LOW CAM POWER SW is inverted by the chatter prevention circuitry and applied as a clock signal to pin 11 of the D-type flip-flop, LC602 (pins 8–13). The D-type flip-flop is connected with the Q output (pin 12) and D input (pin 9) connected. In this configuration, the flip-flop will act as a toggle circuit, and the output Q will change states at every positive-going clock transistion. Therefore, the clock pulse produced when the CAMERA POWER switch is activated produces a high at IC602 (pin 13), which is applied through CN10 (pin 2) and CAM F/F, to the Mode Control CPU. The signal goes through the steering diode D602, making VTR ON high (CN12, pin 2). This signal is applied to the dc-dc converter DR901 to produce 5 Vdc, which is applied to the circuits in the VTR section of the unit (including the Mode Control CPU). The Mode Control CPU, upon receipt of the high CAM F/F, produces a high VTR ON command at CN10 (pin 4).

The VTR ON command is applied through R627, R628, to Q604/B, with a slight delay caused by C617 and C615. If Q604/B, goes high, the power control circuit would reset and turn the power off. This is prevented by the S/S 5 V at CN12 (pin 7), which was produced when DR901 was turned on. This voltage turns Q613 on, grounding the signal to the Q604/B and preventing the reset of the power control circuit. In this way, the circuit automatically shuts the unit down if the systems control did not have the proper voltage because of DR901 not functioning properly.

POWER ON in the VTR mode of operation is accomplished in the same manner. The action of the VTR POWER ON switch is applied through the chattering prevention circuit to toggel the flip-flop IC602 (pins 1 through 7). The high output from this flip-flop produces the VTR F/F command to the microprocessor and is also applied through D601 to produce the VTR ON output at CN12 (pin 2).

Because of the action of the diodes D601 and D602, the VTR ON signal to the dc-dc converter, DR901, is held high, even if one of the inputs is made low. POWER OFF, therefore, must be accomplished by the systems control through a different method.

The mode control CPU constantly monitors CAM F/F and VTR F/F. When one of these signals goes from a high to a low state, indicating that the switch has been operated, a POWER OFF signal is produced by the mode control CPU, as well as a LOW VTR ON signal at CN10 (pin 4). The POWER OFF at CN11 (pin 5) is applied through R612 to the Q604/B. This resets the power control circuit, with both flip-flops producing a low output. With all the inputs to D601 and D602 low, VTR ON at CN12 (pin 2) is also low, and the power is turned off.

The unit can also be turned on by an external source, such as the tuner timer. To do this, EXT POWER ON at CN11 (pin 10) goes high. This is applied to Q603/B, where it is inverted. The low output is inverted by IC601 (pins 5 and 6) and the resultant high is applied through D602 to produce a high VTR ON. This is applied to the dc-dc converter, DR901, and powers the unit.

Q605 is turned on during reset to prevent the unit from being turned on by an external POWER ON command while in the reset mode.

If the UNREG voltage goes above 9 volts, D605 will conduct, placing a high on the Q604/B. This puts the power control circuit in the reset mode and shuts the power off.

CPU POWER CONTROL

The mode control CPU, IC601 on the SK-3 board, is used to control power turn-on together with the power control circuit on the DS-10 board. Refer to Fig. 13-3.

Whenever power is applied to the CCD-V8 unit, even when the power is turned off, SYS 5 V at CN604 (pin 7) is present to maintain memory. This voltage is applied to IC601 (pin 26) and through L601 to IC605 (pin 14) and IC601 (pin 58). This provides power for IC605 and IC601 to maintain memory. However, the microprocessor IC601 does not operate until the power is turned on, as it receives no clock input to process data. The clock for IC601 is a 400 kHz crystal oscillator consisting of IC605 (pins 10 through 13) and associated components. This crystal oscillator applies a clock signal to IC601 (pin 57).

When the power is off, S/S 5 V at CN604 (pin 8) is low. This low voltage is applied to the cathodes of D601. D601 is forward biased, and the input to the crystal oscillator IC605 (pin 13) is grounded, preventing the oscillator from functioning. When power is turned on, the power control circuit in the DS-10 board produces a VTR ON command through the steering diodes and turns S/S 5 V on. When this occurs, the high at Cn604 (pin 8) is applied to the cathodes of D601, reverse biasing them and allowing the oscillator to function.

When S/S 5 V goes high, IC601 must also be reset to ensure proper operation. This is accomplished by the reset circuit consisting of IC604 (pins 5 through 7) and associated components. As S/S 5 V increases, the positive input at IC604 (pin 5) is held low momentarily by C612, producing a low at IC604 (pin 7). This is inverted by

IC605 (pins 8 and 9), and a positive reset pulse is applied to IC601 (pin 56). This positive reset is maintained for the charge time of C612.

The inputs from the power control, either VTR F/F or CAM F/F, are applied through the scan matrix to IC601. The presence of these pulses is sensed by IC601, which produces a HIGH VTR ON or CAM ON at pins 18 or 19.

An additional circuit consisting of Q606 and Q607 is turned on when S/S 5 V goes high. This connects the SYS 5 V and S/S 5 V while the unit is operating. The other functions of IC601 are in reference to the mode control CPU.

SYSTEMS CONTROL CPUs

The systems control in the CCD-V8 contains three CPUs. These three CPUs are all tied together to form the systems control circuit as shown in the Fig. 13-4 block diagram. The first CPU, the mode control CPU, receives the inputs from the key matrix and controls the LCD display. The second CPU is the mechanical control CPU, which controls the mechanisms and servos. It also monitors the mechanical mode switches, RF SW PULSE and CAPSTAN FG to ensure that the mechanism is functioning properly. Because of the interrelation between the commands and the mechanical controls, the three microprocessors must constantly communicate with each other as well as with the individual circuits they monitor or control. This is taken care of by the communications CPU, which synchronizes the two other CPUs with each other and any external circuitry connected to the CCD-V8.

The main control in the CCD-V8 is the mode control CPU, a UPD7503G microprocessor. This CPU monitors a 4×8 input key matrix that detects when a command is received for the unit to execute an operation. This CPU also directly controls the LCD display, and through the inner bus (a series of data communication lines between the microprocessors) controls the digital servo, the LED drivers, and the video/audio switching circuits.

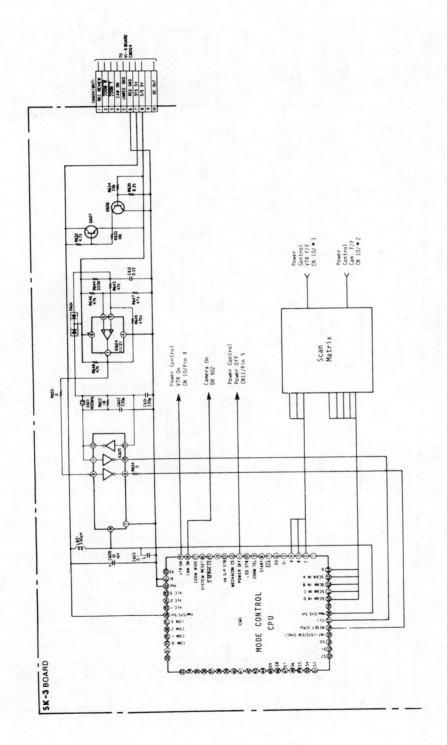

Fig. 13-3. *CPU power control circuit. Courtesy of Sony Corporation of America.*

285

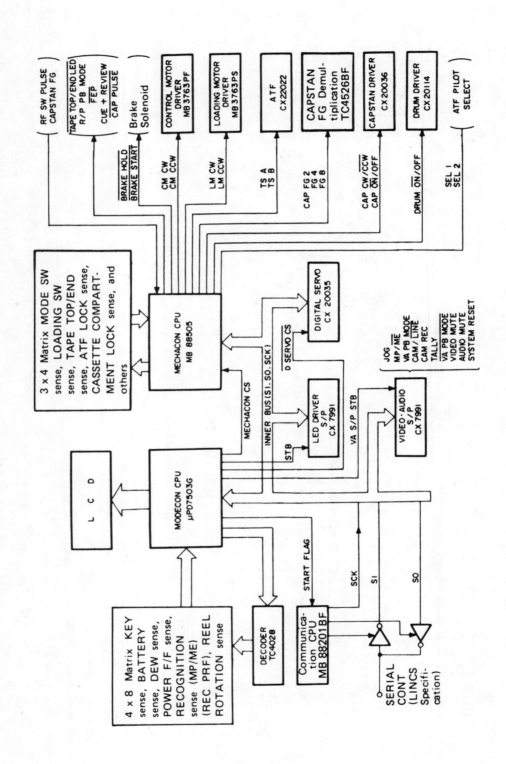

Fig. 13-4. *CPU input/output signals. Courtesy of Sony Corporation of America.*

The second CPU in the CCD-V8 is the mechanical control CPU, a MB88505 microprocessor. This microprocessor, as its name implies, controls the mechanism of the unit, including the capstan/drum servo, capstan driver, capstan FG, ATF servo, the drum driver, loading and control motor drivers, and the brake solenoid. In addition, it controls the tape top/end LED sensors and monitors the RF switching pulse and capstan FG signals to ensure that the motors are correctly operating. In the event of a malfunction of any of the motors, the malfunction is detected by the mechanical control CPU, using these inputs, and the unit will be shut down. The mechanical control CPU monitors the mode sensors with a 3×4 matrix.

The mechanical control CPU communicates to the digital servo and the mode control CPU through the inner bus. In this way, the mechanial control CPU not only receives commands from the mode control CPU but also informs it in the event of a servo malfunction to shut the unit down.

This bi-directional communication between the mode control and mechanical control CPUs is controlled by the third CPU. This is the communications CPU, an MB88201BF microprocessor. Using the START FLAG input, the communications CPU generates a serial clock (SCK) that synchronizes the SERIAL IN (SI) and SERIAL OUT (SO) signals between the two microprocessors. The communications CPU also controls the direction of the SERIAL IN and SERIAL OUTPUT lines to an external data communications line. This data communication line can be used by such accessories as a tuner/timer or remote control units.

INPUT KEY MATRIX

The systems control begins its operation, when the commands are initially sensed by the systems control, it begins its operation in the input matrix of the mode control CPU. Refer to Fig. 13-3.

The mode control CPU uses four scan-in and four scan-out ports to sense the inputs from 26 different sources. In order to do this, an input matrix, which is an array of diodes, transistors, and an integrated circuit is used. These devices are arranged in such a way that a unique input is received for each sensor to be detected.

To achieve this input detection, four scan-out pulses are generated within the mode control CPU at the scan-out ports (pins 2, 3, 4, and 64). These scan-out pulses are in the form of a 4-bit binary word, which is sequenced by the microprocessor. This 4-bit word is applied to a TC4028BP IC. This IC is a binary-to-decimal decoder. It converts the four binary bits from the scan-output to ten outputs, one unique output for each binary word. Of the ten possible binary outputs, only eight are used in the CCD-V8 system.

In order to make the transistor switches in the key input matrix operate, the C scan-output at pin 2 is connected through pull-up resistors to the input key matrix. The sequence of operation that takes place in sensing a command can be considered, assuming the playback button is pressed. The SYSTEMS SYNC at pin 55 of the mode control CPU synchronizes the timing of the key matrix and other functions of the microprocessor. When the systems sync goes high, the binary code for 6 (A = 0, B = 1, C = 1, D = 0), is output at the scan-out to the input key matrix. The binary/decimal decoder converts this to a high at Q6, which is applied to the three reel-sensor transistors at the bottom of the key input matrix. During this time, the input from the reel sensors is detected by the Q6 output, turning on the three transistors. If a reel sensor is active, the transistor will conduct, bringing down the high voltage applied to the scan-inputs during this time. After the initial period, when the Q6 output is high, the mode control CPU changes the scan-output code to indicate a binary 1. This binary 1 is converted by the binary/decimal decoder to a high output for Q1. This Q1 output is applied to the first row of switches, which includes the playback switch. The closed playback switch routes the signal through the isolation diode to the C input at pin 61 of the mode control CPU. Whenever the C input of the mode control CPU goes high during Q1 time, this must be because of a closed playback switch.

DRUM AND CAPSTAN FG CIRCUIT

The signal from the Fg and PG coils in the drum and FG device in the capstan are very small. Note the waveforms shown in Fig. 13-5. To be usable in the digital servo circuit, they must be amplified and shaped into 5 V square waves. This is the function of the drum and capstan FG circuit. The drum FG output is a 50 mV p-p sine wave. The drum PG output is a 1 mV p-p pulse. This signal is so small that it should be considered unmeasurable. The capstan FG signal, being produced by a DME device, is approximately 50 mV p-p.

The capstan FG signal consists of CAP FG (+) and CAP FG (−). These two signals are applied to the inputs of a comparator that produces a shaped output that exits pin 9 as a 960 Hz 5 V square wave. The ×2 logic circuit in IC501 is not used in this unit. The frequency of the output at pin 9 is the same as the capstan FG input.

The bias for the drum PG and FG coils is produced within IC501. Refer to Fig. 13-6 for

IV501 circuit. The bias voltage exits IC501/pin 1 and is applied to the D COM connector of the FG and PG coils. The outputs from these coils are applied to independent preamplifiers and comparators at IC501 (pin 2) and IC501 (pin 4). These pulses are both amplified to 5 V square waves that are in phase with the input signals. The drum PG and drum FG outputs exit IC502 (pin 7) and IC502 (pin 3) and are directly applied to the drum servo IC (refer to drum servo IC502 in Fig. 13-7).

DRUM SERVO

The drum PG and drum FG signals, which have been shaped into 5 V square waves, are applied to the capstan/drum servo. IC502 (pins 6 and 7). Within the IC, these signals are applied to a drum counter. The IC uses the 3.58 MHz input at IC501 (pin 2) as a time reference. This signal is obtained from the color section of the video processing circuit. It is applied to a drum reference generator that produces the reference signal for the drum speed and drum phase servo. During the record mode, the drum reference generator is synchronized to the reference vertical signal entering IC502 (pin 19). This signal is

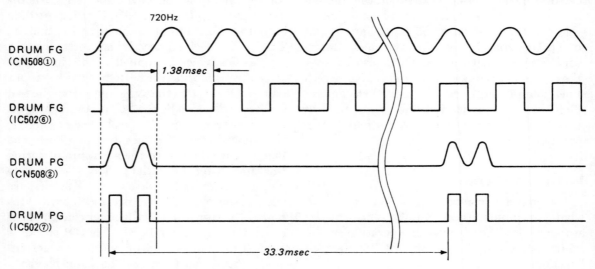

Fig. 13-5. *Drum PG and FG waveforms. Courtesy of Sony Corporation of America.*

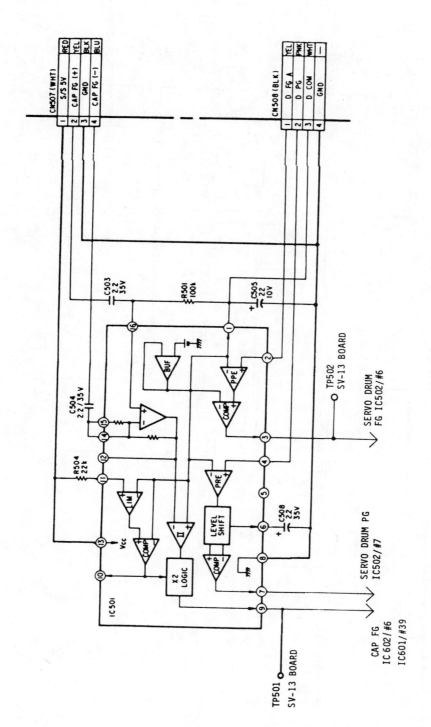

Fig. 13-6. *Drum and capstan FG circuit. Courtesy of Sony Corporation of America.*

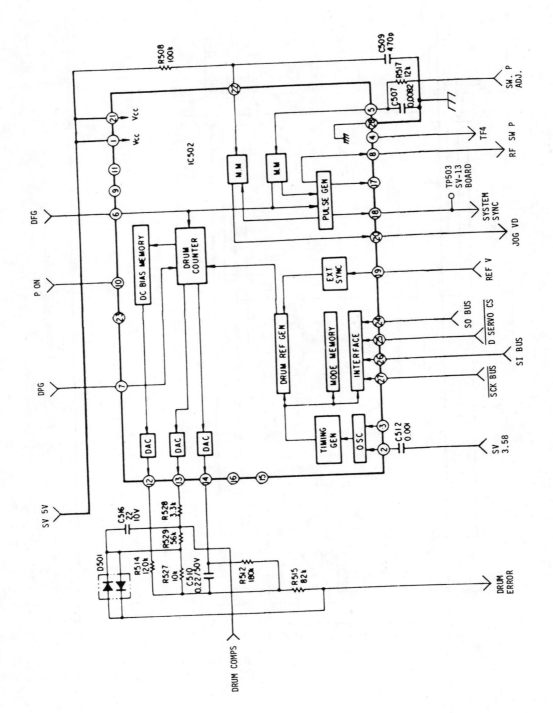

Fig. 13-7. *Drum servo circuit. Courtesy of Sony Corporation of America.*

derived from the video sync of the signal to be recorded.

The switching between modes is accomplished within IC502. The mode control CPU directly controls IC502 via pins 24 through 27. These pins are the inputs from the system control. Pin 27 is a serial clock (SCK BUS), pin 26 is serial data from the serial input bus (SI BUS), pin 24 is connected to the serial output bus but is also data entering IC502 (SO BUS), and pin 25 is the chip enable input (D SERVO CS), which allows the IC to read the data entering the other 3 pins when it is low.

In normal mode of operation, data is always present on all four of these lines and repeats at a rate of 60 Hz. The data (SO) occurs when the chip select input is low. This data is read into the memory by five groups of clock pulses that occur when the chip select signal is low. Each group of clock pulses contains eight negative clock pulses. These clock pulses are always present and do not change from mode to mode. The SI signal at pin 26 also provides data to IC502.

DRUM SERVO INPUT TROUBLESHOOTING

When troubleshooting these inputs, it is important to observe their presence. If pulses are present (note that in the record mode, no pulses are present on the SI line), the system control, which produces these pulses, can be assumed to be operating properly.

The output from the drum counter, as well as the output from the dc bias memory, are applied to the motor driver IC to control the speed. The dc bias memory produces a dc output at pin 12. This output voltage changes from mode to mode and is controlled by the serial data from the system control. Once the mode has been entered, no change should be observed in the dc voltage at pin 12. The voltages at pins 13 and 14 are dc voltages produced by digital-to-analog converters within IC502. These voltages are the speed and phase servo outputs at pins 13 and 14, respectively. They change the dc level according to the speed and phase of the motor. The three outputs from the IC are combined in a resistor

and capacitor matrix to produce the drum error voltage that is applied to the drum and controls the speed of the motor through the drum driver circuit. A diode D501 is incorporated into the circuit so that if the speed servo voltage becomes very high, indicating a severe difference in speed, the speed servo will bypass the resistor capacitor network and directly control the drum error and the motor speed to quickly lock up the motor.

The drum FG input (DFG) is also used to produce the RF switching pulse. The DFG is applied to a multivibrator within IC502. The positive transition of this multivibrator is controlled by the capacitor at IC502 (pin 5). This is the RF switching pulse adjustment. By adjusting the resistance of the switching pulse adjustment, the time constant attached to pin 5 will change. This controls the slope of the signal at pin 5 and, therefore, the position of the RF switching pulse that exits at pin 8.

CAPSTAN SERVO

IC502 contains both the drum and capstan servo circuitry. The capstan FG, from the FG amplifier, is applied to IC502 (pin 23) through IC603, a programmable divider. In the play mode, this programmable divider is set to divide capstan FG frequency by one. Refer to circuit in Fig. 13-8.

In the CCD-V8, the capstan drives the tape in the cue and review modes of operation. The capstan servo is also used during this mode, but to speed up the tape, a programmable divider, IC602, is incorporated in the capstan FG input circuit. In the cue mode, commands from the mechanism control CPU place a high FG8 input at IC602 (pin 2). This changes the programmable divider from a divide-by-1 to a divide-by-9 circuit. Because the capstan FG has been divided by 9, the servo within IC502 will automatically increase the output voltage to restore the capstan FG to the proper frequency. This results in speeding up the tape for the cue mode.

In the review mode, high FG2 and FG4 commands from the mechanism control CPU

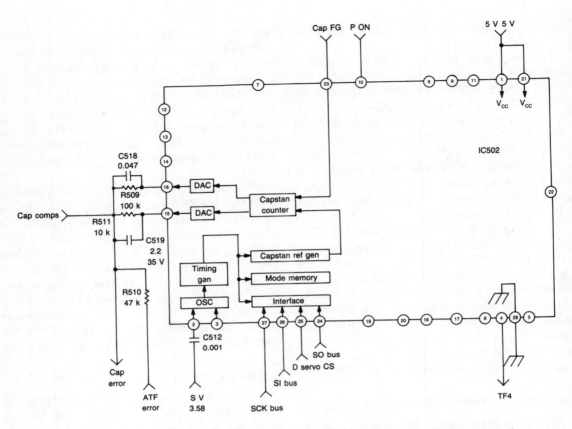

Fig. 13-8. *Capstan servo found on MC-4 board. Courtesy of Sony Corporation of America.*

change the programmable divider to divide by 7, and the speed of the capstan is changed accordingly by the capstan servo circuit. The capstan motor also drives the reel tables through a timing belt.

CAPSTAN FREE-SPEED COMPENSATION

The capstan error voltage, which was produced in the capstan servo circuit by combining the capstan speed, phase and AFT servo outputs, is amplified by IC504 (pins 1 through 3) and applied to the non-inverting input of an amplifier within IC502 (the capstan motor driver IC). IC502 compares the capstan error voltage at pin 23 with the capstan bias at pin 24. The capstan bias is produced in a resistor divider network consisting of R575, R576, RV504, RV503,

and RV574. This voltage divider produces a dc voltage that is buffered by IC504 (pins 5 through 7) and applied to IC502 (pin 24). The capstan bias voltage is adjusted so that the capstan motor turns at the correct speed if the ATF servo output is missing. This is essential to allow the ATF servo the ability to correct in either direction for a relatively large range to ensure proper servo lock-up.

AUTOFOCUS

One way to have autofocus relies on the camera emitting a light beam to place a spot at the center of the scene to be picked up by the camera. This is accomplished by a sensor within the camera. The sensor consists of two photo sensors that are only sensitive to the light beam projected by the camera. If the image is in focus, the outputs of these two sensors is equal (because of the mechanical coupling between the light-emitting diode and the sensors). The focus servo tries to adjust the position of the lens to achieve this desired output.

In the real world, several problems arise that must be elminated by the autofocus circuit for reliable focus operation. One of these is caused by random light sources interfering with the beam emitted by the camera. Another problem is the varying intensity of the light reflected back to the camera. Light intensity changes with distance from the camera to the object, as well as with differing colors and reflectivities of the image being picked up by the camera. Another problem is that of moving or vibrating objects. These three problems are eliminated by the autofocus circuitry.

Autofocus Operation

The output of the IRED (infrared-emitting diode) is controlled by IC12 (pins 32 and 33), the autofocus controller. Refer to the block diagram in Fig. 13-9. The IRED is turned on and off at a rate of 12.5 kHz by IC12 to enable the sensors to distinguish between the light coming from the IRED and the background light. The light emitted by the IRED is focussed by a lens onto the object that reflects it back into the sensors.

The two sensors detect the light reflected from the object and apply two signals to the preamplifiers located on the PA-7 board. The output from the preamplifiers are chopped by Q3 and Q4. Transistors Q3 and Q4 are controlled by SYNC SIG generated by IC12. This signal is 12.5 kHz and is in phase with the IRED drive. Therefore, it turns Q3 and Q4 on at exactly the same time that the IRED is turned on, so only the light generated by the IRED is allowed to enter the autofocus circuitry. The outputs of Q3 and Q4 are applied to integrators that are controlled by RESET SIG. This signal is also generated within IC12. When RESET SIG is low, Q5 and Q6 are turned off, and the amps function as integrators.

The signals from the sensors are integrated, i.e., many pulses are added to produce ramp outputs AF SIG 1 and AF SIG 2, representing the signals received by the sensors from the IRED. In this way, minor fluctuations in the signal returned from the IRED are removed as several hundred pulses are averaged together to produce the autofocus signal. This removes the problem of random noise as well as movement of the subject in the camera. After a short period of time, the IRED and SYNC SIG pulses stop, and the level of the ramp is maintained. It is this level that is used by IC12 to produce the autofocus control. After the signal is used, RESET SIG goes high to reset the integrators, readying the autofocus circuit for the next group of pulses from the IRED.

Power for the autofocus circuit is derived from camera unregulated input. This voltage UNREG is the battery voltage. It is applied through THP001 (a positive-temperature-coefficient resistor) to the IRED drive circuit and through PS001 to the autofocus power circuit. Because UNREG is always high, it must be turned on and off by the P5 VOLT input at CN001 (pin 2). When this goes high, Q15 is turned on, turning Q14 on, which applies 5 volts to the autofocus circuit. Because autofocus might be undesirable at times, an AUTOFOCUS switch is incorporated on the RR-8 board that allows the autofocus circuit to be defeated.

CCD/TRINICON COMPARISON

The CCD-V8 achieves its small size through the use of 8 mm video tape and a CCD (charge coupled device) camera. A CCD camera employs a solid-state device in place of a conventional glass pickup tube. Thus, it has many advantages that a transistor has over a vacuum tube, including long life, reduced sensitivity to vibrations and shocks, no warm-up time, small size and weight, not influenced by magnetic fields, and very low power consumption. In addition, because of the solid-state structure, there is no lag or blooming in the CCD device.

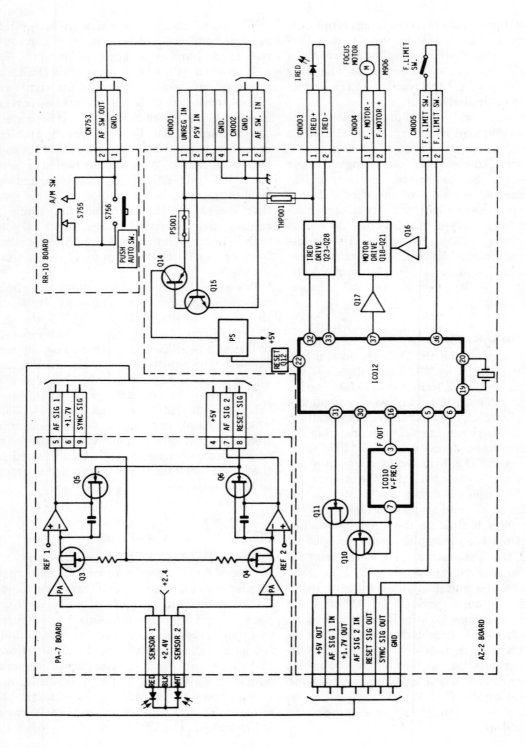

Fig. 13-9. *CCD-V8AF auto focus circuit. Courtesy of Sony Corporation of America.*

In the Trinicon, however, lag and blooming are characteristic of the photoconductor that senses the light. This is because very bright images produce leakage across the surface of the photoconductor, which is impossible in the CCD device. Note the comparison chart in Fig. 13-10.

An additional advantage of the CCD device is that it is not prone to damage by the customer. The CCD-V8 is much less sensitive to image burns. This is because the CCD element is a single piece of metallic circuit. If a bright light is focused on the surface of a photoelectric film in a Trinicon, the heat generated could damage it. On the other hand, the CCD chip easily dissi-

pates any heat because it is made of metallic silicon and is very difficult to damage in normal use.

CCD IMAGER

The heart of the camera in the CCD-V8 is the pickup element itself. This is a single chip of silicon that is constructed to be light-sensitive and to be able to sequentially draw information from this device to produce the video signal.

The CCD element is divided into discrete elements that can be individually accessed. This is similar in concept to the sampling of a digital

	CCD (Charge Coupled Device)	Pickup Tube
1. Life & Reliability	Long-life	Heaters progressively weaken from beam radiation.
2. After Image & Scorch	No after image. No scorch with lengthy photographing of the same object or with strong light.	Characteristics of photoelectric film make this unavoidable.
3. No Figure Distortion	As picture elements are arranged regularly and they also perform selfscanning, exact geometrical figure can be obtained.	With beam-scanning, it is difficult to inspect accurately both the center and perimeter by scanning.
4. Vibration Proof & Impulse Proof	Strong because it uses the semiconductor chip.	Weak because it uses glass tube, filament and socket.
5. Picturing Time	Fast because it uses no heater.	Needs time for heater warm up
6. Size & Weight	Small and light.	Needs length for emitting beams and also space of coil for focussing and deflection.
7. Use in an Electro-magnetic Field	No influence.	Easily influenced by electronic beams.
8. Electric Power Consumption	Low power consumption because it is composed of semiconductor.	Large amounts because the heater and coil, etc. are used with several hundreds of voltages.

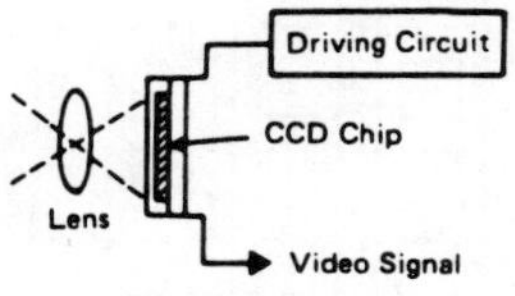

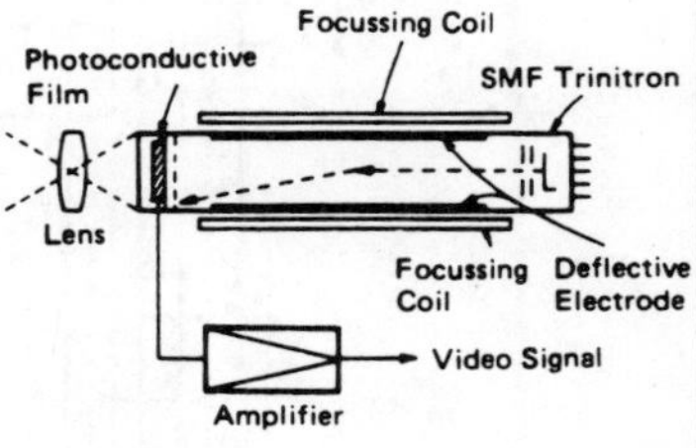

Fig. 13-10. *Comparison chart of CCD and pickup tube. Courtesy of Sony Corporation of America.*

audio signal where the information is sampled at specific points. In the CCD imager, the picture is sampled at individual points both horizontally and vertically. Refer to Figs. 13-11 and 13-12. In the CCD-V8, there are 532 columns of elements and 504 rows of elements, resulting in a total of over 250,000 elements. These elements are called *pixels* and are the smallest amount of detail that can be resolved by the CCD pickup. The CCD imager used in the CCD-V8 contains 60,000 pixels — more than the pickups used in

previous CCD cameras — resulting in much higher resolution of detail. In the CCD-V8, several rows and columns of the picture elements are not used. These elements are black-masked and form the black border at the top, beginning, and end of each line. This leaves an array of 492 by 510 active picture elements in the CCD-V8.

In order to produce color, each pixel is placed behind a color filter. The color alternates as shown in Fig. 13-12. In the first two rows, the elements are colored green and red. In the third and forth rows, the elements are green and blue; and in the fifth and sixth rows, the elements again go to green and red. This results in an optimal pattern for both high-resolution luminance and color signals. The two identical rows are alternated between fields in order to perform odd/even line interlacing. The individual light receptors are addressed using a series of pulses to move the information from the light receptor to the vertical register, down the vertical register, and ultimately out of the device.

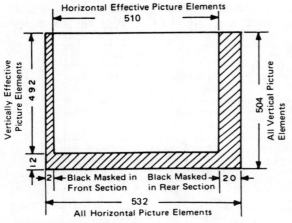

Fig. 13-11. *Effective picture elements. Courtesy of Sony Corporation of America.*

Charge Transfer in the CCD Imager

The CCD imager must transfer the signal produced by the light across the device to the output lead. A principle of charge coupled devices is employed. These devices are very similar

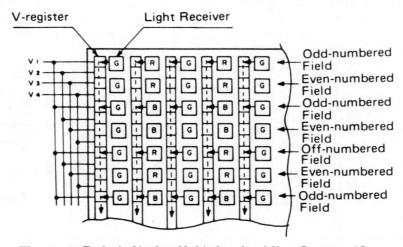

Fig. 13-12. *Each pixel is placed behind a colored filter. Courtesy of Sony Corporation of America.*

to a bucket brigade. In that, the charge from one device is passed to the adjacent device by varying the voltage on both devices. From the second device, the charge can be transferred to a third device, without loss, by again varying the voltage in the proper manner. The individual sensors are MOS transistors constructed in such a way that in the presence of light, electrons flow from the gate to the sensor, where they are accumulated. For this device to produce a video output signal, the charge from each sensor must be moved to the output port at the correct time.

Because of an inversion that takes place in the lens, the top of the image is seen at the bottom of the diagram. The charge from the sensors must be moved to the output from the bottom row first, followed by the next row above it and the row above that, and ultimately to the top row of the diagram shown in Fig. 13-13.

The last step in the CCD process involves moving the charge from the last horizontal register to an output buffer and at the same time converting the charge to a voltage. In the diagram of Fig. 13-14 at time A3, H1 and PG are high. This causes the switch to close, which charges the capacitor on the gate of the output MOS to 18 Vdc. CCD output at this time is high.

During time B3, H1 and PG go low. When this happens, the switch opens, and the charge in the H1 register starts discharging the capacitor. The amount that the capacitor discharges is exactly proportional to the amount of charge in the H1 register and will determine the final voltage on the capacitor. This voltage is then buffered by the output of the CCD. The output is proportional to the intensity of the light striking a particular sensor and represents the video level. At time C3, H1 and PG go high, and the process repeats. The resultant output is a series of amplitude-modulated pulses that are very similar to a PAM output seen in digital audio. The signal take-out circuit is shown in Fig. 13-15.

The camera system block diagram is shown in Fig. 13-16. The Sony model CCD-V9 camcorder is shown in Fig. 13-17, and the CCD-V8 camcorder is shown in Fig. 13-18.

ZENITH VM6200 CAMCORDER

The Zenith VM6200 is a high-picture-quality, high-performance camcorder (see Fig. 13-19). It is one of the smallest units that features both record and playback modes. The following in-

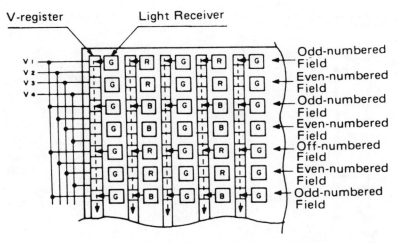

Fig. 13-13. *The two identical rows are altered between fields to perform odd/even line interlacing. Courtesy of Sony Corporation of America.*

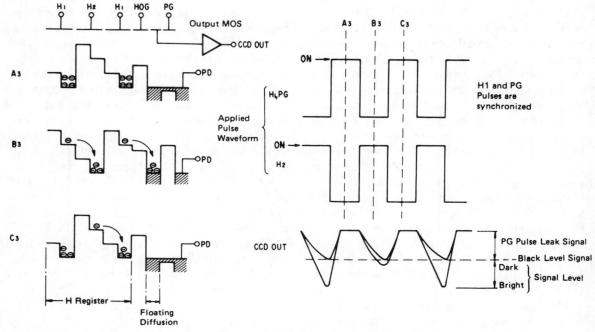

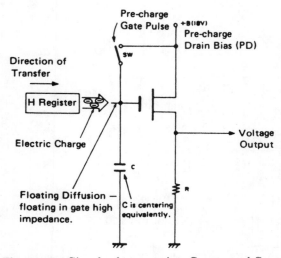

Fig. 13-14. *Waveform transfer mechanism from H register to output part. Courtesy of Sony Corporation of America.*

Fig. 13-15. *Signal take-out point. Courtesy of Sony Corporation of America.*

formation and illustrations are courtesy of Zenith Electronics Corporation.

This camcorder uses the newly developed ½-inch CCD (charge coupled device) pickup element and HQ (high quality) VHS picture improvement technology that makes 1-hour recordings possible. The unit has full automatic operation that includes auto focus, auto white balance, auto filter, and auto iris. For convenience, the auto focus and auto filter circuits are switchable from auto to manual.

The VM6200 has an SP-EP switch. This allows a full hour of recording in the EP mode using a TC-20 tape. The unit comes with the VAC 415, a 1000 mAh battery. The weight of the unit with the battery is approximately 3.9 pounds.

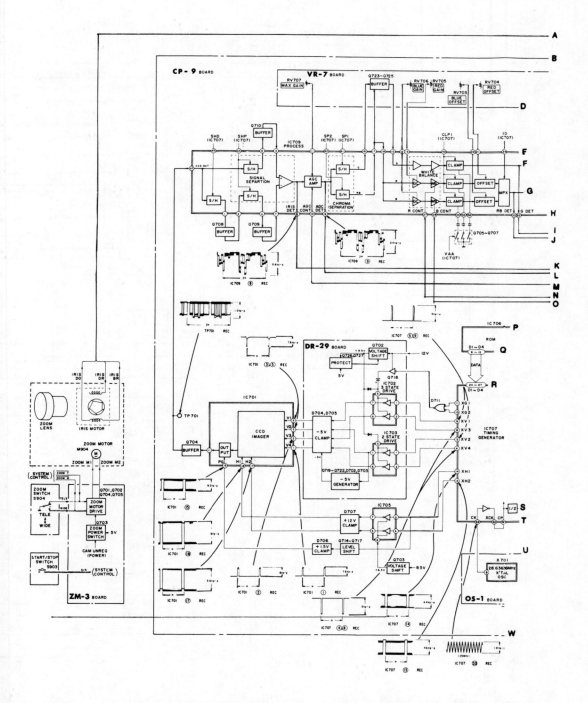

Fig. 13-16. *Sony CCD-V8 camera system block diagram. Courtesy of Sony Corporation of America.*

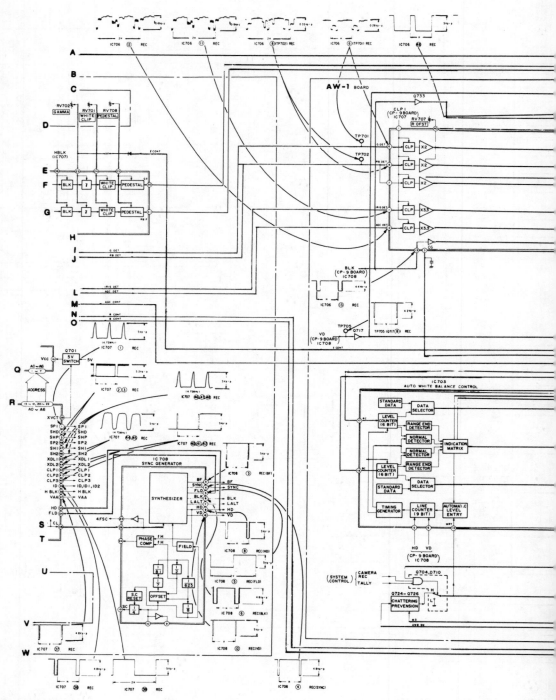

Fig. 13-16. *(continued)*

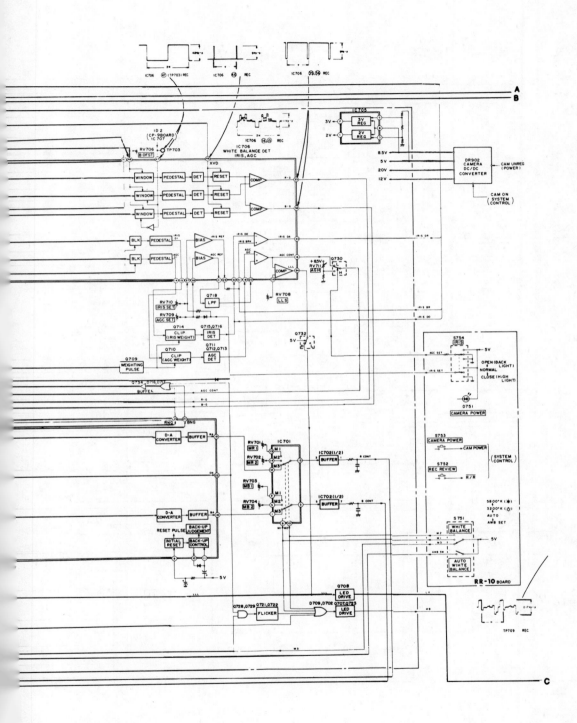

Fig. 13-16. *(continued)*

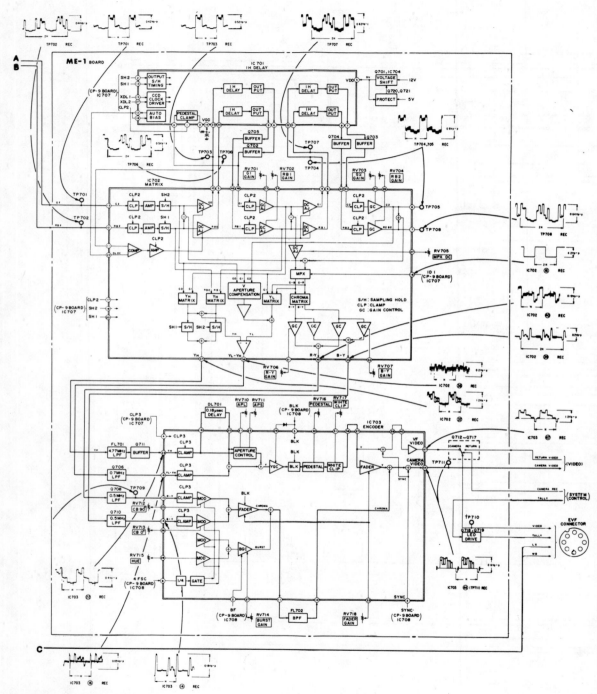

Fig. 13-16. *(continued)*

Fig. 13-17. *Photo of Sony CCD-V9 camcorder. Courtesy of Sony Corporation of America.*

Fig. 13-18. *Photo of Sony CCD-V8AF camcorder. Courtesy of Sony Corporation of America.*

Fig. 13-19. *Zenith VM6200 VHS-C camcorder. Courtesy of Zenith Electronics Corporation.*

Recording Color Signal Flow

The color circuits block diagram is shown in Fig. 13-20. IC4 selects the color signal input for either external or camera section. The external video signal input from the A/V cable is applied via Y/C board CN3 (pin 6) to IC4 (pin 10).

The signal goes through switch SW1, and from IC4 (pin 5) to bandpass filter BPF1. This yields the color component, which goes to IC4 (pin 3). This signal is applied through SW2 and IC4 (pin 8) to IC5 (pin 1).

Playback Color Signal Flow

Low-pass filter LPF3 separates the down-converted color component from the signal played back from the tape and supplies it to IC5 (pin 20). R56 adjusts for a suitable level.

Through the electronic switch and ACC of IC5, the playback color signal is applied to the main converter. In the same manner as recording, 3.58 MHz + 40 fH is obtained from the sub converter, and the output becomes the 3.58 MHz color signal.

After passing through various circuits, the signal is applied to IC5 (pin 1) where the burst level is reduced by 6 dB (reversing the 6 dB increase applied during recording). The result is sent via the playback amplifier and killer switch of IC5 to the pin 7 output.

Color level control R33 adjusts the playback color level, while the trap circuit of L8 and C35 attenuates main converter leakage. The resulting signal is sent to IC8 (pin 8).

COLOR CIRCUIT SYSTEM

Hybrid IC5 has been specially developed for performing the main color circuit functions of this camcorder. In addition to contributing toward circuit minaturization and simplification, the reduction in the number of discrete parts considerably reduces power consumption. Further cost and space reductions are achieved by using the same delay line for both the camera and recorder sections.

THE AUDIO CIRCUITS

Refer to the audio system block diagram shown in Fig. 13-21.

Recording Audio Signal Flow

Audio signals supplied to pins 2 and 4 of IC303 are imput to the ALC circuit after they are switched by IC303 pin 1's voltage.

Level is adjusted at the line amplifier according to the control voltage from the ALC (automatic level control) detector. Resistors R338 and R336 determine the ALC operating level, while the response characteristics are determined by R339, C325, and C326.

A low-pass filter (LPF) removes high-frequency noise (above 12 kHz approximately), after which the signal goes in two lines. One of these is amplified by the monitor amplifier and supplied as the E-E signal to the earphone and 8-pin A/V connector (this is also sent to the ADC detector). The other signal line goes via R314 (REC LEVEL) to the recording amplifier. The REC amp serves to compensate for high-frequency loss in the tape recording and playback processes. From this point, the signal is mixed with ac bias and supplied to the audio head for recording.

Playback Audio Signal Flow

The signal played back by the audio head is supplied to the playback equalizer amplifier, which functions for low-frequency level compensation. This output goes via the PB LEVEL control to the line amplifier. Afterwards, signal flow is the same as the E-E output.

MODE CONTROL

The four operating modes (REC, E-E, PB and PB MUTE) of the audio circuit are switched by the mechacon circuit. The modes of IC301 are determined by the sates at pins 15 and 16.

Mechacon

The mechacon section of the VM6200 has quite a few features, such as a feature that monitors the functioning of the mechanical operations and provides necessary signals and voltages to carry out these operations.

One feature of the mechacon circuit is an expanded "Emergency Mode." This feature indicates where a failure has occurred by causing different LEDs to flash when the unit goes into auto-stop. In the VM6200, there are 27 possible emergency modes. The expansion of the emergency mode gives the service technician a more accurate answer to the cause of a certain failure. These 27 modes are indicated in the chart shown in (Fig. 13-22).

The VM6200 also has expanded "Test Mode" capabilities. If a failure occurs during one of these test modes, the unit enters one of the emergency modes. There is a special switch to activate these test modes.

The mechacon circuit in the VM6200 also has a feature that allows it to use less battery power in the off mode. The previous camcorders used 4 mA/h and 2 mA/h respectively when off. The VM6200 uses only 100 μA/h when off. This means that while the VM6200 is between recording segments, it uses less battery power than its predecessors. (However, it is still recommended that the battery be removed for extended storage.)

SERVO SECTION

Because of the addition of the EP mode of operation, there are some changes in the servo

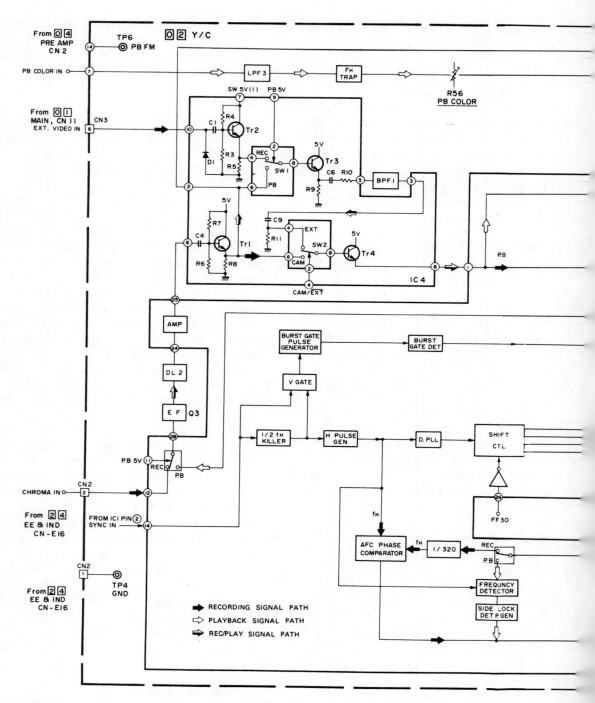

Fig. 13-20. *Portion of color system found in Zenith model VM6200 camcorder. Courtesy of Zenith Electronics Corporation.*

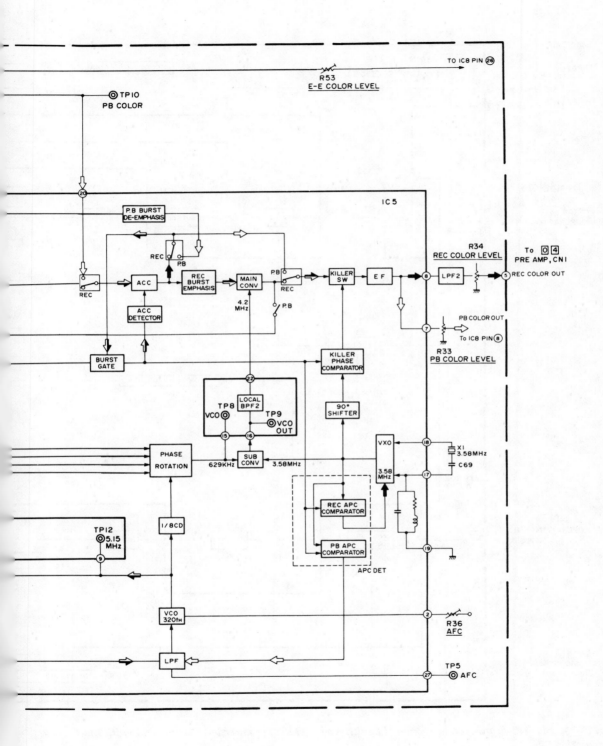

307

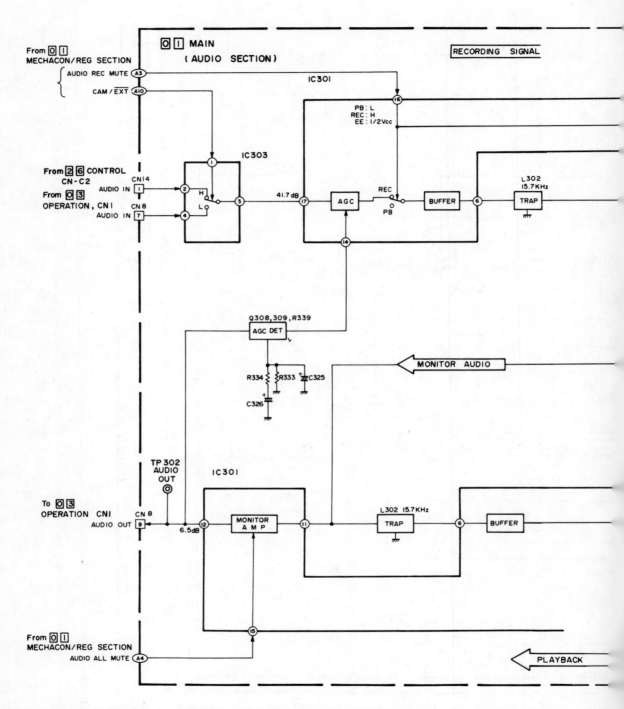

Fig. 13-21. *Block diagram of audio section of Zenith model VM6200 camcorder. Courtesy of Zenith Electronics Corporation.*

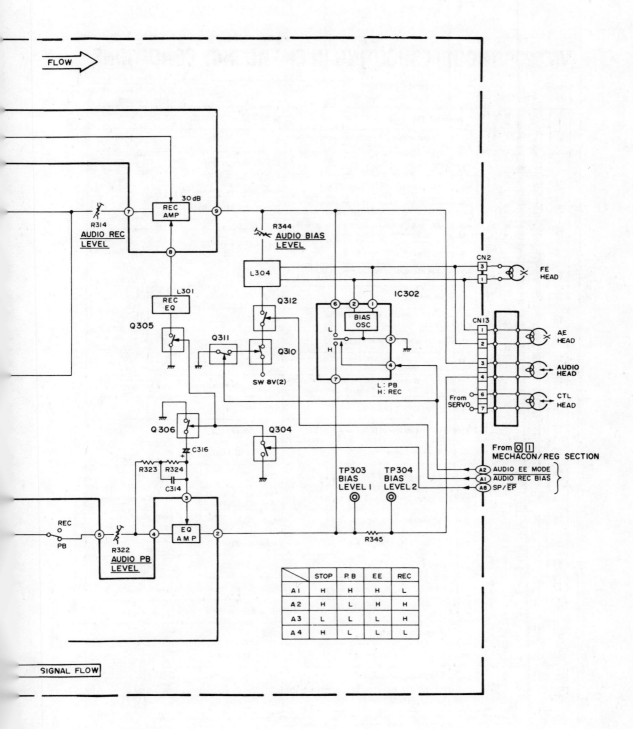

	STOP	P.B	EE	REC
A1	H	H	H	L
A2	H	L	H	L
A3	L	L	L	H
A4	H	L	L	L

309

VM6200 TROUBLESHOOTING IN EMERGENCY CONDITIONS

*Specifications and appearrance of this unit are subject to change without prior notice.

No.	Emergency Indication	Cause and Description	Check Point	Normal Waveform & Voltage
1		Defect in 8 V SW	CP3(⊗①), IC1, Q1(②⑧)	M3 DC 8 V ✱
2		SW REG's input voltage for the drum/capstan motor is abnormal.	CP2	M2 DC 9.6 V ✱
3		No input of P.B. CTL pulse	CTL signalling system	S11 ⎍⎍ 33.3msec DC 5V / DC 0V ✱
4		Unstable input voltage of IC406 pins 27 − 31	IC406 pins 27 − 33 (⊗①) (in other modes than (TEST) No chattering must be confirmed	DC 5V or D.F.F. ⎍⎍ DC 5V
5		Abnormal CAM SW's input in the condition that the loading motor is rotating in the direction of loading.	CAM SW, CN19(⊗①)	CN19 H:5 V L:GND
6		Abnormal CAM SW's input in the condition that the loading motor is rotating in the direction of unloading.	Same as above	Same as above
7		Abnormal inputs of IC406 pins 2 − 9	IC403~405 ⎍⎍⎍ DC 5V / DC 0V High level = 5 V approx.	Port A input check.
8		No CAP. FW FG input at back-spacing in QUICK REV. mode.	CN16, IC411(⊗①)	S12 ⎍⎍⎍⎍⎍ DC 5V / DC 0V ✱
9		Dew sensor is faulty.	CN21, IC407(⊗①)	N1(CN21-⑤) NORMAL:L DEW :2.5 V
10		No CAP. FW FG input at REC start in the assembly mode.	CN16, IC411(⊗①)	S12 ⎍⎍⎍⎍⎍ DC 5V / DC 0V ✱
11		Abnormal clock pulse of IC401 pin 6	IC401(⊗①)	CN9-①, R487 DC 5V OSC RANG 0.1msec
12		The set does not enter the next mode 10 sec after the loading motor has started rotation in the loading direction.	CAM SW, IC403(⊗①)	CN19-②,③,④ H:5 V L:GND
13		The set does not enter the next mode 10 sec after the loading motor has started rotation in the unloading direction.	Same as above	Same as above
14		Defect in the CASSETTE SW	CN4-①(⊗①)	ON :L OFF:5 V
15		Abnormal input voltage of IC401 pins 27 − 32	IC401 pins 27 − 32	H:5 V
16		SW 5 V (01) level is abnormal.	CP1, SW REG BLOCK	IC405 pin 2 DC 5 V (at POWER ON)
17		Abnormal level of LOAD 9.6 V	CP5	IC405 pin 8 DC 9.6 V (at POWER ON)
18		KEY data inputs (IC401 pins 2 − 9) are unstable.	Pins 3, 5, 7, 9, 12, 14, 16, 18 of IC403, 404 or 405	DC 5 V in Stop mode
19		There is SERVO EMERGENCY input.	Servo circuit	IC405 pin 17 = 5 V at normal
20		There is AUDIO EMERGENCY input.	Audio circuit	IC405 pin 17 = 5 V at normal
21		No DRUM FF input	IC401-②(⊗①)	⎍⎍⎏ DC 5V
22		No CAP. FW FG input in PB quick review	Refer to Item No. 8.	Refer to Item No. 8.
23		No CAP. FW FG input at back-spacing in REC PAUSE mode.	Same as above	Same as above
24		Defect in REC SAFETY SW	CN3-①	ON :L OPEN:5 V
25		Abnormal H level at IC401 pin 60	IC411, 410, 408	H:5 V L:GND
26		No TU REEL SENSOR INPUT	CN20, IC409, IC403	Set the oscilloscope's range to 20 ms, and confirm that there are periodic SCAN inputs to IC403 pins 3 and 12.
27		No SUP. REEL SENSOR INPUT		⎍⎍⎍⎍ DC 5V

✱ : "M" and "S" in the right column show checker lands on circuit boards. Check them referring to pattern figures.

Fig. 13-22. *Troubleshooting guide for Zenith VM6200. Courtesy of Zenith Electronics Corporation.*

circuit. The changes are mostly within the servo IC and are switched by the mechacon circuit.

RCA MODEL CC011 CAMCORDER

A brief circuit description is given on some of the CC011 camcorder systems in the following section of this chapter. The operating controls for the CC011 are shown in Fig. 13-23. The following information and illustrations are courtesy RCA Corporation.

COLOR MULTIPLEXING SYSTEM USING A SINGLE NEWVICON

The incoming light to the camera passes through the lens, the optical system (in which the color temperature conversion filter is installed), the infrared (IR) cut filter, and the crystal filter (where it is imaged on the surface of a striped filter). Note the block diagram in Fig. 13-24.

COLOR TEMPERATURE CONVERSION FILTER

Generally, the human eye is sensitive to electromagnetic waves from 380 nm to 780 nm in wavelength (the visible region). In addition, the human eye also discerns the wavelength difference as a color difference. The human eye responds to the light between 400 nm and 500 nm wavelengths as predominantly blue information, the light between 500 nm to 600 nm as green information, and the light between 600 nm and 700 nm as red information.

HORIZONTAL SAWTOOTH/PARABOLA GENERATOR CIRCUIT

This circuit (see Fig. 13-25) generates the horizontal sawtooth/parabola waveforms required in the bias light shading correction circuit and the dynamic focus correction circuit. The horizontal sawtooth/parabola signals are also supplied to the process circuit board for signal shading correction.

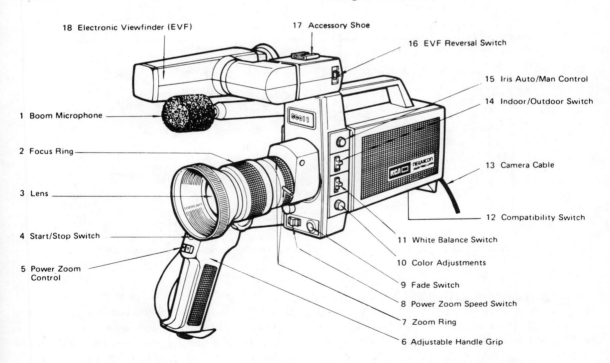

Fig. 13-23. *Operating controls for RCA model CC011 camcorder. Courtesy of RCA.*

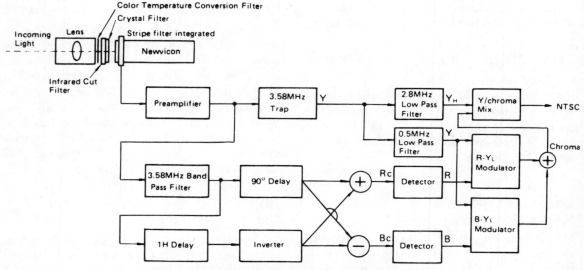

Fig. 13-24. *Block diagram of multiplexing system using a single tube. Courtesy of RCA.*

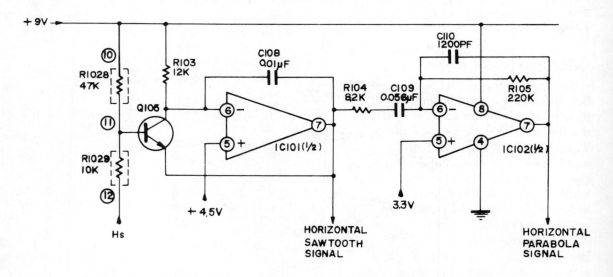

Fig. 13-25. *Horizontal sawtooth/parabola generator circuit. Courtesy of RCA.*

BIAS LIGHT SHADING CORRECTION CIRCUIT

The light intensity of the bias light LEDs is not uniform along the entire target surface but has a shading. The red and blue signals are obtained by detecting the modulated signal from the preamplifier output signal with a bandpass filter; therefore, the red and blue signals are free from bias light shading. However, the luminance signal is obtained by removing the modulated signal with a trap circuit. Thus, the luminance signal receives the adverse effect of the bias light

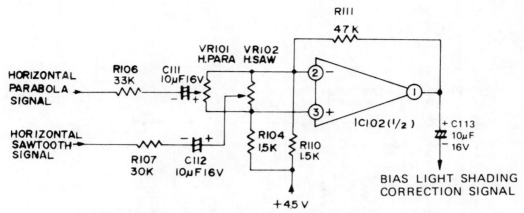

Fig. 13-26. *Bias light shading correction circuit. Courtesy of RCA.*

shading. The bias light shading correction circuit supplies horizontal sawtooth and parabola signals "B" to the process circuit board to correct color shading at low illumination levels. Refer to circuit diagram in Fig. 13-26.

DYNAMIC FOCUS CORRECTION CIRCUIT

The level of focus voltages that produces the best focus of the beam in the center part of the pickup tube is different than that which produces best focus at the edges of the pickup tube. Therefore, the modulation depth for the center and the edges of the pickup tube differ. When the camera is directed at an evenly illuminated white object, the red and blue signals modulated at 3.58 MHz do not have a uniform level, and as a result, color shading appears. Note Fig. 13-27.

The dynamic focus correction circuit supplies horizontal and vertical sawtooth and parabola signals to the focus grid (grid-4). These signals, together with the dc focus voltage from the high-voltage circuit, focus the electron beam along the entire scanning area to correct the unevenness of modulation.

Note: Dynamic focus should be adjusted when color nonuniformity is seen in the picture of a white card after bias light shading is properly adjusted and electrical focus and beam alignment are properly set.

FADE CIRCUIT

The fade circuit generates a triangle pulse that is applied to IC 207 (pin 4). Here it controls the video signal, creating the "fade in" or "fade out" effect. Refer to Fig. 13-28 for the fade timing chart waveforms.

SHADING CORRECTION CIRCUIT

Even when dynamic focus is applied to grid-4 (G4), the non-uniformity of modulation (due to uneven focus on the target) cannot be completely eliminated. The shading correction circuit compensates for residual picture shading using horizontal (H) and vertical (V) sawtooth and parabola signals that are generated in the deflection circuit. Note the block diagram in Fig. 13-29.

CHROMA ENCODING CIRCUIT

This circuit contains the color reproduction correction, gamma correction, chroma signal generator, and the chroma clip/suppression circuits. The chroma encoding block diagram is shown in Fig. 13-30.

GAMMA CORRECTION CIRCUIT

This circuit receives the red, blue, and yellow signals from the color reproduction circuit.

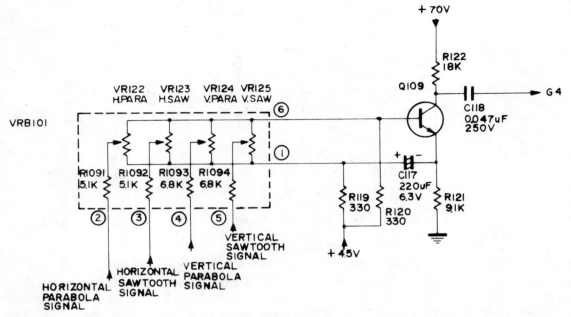

Fig. 13-27. *Dynamic focus correction circuit. Courtesy of RCA.*

These signals, after gamma correction, are supplied to the chroma signal generator circuit.

CHROMA SIGNAL GENERATOR CIRCUIT

This circuit matrixes the R, B, and YL signals with the color reproduction correction signal to generate the R-YL and B-YL color difference signals. It also modulates these baseband color difference signals on two subcarriers that differ in phase by 90 degrees to generate a chroma signal.

CHROMA CLIP/SUPPRESSION CIRCUIT

This circuit contains the high Y level chroma clip circuit and low-level chroma suppression circuit. This circuit clips the green color that results when an excessively bright object is shot with the camera. This is caused by the lack of

beam in the newvicon. When the beam of the newvicon is insufficient to discharge the target due to excessive light reflected from a brightly illuminated object, the red (R)/blue (B) signal component modulated and riding on top of the green (G) signal is lost. Thus, the bright picture parts turn into an unnatural green color.

Hence, if this circuit is faulty, the bright picture parts would turn into an unnatural green color, and the chroma signal would appear on the blanking period of the NTSC signal.

The low-level chroma suppression circuit enhances color reproduction in a low luminance condition by improving the white balances.

NTSC SIGNAL PROCESSING CIRCUIT

In the circuit shown in (Fig. 13-31), the chroma/luminance mixed signal, burst signal and the sync signal are mixed to produce the NTSC signal. The NTSC signal is applied di-

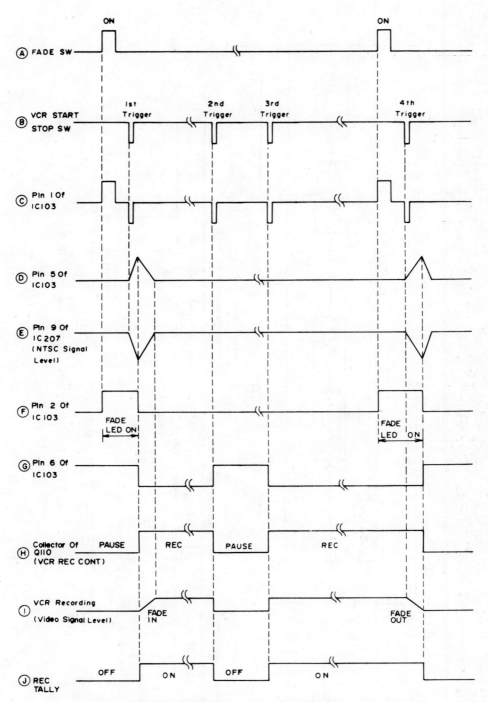

Fig. 13-28. *Fade timing chart waveforms. Courtesy of RCA.*

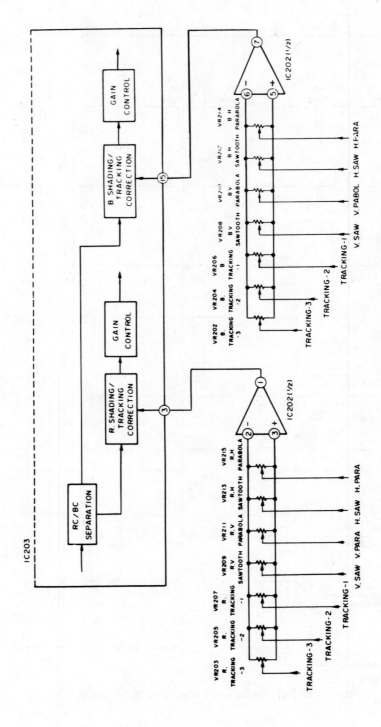

Fig. 13-29. *Shading correction circuit. Courtesy of RCA.*

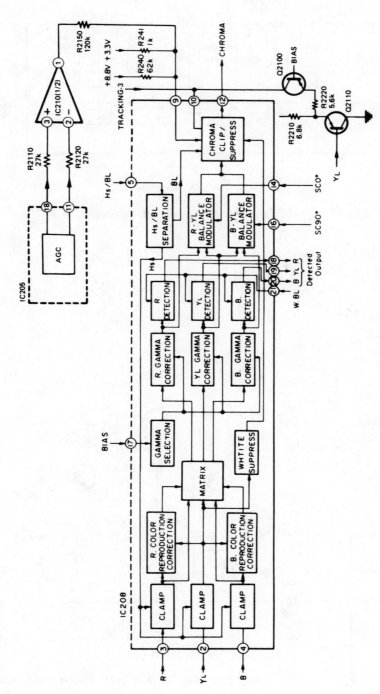

Fig. 13-30. *Chroma encoding circuit. Courtesy of RCA.*

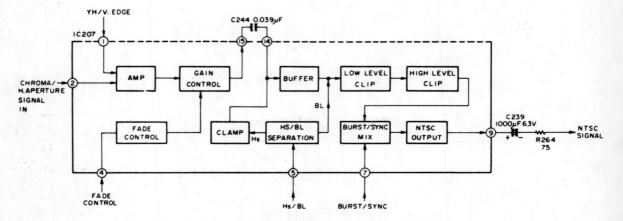

Fig. 13-31. *NTSC signal processing circuit. Courtesy of RCA.*

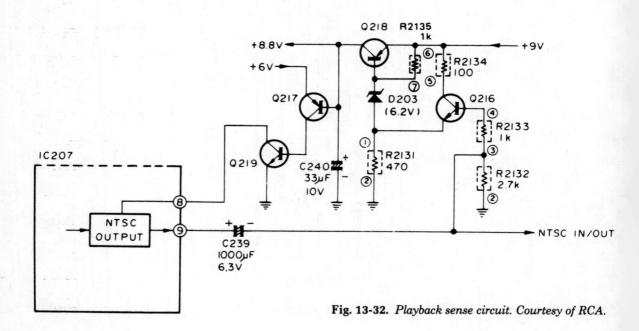

Fig. 13-32. *Playback sense circuit. Courtesy of RCA.*

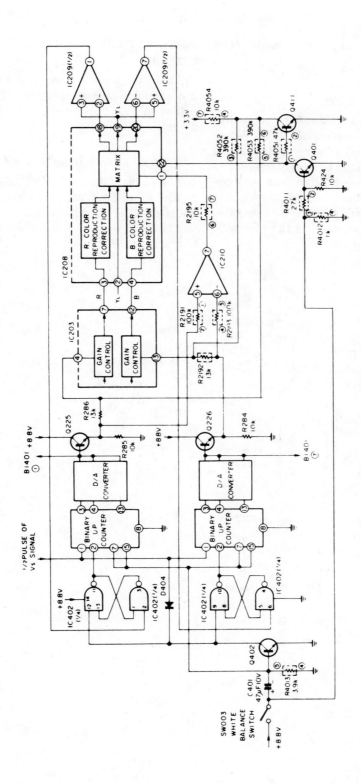

Fig. 13-33. *Automatic white balance setting circuit. Courtesy of RCA.*

rectly to the portable VCR or a table model VCR via the power supply.

PLAYBACK SENSE CIRCUIT

This circuit (Fig. 13-32) senses the recording or playback mode of the portable VCR. When the portable VCR is set in the recording mode, this circuit sets the camera to supply the NTSC output signal to the viewfinder and the portable VCR for recording. When the portable VCR is set in the playback mode, this circuit shuts the camera function off and supplies the playback signal from the VCR to the EVF.

AUTOMATIC WHITE BALANCE SETTING CIRCUIT

This circuit automatically sets the gain of the red and blue signals to match the gain of the yellow signal. Refer to Fig. 13-33 for this circuit diagram.

WHITE BALANCE INDICATOR CIRCUIT

The circuit shown in Fig. 13-34 supplies a 15 Hz signal (¼ pulse of Vs signal) to the indoor/white balance indicator D112. This LED will flash on and off if the white balance is improperly set.

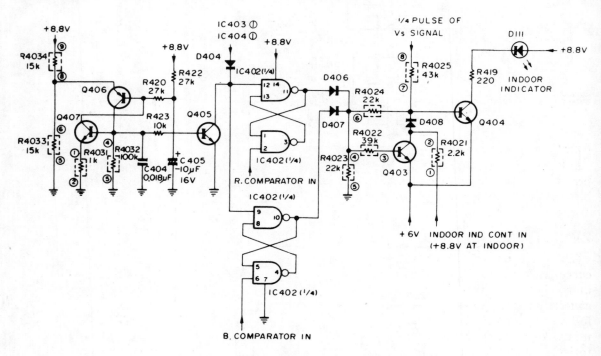

Fig. 13-34. *White balance indicator circuit. Courtesy of RCA.*

14

Case Histories: VCR Problems and Solutions

This chapter features some actual VCR case history problems and solutions for various brands of video cassette recorders. These troubles cover old and newer model machines.

ZENITH MODEL JR9000

This section provides the playback and record checklists for verifying proper operation of this unit. Then a list of questions on helping you collect symptoms of various faults is included, followed by a note of caution for servicing these units. This section then concludes with case histories specifically pertinant to the Zenith JR9000.

VCR PLAYBACK CHECKLIST

This checklist shows you how to check a VCR with a test tape. Note the JR9000 set-up in Fig. 14-1.

- Install alignment tape (Zenith part number 868-27).

- Push the cassette lift assembly down. The stand-by lamp should light up, and tape threading should be completed in 3 seconds or less.
- Depress the play button. The TV monitor screen should blank video and mute audio, followed by the presentation of picture and sound.
- Depress the stop button, then depress the fast-forward. The VCR should stop at the end of the tape.
- Depress the rewind button. The VCR should stop at the end of the tape.
- Depress the eject button. The tape should unthread and the cassette lift assembly should pop up with in 5 seconds. During the unthreading, the stand-by light should be on.
- Reinstall the tape, and locate the beginning of the 1-hour playback portion. Depress the play button with the tape in the 1-hour position. The VCR should blank the video and mute the audio, followed by a normal picture and sound.
- As the 2-hour portion passes the heads, the

Fig. 14-1. *Photo of test equipment on authors VCR test bench.*

video should blank and the audio mute. Next, the VCR should automatically change capstan speed, followed by a normal video and sound reproduction.

- Check the memory mode of operation.

VCR RECORD CHECK LIST

- Insert a blank cassette.
- Select a program to be recorded.
- Record a portion in the 1-hour mode.
- Switch to the 2-hour mode and continue to record at least five counts on the counter.
- Play back the tape to determine which modes are normal. This procedure is not necessary each time, but it does help to isolate problems more accurately. These checklists should also be followed after completing repairs. The machine being serviced should be operated in both the 1- and 2-hour modes.

Another important consideration involves the functioning of the chroma and luminance systems. If no chroma is present in the playback of a recorded signal that *is* in color, then the

chroma processing section of the VCR must be checked out.

Because the JR9000 VCR contains a combination of mechanical and electronic systems, another step is in the type of fault — mechanical, electronic, or both.

COMPILING SYMPTOMS

- Is the record mode operating properly?
- Is the play mode operating properly?
- Is the 1-hour mode of record/playback functioning correctly?
- Is the 2-hour mode functioning correctly?
- If the recorded signal includes chroma information, does this chroma appear to be normal at the output?
- Is the malfunction a mechanical or an electronic problem?

NOTE OF CAUTION FOR BENCH SERVICING

When placing the VCR on its left side (the RS-L board side) with the cover removed, damage can occur to the P-7 board if it is done improperly. The P-7 board, which supplies the regulated +12 volts to most sections of the recorder, is located on the left front corner of the machine. Should you *roll* the machine from its normal horizontal position onto its left side, the P-7 board could crack. This crack has been known to open the conductive path that leads to the output pins of CN5002; this cuts off the +12 V output.

To avoid this damage, the machine should be *lifted,* not *rolled,* from one position to another. Always take this precaution when servicing any VCR with the covers removed.

ZENITH JR9000 CASE HISTORIES

This section describes actual problems and solutions that were encountered with this particular model. Some symptoms could pertain only to this model, while others might be more universal. Each symptom is listed, followed by the corresponding service technique and actual cause of the problem, plus perhaps some background information for clarification.

Symptom

The standby lamp did not light during the threading and unthreading (eject) operations. Other functions of the machine were normal.

Background Info. The threading/unthreading operations are functions of the SY-2 board, as is the standby lamp circuit. On the SY-2 board, transistor Q9 is the device that operates the lamp. The lamp is in the collector circuit of Q9, with one side of the filament connected to 12 V (B+).

Service Techniques. The standby lamp could fail to operate because of several conditions: (1) loss of the 12 V; (2) a defective lamp; (3) a defective Q9 transistor. A voltmeter check of the collector of Q9 can determine which of the above conditions exists. This voltage point was measured during the threading and unthreading modes. The reading indicated a high voltage, meaning that the B+ was present, and the lamp filament was good. A clip lead to ground from the collector lit the lamp.

The Cause. Transistor Q9 was defective (open).

Trouble Symptom

The eject function did not operate. The threading function was normal. The dc threading motor turned continuously as long as the unit was turned on.

Service Technique. The clue here is the constant operation of the threading motor. The SY-2 board controls the threading and unthreading functions. The current through transistor Q10 is the current through the threading motor. If this motor is constantly running, Q10 is always conducting. The operation of Q10 is controlled by transistor Q3, which is only cut-off during normal threading and unthreading. This causes a high voltage on its collector and on the base of Q10, turning on Q10 and operating the motor. Voltage measurement indicated that this was the case. After threading, Q3 should turn on and Q10 off.

The Cause. Transistor Q3 was defective (open).

Trouble Symptom

The VCR would not unthread. Depressing a function button put the recorder into auto-stop. The ac motor would not turn.

Service Techniques. The cassette must be in and the threading switches must be closed to deliver the 12 V (B+) to the proper circuits. The voltage level on pin 19 of IC4001 on the SY-2 board must be high, and pin 21 must be low. Transistor Q3 must be off, and Q10 must be on. Switch 4004 must be in the non-eject position. An obvious first measurement would be the +12 V beyond the cassette in switch (normally closed with a cassette pushed down into the operating position). This measurement point is pin 1 of connector CN4001. The +12 V did not appear at this point but was found at pin 10 of the same connector. Pin 10 connects to the other side of the cassette in switch.

The Cause. The mechanical movement of the cassette IN plunger did not actuate the cassette in microswitch. A slight re-adjustment of the position of the microswitch solved this problem.

Trouble Symptom

The VCR would not record or play in the 2-hour mode. The recorder operated normally in the 1-hour mode. A known, good tape cassette, recorded in the 2-hour mode on another VCR, played back properly on the defective machine. Thus, the actual symptom was that there was no operation of the 2-hour record mode.

Background Info. If you check the pre-emphasis route of the luminance signal during the 2-hour record operation, note that it includes the "outside" circuitry between pins 21 and 6 of IC1001 on the YC-L board.

Following the luminance/video signal with a scope through these stages showed a proper waveform up to and including at the base of Q36. The collector and emitter of this transistor showed no signal. Transistor Q36 was cut off because of a faulty bias.

The Cause. Transistor Q49, the thermal compensator in the emitter circuit of Q36, was open.

Trouble Symptom

There was no chroma and/or luminance during playback of a good recorded tape. This indicated the fault was in the VCR record system. In addition, the E-E signal, without any button activation, was defective.

Service Technique. Block diagrams in service manuals usually show the paths of the luminance, chroma, and E-E signals from the input to the output of the VCR. It is always good practice to check inputs and outputs of these blocks (in this case the YC-L board) first before looking inside the circuitry of the blocks.

A scope look of the composite video input signal at pin 6 (CN1001 connector) revealed an improper video signal. The source of this signal is the PJ-7 board, incorporating the video in/out jacks and the output of the IF-1 board. Next, these sources were checked.

The Cause. The shielded leads going to and from the video in/out jacks at the rear of the VCR had been pinched under the rear cover mounting stud by the cover when it was installed. This resulted in a shorted coax cable and a loss of signal.

Trouble Symptom

No playback was possible in any mode. A known, good recorded cassette in the machine did not produce a picture on the TV screen. Also, the machine did not record properly. Thus, this fault was common to both the record and playback functions.

Service Techniques. Most of the record and playback circuitry is located on the YC-L board. The first step is to check the dc power to the board, in this case +12 V. No dc voltage was present. Fuse F5001 on the P-7 board was open. An ohmmeter check indicated a short to ground for the 12 V supply, which was traced to the YC-L board. The YC-L board has two 12 V circuits. One is switched on all the time the recorder is on. The latter 12 V potential was the one that was shorted. Locating a short to ground for the 12 V B+ conductive path on a PC board can be difficult.

There is some help available on the YC-L board, however. There are several jumper wires that connect sections of the B+ PC patterns together. Thus, unsoldering a jumper can isolate the sections of the B+ line for resistance checks. This technique centered around the IC5 area of the YC-L board.

The Cause. There was a short between the primary (connected to ground) and the secondary (connected to the 12 V line) of transformer T13.

Trouble Symptom

Playback of tapes recorded in the 2-hour mode was not possible. All other functions were normal.

Service Techniques. The problem was isolated to the deemphasis process circuitry of the 2-hour playback of the luminance signal. The luminance signal could be traced through the system up to the Q1029 switching transistor, but not beyond.

The Cause. The switching transistor was defective.

Co-channel Cable Interference

Sometimes on cable TV systems, a beat can develop between a channel 3 or channel 4 signal on the cable and the corresponding output frequency of the VCR's RF modulator. Two conditions appear to be required for this co-channel interference to occur: (1) an unusually high (80 mV or more) cable signal level; (2) a mis-termination of the cable input.

AC Motor Failures

Some AC motors have failed in the Zenith JR9000 units. The failures were traced to the thermal protective device within the motor, which was modified on later production model runs.

Audio/Control Head Failures

This failure has occurred when the leads going to the head assembly broke. Two methods of connecting these leads have been used in production. Automatic lead connection sometimes did not leave any slack in the wires, causing breakage later from mechanical vibration. Manual connection of the head leads provides more slack and is used in all later production of these models.

Trouble Symptom

The tracking control seemed to drift in and out of adjustment intermittently. With the case removed, the recorder operated perfectly. The problem looked the same as when the head drum brakes were unplugged from the CS board.

Service Techniques. All scope waveforms appeared to be normal. All adjustments were set properly. The recorder operated perfectly with the covers removed. When the recorder was turned on end to disassemble it (tuner end), the 3.15 amp fuse blew on the P-5 board. The short was traced to the CS board.

The Cause. Component leads protruding from the foil side of the CS board were touching the metal shield on the AD-L board, causing a partial loss of the drum braking, and with the recorder on end, the additional pressure caused the B+ to short, as several other leads were grounded to the shield.

The UHF tuner bracket and frame were straightened. The rear of the UHF tuner had been pressed against the component side of the CS board, causing it to swing in toward the AD-L board.

Trouble Symptom

There was no audio during playback of a tape recorded on this VCR. A known, good tape recorded on another machine produced normal sound when played back on the defective VCR. Hence, the problem was in the audio record section of the VCR.

Service Technique. The audio record system includes the audio heads, the AD-L board, the tuners and IF, the MC-4, and PJ-7 boards. The heads are not the problem, because the playback with a good tape was normal. During record, the IF audio output was normal. It was also okay at pin 4 of connector CN7010 and at the input to the AD-L board (pin 1 of CN3603). No audio signal was present at the AD-L board output (pin 1, CN3604) to the audio heads.

The Cause. Signal tracing from the AD-L board input into the board quickly determined that transistor Q611 and Q612 were defective. They are the active devices in the first two stages of amplification during the record mode. Replacing these transistors solved the problem.

Trouble Symptom

Snow was present on channel 4, but not on other channels.

Service Technique. These symptoms indicate an RF problem. Thus, the function blocks that are part of the RF signal path were checked. This included antenna input, VHF splitter (inside the VCR), and the VHF tuner. Direct connection of the antenna lead into the TV set's antenna terminal gave a good picture.

The Cause. The VHF splitter in the VCR machine was replaced. The best way to check the splitter is to bypass the RF signal around it.

Trouble Symptom

On playback, the picture on the TV set screen broke up. This symptom occurred during the playback of a known, good recorded video tape. A tape recorded on this VCR played back normally on a good machine. Thus, the fault would probably be in the VCR playback system.

Service Technique. The RS-L board, the YC-L board, and the RF modulator are included in the playback system. The output of the RS-L board is at the "center" of this playback signal path. Checking with a scope on this output at CN2004 (pin 5) revealed an abnormal signal. A similar condition was present at pins 16 and 17 of IC2001 on the RS-L board. The input signal to IC2001, at pins 23 and 24, was found to be normal.

The Cause. A close, visual examination of the board area near IC2001 showed a short between pins 17 and 18. Scraping the space between the board conductors leading to the pins cleared the short and restored the VCR to normal operation.

Trouble Symptom

There is "wow" in the reproduced sound. Video is normal in the playback mode. A known, good recorded tape gave the same symptoms. Thus, the fault is in the playback circuits.

Service Technique. Characteristically, the "wow" condition of sound is directly related to the movement or the tracking of the tape. The tape movement is a mechanical operation,

whereas the tracking is mostly an electronic operation.

The Cause. Replacement of the CS board solved the problem, thus the electronics of the tracking were involved. In this particular case, Q4631 of the tracking chain circuits was defective. When defective, it gives a "wow" or "warbling" sound during playback.

Trouble Symptom

The VCR would not operate or unthread the tape. The tape would not move.

Service Technique. Tape movement for the various operating modes is mechanically controlled by the operation of the capstan and pinch roller. The capstan motor was not rotating. An ohmmeter test at the CM board indicated a shorted winding condition. The resistance between the orange and white leads measured 12 to 20 ohms instead of the usual 200 ohms.

The Cause. The capstan motor was defective. A new motor solved the problem.

Trouble Symptoms

There was buss or hum in the sound when a local UHF channel (20) was recorded. This condition only ocurred in one large area of a city. All the recorders used had the same problem. The following conditions were observed:

- Very strong channel 20 signal.
- Adjusting recorder AGC did not remove buzz.
- Using various TV sets did not eliminate buzz.
- Buzz was not from the power supply, but it appeared to be a video signal in the audio, because buzz level varied with picture contrast.
- Attenuating the channel 20 signal input removed the buzz in the E-E mode, but it still appeared in the playback of the recorded tape.
- The attenuated channel 20 recording exhibited the buzz when played back on a different VCR.

Service Technique. Analysis of the composite video waveforms out of the recorder IF only proved that the signal was exceptionally strong and that the UHF channels were not controlled quite as effectively as VHF by the AGC circuits. No distortion or video was present in the sound. Sync level in the composite signal appeared to be about 28 percent for channel 20 as compared to 30 percent for other UHF signals. A 2 percent difference is not much change in amplitude with a service type scope, but at least there was some reduction in sync level of channel 20 as compared to the other stations.

Because the buzz was evident when a recorded tape was played in different machines, it was assumed that the buzz was recorded on the tape. However, the audio was clean at the audio output jack of recorder.

Analysis of the signals and voltages fed to the RF modulator proved that an excessive composite video signal fed to the RF modulator was causing overmodulation of the picture carrier and resulting in video intrusion in the sound.

The Cause. Following the adjustment procedures from the modulator backward through the playback circuitry in search of the excessive video resulted in these findings:

- All levels were high, but apparently the recorder was able to handle the strong signal and diminished sync in the record mode.
- In playback, the recorder was again able to cope with the conditions, but signal levels gradually increased from the recommended levels until the RF modulator input equaled 2.0 V p-p video.
- The most effective means of controlling the video signal proved to be the adjustment of VR21 (on the YC-L board) to the specified 1.3 Volt p-p at the emitter of Q25. Original level was about 1.9 V p-p before adjustment on channel 20. The original level was normal on any other channel. After adjustment to 1.3 V p-p for UHF 20, other channels read as low as 1.0 V p-p. The only noticeable change in the picture on other channels was a slight change in brightness. Originally, all the UHF channels seemed to playback slightly brighter than

VHF channels, probably because of the stronger (local) signals.

Note: Do not attempt to compensate by adjusting VR24 (1 V p-p at TP3) for the proper video signal to the modulator. It would not reduce the signal sufficiently, and it caused deterioration in the playback of the other channels.

The recorder is able to handle a non-standard signal and record properly with this adjustment. All of the tapes now play good, with the video playback signal reduced.

Trouble Symptom

After 15 minutes of playing time, lines appeared in the picture on the TV set. This effect was seen on all channels. Direct connection of the antenna leads to the TV set's tuner input terminals did not display any lines. Thus, the fault was in the recorder.

Service Techniques. The location of this fault could be in the VCR RF input, the VHF splitter, the IF-1 board, or in the RF modulator. The RF modulator is the easiest unit to change, so this is a good place to start.

The Cause. The RF modulator was changed, and the lines in the picture cleared up. The RF modulator was defective.

Trouble Symptom

On playback, no signal was visible on the TV screen (the raster was blanked). A known good tape was then played back on the faulty machine with the same results. Thus, it appeared that the fault lay in both record and the playback. An additional symptom was the absence of sound (audio muted).

Service Procedure. The functional blocks that are common to both record and playback are as follows:

- video tape
- audio head
- video heads
- control heads

The chances of a defective video tape were remote. If the audio head had been open, there would have been some noise in the speaker instead of the muted sound condition.

If the video heads had been open, there would have been a visible TV raster, without video, instead of a blanked (black) raster.

The control head could have been open, because both record and playback were affected. An open control head would result in no control track on the tape during recording.

The Cause. The clue here lay in the fact that both the audio was muted and the video was blanked out on playback. This condition occurs if the CS board circuitry does not sense a good, steady control track signal. Thus, it appeared that the control signal system should be checked first (both record and playback).

During the record operation, a scope check at pin 1 of CN2005 (RLS board) showed a good 30 Hz signal going to the control head. During playback, no recovered control track signal was present at this point or at pin 12 of IC502. The control head was then suspected to be faulty. It was discovered to be open. Replacement with a new audio/control/erase head assembly returned this VCR to normal operation.

Trouble Symptom

The VCR would not complete the threading operation and the standby lamp remained lit. Hence, no other function could be activated. The threading ring did not complete its cycle. Electronic circuitry is part of the threading system, but the mechanical action might be easier to check first.

Service Procedure. A careful visual check of all significant parts of the threading mechanism disclosed and broken tip on the threading arm assembly.

The Cause. The arm assembly tip was replaced, and the threading operation then worked properly. The broken part of the threading arm assembly allowed the arm to fold in the wrong direction, preventing the completion of the threading cycle.

Trouble Symptom

After being used for several hours, the JR9000 VCR would exhibit a tracking problem. Recordings made on a known good machine would play back improperly. Visible symptoms on the TV screen resembled symptoms that appear when the drum brake cable is disconnected from the CS board. The picture would alternately clear up and then break up as if the tracking control was being rotated back and forth. As the recorder continued to heat up, the symptoms became more severe.

Service Procedure. The symptoms indicated a temperature warm-up situation, and the evident tracking problem pointed the way to the CS board (where the tracking control circuitry is).

The recorder was allowed to cool down to room temperature. It was then turned on, placed in the playback mode, and operated normally. Applying a heat lamp to portions of the CS board localized the defect to the area of IC4601, which was causing the tracking problem. Coolant applied to IC4601 restored the recorder to normal operation. Replacement of the chip solved the problem.

Trouble Symptom

The cassette would not eject from the VCR.

The Cause. A broken eject spring, part of the slide lever assembly, was the trouble.

Trouble Symptom

Intermittent color playback on VCR.

The Cause. An intermittent delay line was found on the YC-L PC board.

Trouble Symptom

The tape cassette would not load into the machine.

The Cause. One section of the functions switch was open. This switch can also have intermittent contact problems.

Trouble Symptom

A situation of poor playback performance was noted when recording and playing back programs received over cable systems that delivered very "short" sync signals. On playback, the pictures showed white flaring, and a buzz was audible. A scope check at the TV receiver's video detector showed the white going to zero carrier level, indicating that the VCR modulator was being overmodulated. The emitter of Q25 on the YC-L board (JR) had a peak-to-peak signal of 2.2 V when the recorded cable program was played back. When the alignment tape was played back, the Q25 emitter signal read the specified 1.3 V peak-to-peak.

Service Procedure. On the JR9000 VCR, two adjustments can be made to solve this problem, one for each record and playback. In the record mode, while observing the E-E level, adjust (VR5) on the YC-L board slightly to remove the unwanted video/audio symptoms.

In the playback mode of a tape recorded on the VCR being serviced, adjust VR21 (YC-L board) while observing the signal on the TV screen. Adjusting the control only enough to clear the buzz and sync compression symptoms.

Trouble Symptom

Would not record. Capstan slows down during record mode and then stops.

The Cause. Dirty contacts on switch S601 on the AD-L board.

Trouble Symptom

Intermittent, slow tape movement on both record and playback modes. When tape is running too slowly, sync and picture breakup problems occur.

The Cause. Intermittent open winding on the supply reel brake solenoid. Replace the brake solenoid.

Trouble Symptom

No response from any of the control functions that relate to the SRP board, such as partial threading or unthreading, partial rewinding, or non-functioning of the counter memory circuit.

The Cause. Defective IC1 (CX141) on the SRP board (Zenith part number 905-103).

Trouble Symptom

Cross modulation of a strong channel 3 on a much weaker channel 2 when a tape recorded on the machine was played back.

The Cause. Too much difference in signal strengths of adjacent channels. Adjust AGC control located in the IF-4 board.

Trouble Symptom

No eject function.

The Cause. Spring leaf broken on the eject switch lever.

Trouble Symptom

No eject function. Dc threading motor continually turned. Machine went through threading cycle without tape cartridge installed.

The Cause. Transistor Q3 on SY-2 board open.

Trouble Symptom

No 2-hour playback operation on the JR9000 machine.

The Cause. Defective Q1029 transistor located on the YC-L board.

Trouble Symptom

No chroma in playback operation.

The Cause. Replace IC4 on the YC-L board.

Trouble Symptom

No chroma on playback or record.

The Cause. Zenith JR9000 — T6 and IC2 on the YC-L board. Zenith KR9000 — T6 and IC2 on the YC-2 board.

Trouble Symptom

Sound on playback warbled, 2-hour free speed would not lock in, and picture was flashing at about 2 Hz interval intermittently.

The Cause. Transistor Q4631 on the CS board defective.

Trouble Symptom

No record on playback.

The Cause. JR9000 — Shorted T13 on the YC-L board. KR9000 — Shorted T13 on the YC-2 board.

Trouble Symptoms

Playback and record buttons would not lock. Machine went immediately into auto-stop. The threading and unthreading cycles seem too fast.

The Cause. Defective Q5501 transistor on P3 board. With this fault, the 12 V B+ will measure about 18 V.

Trouble Symptoms

Machine would not unthread. All control buttons put the machine into auto-stop. The ac motor would not turn.

The Cause. Defective ac motor caused by shorted windings.

Trouble Symptom

The VCR made poor recording. When played back, the recording on this tape caused the machine to go into mute (no video or audio) and the set has a blank raster.

The Cause. Transistor Q501 is shorted on the RS-L board (for JR9000) or ARS board for (KR9000) causing loss of control track pulses.

Trouble Symptom

No chroma on playback. Loss of 4.27 MHz signal.

The Cause. Open transformer T13 on the YC-L board.

Trouble Symptom

Intermittent loss of chroma on playback.

Servicing Procedure. Touch-up ACK level adjustment on YC-L board.

Trouble Symptom

Snowy picture.

Servicing Procedure. Re-adjust the REC current and F-Chara adjustments on the ARS board.

Trouble Symptom

Cassette lift assembly would bind up halfway down in position.

The Cause. Tape-up sensor bracket pushed against cassette, causing binding. Cassette cartridge not pushed all the way into the lift assembly. Cassette guide assembly too far toward front of the machine.

Trouble Symptom

Machine would not thread. Loss of dc (12 V) to the threading motor.

The Cause. A broken P7 board feed-through connector was found. This caused a loss of the 12 Vdc to the motor.

Trouble Symptom

Machine would not rewind the tape.

The Cause. Defective ac motor (part no. 941-104).

Trouble Symptom

Played back pre-recorded tapes normally, but no TV reception.

The Cause. Fuse F5301 (0.5 amp) was open on the P-8 board. A short circuit was caused by the AD-L board being loose and not locked into the nylon clips. It was laying against the 18 V connection pin. To correct, clamp the ADL board into its nylon clips and replace the fuse.

Trouble Symptom

No playback on pre-recorded (on other working VCR) or recorded (on faulty machine) video tapes.

The Cause. Defective audio/control head assembly (part no. 949-102).

Trouble Symptom

Picture breaks up and loses color on playback.

The Cause. Defective playback/receive switch.

Trouble Symptom

Picture breaks up in a symptom similar to that occurring when the control head is defective.

The Cause. Defective IC501 chip on the RS-L board.

Trouble Symptom

Noisy picture on playback.

The Cause. AGC delay (in TV receiver) control is set incorrectly. Readjust the AGC delay control.

Trouble Symptom

No tracking in the playback mode.

The Cause. Defective IC601 (CX143A) and Q629 transistor located on the CS board.

Trouble Symptom

Machine would not play back previously recorded tape. Tape recorded on defective machine played back on a good machine. Tapes played on the faulty machine were erased.

The Cause. Inspection of the SRP board revealed a shorted diode (D4009). This allowed 10.2 V to be applied to the record circuit during playback. Replacement of this diode solved the playback problem.

Trouble Symptom

Auto-stop would engage regardless of the function selected.

Service Procedure. The waveform at TP-1 on the SRP board was not correct. Adjustment of VR4001 produced the proper waveform. The machine was aligned to specifications and now operates normally.

Trouble Symptom

Fast-forward mode would not function. Tape would not eject.

Service Procedure. The 12 V was not present. A check of the SRP board verified this. The power transformer did not supply ac to the board, but voltage was not found at the diode bridge.

The Cause. Connector CN4005 (pins 1 and 2) were not making proper contact. The terminals were discolored and the connector was replaced.

Trouble Symptom

In this case, the faulty machine would not record color but played back color from a pre-recorded tape.

Service Procedure. Pin 12 of IC2 (ACC on YC-A board) should measure 3.4 V during record and 0 V during playback. The voltage was 0 V in both modes.

The Cause. Capacitor C39 (0.01 μF), from pin 12 of IC2 to ground was shorted. A replacement capacitor solved the problem.

Trouble Symptom

Audio recorded by this VCR would have a warble (or wow) for 2 to 4 seconds after the pause lever was released. There was no evidence of picture break-up.

Service Procedure. This machine had over 1000 hours of operating time and a visual inspection of the capstan during playback revealed possible binding.

The Cause. The capstan and flywheel were

disassembled, cleaned and the shaft and bearings were lubricated. This solved the problem.

Trouble Symptom

No chroma or occasionally weak chroma during playback. The black-and-white picture was normal.

Service Procedure. It was found on the YC-L board that the playback ACK was 3.2 V and not the required 4 V. On the RS-L board, pin 12 of IC1 measured 2.7 V and should be 3.9 V. *Note:* pin 12 connects to the mono/chroma switch inside the IC1 chip.

The Cause. IC1 on the RS-L board was defective and replaced.

Trouble Symptom

Intermittent blanking of picture during playback.

Service Procedure. A check of the RS-L board playback circuitry revealed a variance in amplitude of the waveform at TP-505 (pin 9 of IC502 servo).

The Cause. Connector CN2005 from the control head to the RS-L board was making intermittent contact. A replacement connector solved this problem.

Trouble Symptom

When the tape was put in rewind mode, it sometimes slowed down or stopped.

Service Procedure. Check for evidence of wear or binding in tape pulleys or tape path.

ZENITH VR8910W

This VCR utilizes a microprocessor system control to operate all of the machine's controls. Refer to (Fig. 14-2) for the block diagram of this control system, which is located on the SS-9 board.

Before looking at any problems in this microprocessor control system, let's take a brief look at the record and playback circuit operation. See Fig. 14-3.

PLAYBACK OPERATION

When the play button is pushed, pin 5 (PLAY IN) of IC501 (the microprocessor) changes over from a high to a low. By this operation, pin 26 (PLAY OUT) produces an output, and 12 V is output from Q502 and Q509.

At the same time, pin 15 (MUTE OUT) outputs a muting signal for about 2.8 seconds. Also, pin 32 (ANT SEL OUT) outputs, and a signal is produced to change over TV/VTR (the changeover electronic switch) in the antenna input section. The VTR is then output by Q508 and Q514.

Control output signals are also sent to other solenoids to complete the play operation. Motor control output is generated from pin 12 (CAPSTAN), pin 13 (DRUM), and pin 14 for the reel. These solenoids and motor control outputs are sent to the DR-1 substrate to drive each solenoid and motor.

RECORD OPERATION

When the REC button is depressed, IC501, pin 8 (REC IN), changes over from high to low. By this operation, pin 27 (REC OUT) outputs and REC 12 V is generated from the Q503 buffer transistor and Q510. Solenoid and motor control outputs are generated in exactly the same way as they are for the playback mode.

Trouble Symptom

The problem with this machine was that it would not play back a pre-recorded tape. However, it would produce a picture in the fast-forward (FF) scan mode.

Also, when the machine was turned on, the red record LED was on. The VCR could record properly.

The Cause. The problem was in the microprocessor SS-9 control board. The Q503 buffer record control transistor was showing leakage. Note that the same problem could have developed in any of the other microprocessor control mode functions.)

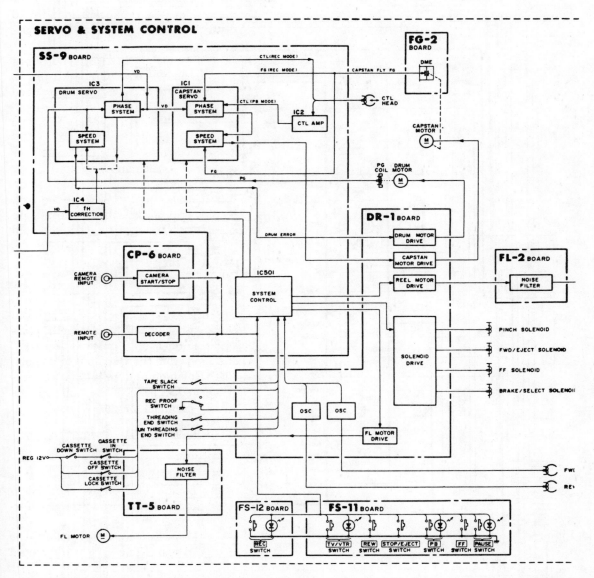

Fig. 14-2. *Block diagram of Zenith VR8910W VCR Servo Control system. Courtesy of Zenith Electronics Corporations.*

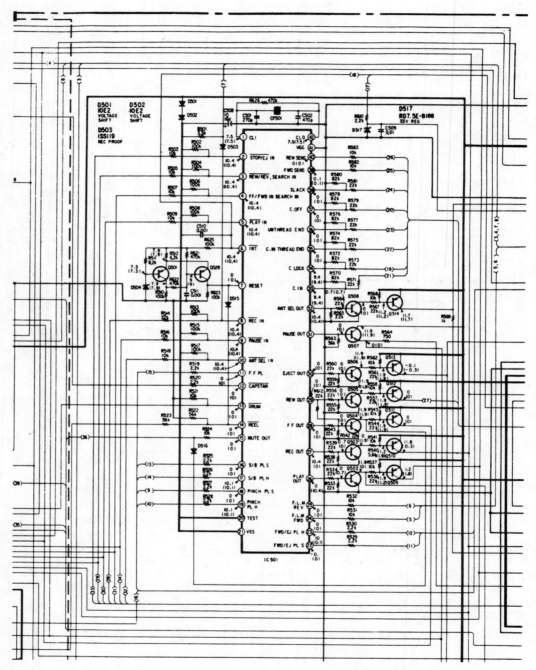

Fig. 14-3. *Microprocessor circuit located on the SS-9 board of the Zenith VR8910W VCR. Courtesy of Zenith Electronics Corporations.*

Q501 2SA1175 INITAL RESET	Q528 2SA1175 INITAL RESET	IC501 µPD553C-149 SYSTEM CONTROL	Q502 2SC2785 PLAY OUT DRIVE	Q505 2SC2785 REW OUT DRIVE	Q508 2SC2785 ANTENNA SELECTOR	Q511 2SA1048 FF OUT SWITCH
D504 RD7.5E-B1 RESET CLEAR	D516 1SS119 FF MUTE	D515 1SS119 RESET DET	Q503 2SC2785 REC OUT DRIVE	Q506 2SC2785 EJECT OUT DRIVE	Q509 2SA1048 PLAY OUT SWITCH	Q512 2SA1048 REW OUT SWITCH
			Q504 2SC2785 FF OUT DRIVE	Q507 2SC2785 PAUSE OUT SWITCH	Q510 2SB740 REC OUT SWITCH	Q513 2SA1048 EJECT OUT SWITCH

Transistors for Fig. 14-3.

ZENITH VR9700 AND VR9000 BETA MACHINES

Trouble Symptoms (System Control)

You could have intermittent shut-down in forward functions (play, record, and fast-forward) or no play or fast-forward operation (when these buttons return to the stop mode immediately). There could also be an intermittent rewind action, or the unit shuts down before the end of the tape during rewind mode.

Service Checks. The sensor circuit IC (see Fig. 14-4) contains an oscillator that has the tank circuit components (sensor coil) connected to pin 2 and a capacitor external to the IC between pin 2 and pin 4 (ground). The forward sensor oscillator in IC3 is on at all times, but the rewind sensor of IC2 is turned on only when the VCR is in the rewind mode. When the sensor foil at the end of the tape passes the sensor coil, the foil detunes the oscillator tank circuit, disabling the oscillator. This in turn biases the last stage in the IC (between pins 7 and 8) off, causing pin 8 to go high. This high causes IC1 to set up the stop mode. The coil at pin 7 of IC2 or IC3 is the dc return path for the base circuit of the stage between pins 7 and 8. When pin 7 is open, pin 8 goes high just the same as if the end of the tape had been sensed.

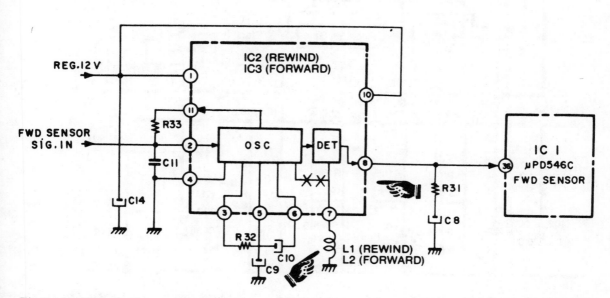

Fig. 14-4. *Block diagram of the Rewind/Forward Sensor circuit. Courtesy of Zenith Electronics Corporations.*

ZENITH VR8500 VCR

Trouble Symptom

The drum servo on this machine lost its lock as the VCR warmed up.

Service Checks. When the VCR was first put into operation, the drum free-speed control (RV9) located on the SS-9 board (see Fig. 14-5) could be adjusted for proper free speed. However, the drum servo would lose lock as the VCR warmed up. Re-adjustment of RV9 would not maintain servo lock. The voltage at test point TP2 should be adjusted for 5.2 V, but it would not go any higher than 3 V when the machine warmed up. The voltage and waveform found at pin 22 of IC3 were not correct. When a circuit coolant was sprayed on capacitor C54 (con-

nected to pin 22), the voltage at TP2 increased to near 10 V.

The Cause. Capacitor C54 and R105 at pin 22 of IC3 are part of a slope (ramp) circuit. The output of this circuit is later sampled to determine the drum speed error voltage. Capacitor C54 was leaky and affected the adjustment range of the drum free-speed control RV9.

Capacitor C54 was replaced at pin 22 of IC3. The voltage at TP2 was then adjusted with control RV9 for a stable 5.2 V. A normal scope waveform then appeared at pin 22 of the IC3 chip.

SONY VCRs

Trouble Symptom

The unthreading process requires the operation of the ac motor, which turns the take-up

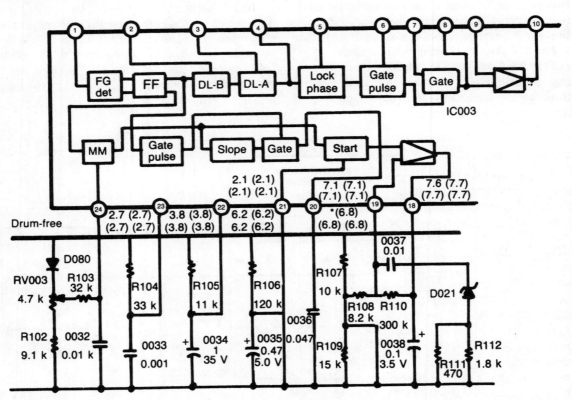

Fig. 14-5. *Drum Free Speed circuit found in Zenith Model VR8500 VCR recorder. Courtesy of Zenith Electronics Corporations.*

reel to pull the tape back into the cassette. The ac motor also rotates the video head disc assembly, developing the 30 PG pulses that are converted to the RF switching pulse. If this pulse in not present at pin 22 of IC4001 on the SY-2 board, the auto-stop function is activated. Hence, this machine would quickly shut down when a function button was pressed.

The Cause. The ac motor was defective due to shorted windings.

Trouble Symptom

The eject function would not operate as it does during threading. To perform the unthreading operation, the motor must rotate in the opposite direction. This reversal is normally accomplished when the threading switch moves to the eject position when the eject button is depressed. Investigation showed that the threading switch was not moving to the eject position.

The Cause. The spring leaf on the eject switch lever was broken. This lever, connected mechanically to the eject button, activates the threading switch on the SY-2 board. Replace the spring leaf switch assembly.

Trouble Symptom

No playback was possible in any mode. A tape attempted to be recorded in the machine did not play back in a known good machine.

Background Information. An important clue to this problem is the ever-present blanking of the video. This condition occurs whenever the apparent control signal, as "seen" by the CS board, is not the normal 30 Hz waveform. In this machine, the automatic tape speed circuit on the CS board determines the operation of the blanking/mute signal. The input of this circuit comes from pin 14 of IC2502 (on the RS-L board). This input for normal operation is the playback control track signal. An oscilloscope connected to this point indicated that the CTL signal was not present.

The Cause. The control head was defective

(open). Thus, no CTL signal could be recorded or played back.

Trouble Symptom

A cassette tape recorded on the VCR would not playback. Video blanking and audio mute are on continuously. Also, the recorded tape would not play back on a known, good machine. Thus, the basic symptom indicates a record function fault.

Service Technique. The symptoms are the same as in the problem above and so is the technique used to find the source of the problem — the lack of a recorded control track. In this case, the faulty component was not the control head. Scope checks revealed that the CTL signal was not reaching the head.

The Cause. Transistor Q2501, the record CTL amplifier, was defective (shorted).

Trouble Symptom

The luminance record and playback functions were normal. Chroma was not visible in the playback signal. A prerecorded video tape did not produce chroma when played back on another machine. Thus, the problem was no chroma on record or playback.

Service Technique. In this machine, one circuit that is common to both the record and playback system is the 3.58 MHz crystal filter. This is the source of a signal that develops the ACK (automatic color-killer) voltage of $+4$ V when chroma is present in the composite video signal. Without this voltage, chroma in record and playback are not possible. Thus, this circuitry was checked and revealed a loss of the 3.58 MHz CW signal.

The Cause. Transformer T6, part of the 3.58 MHz filter, was defective (opening winding).

Trouble Symptom

There was an intermittent shut-down of all functions in the VCR. The machine was thus intermittently in the auto-stop mode.

Service Technique. There are several conditions that can cause activation of the auto-stop mode, for example, end of tape (during rewind or fast forward FF) or when the drum head assembly stops rotating. In this particular VCR, the symptoms were not related to the end-of-tape situation, and the slack sensor was not activating. Thus, attention was directed toward the drum assembly.

Visual observation indicated that the cylinder head drum was rotating steadily during the intermittent shut-down condition. When the cylinder head rotates, the two magnets on the bottom of the disc generate pulses as they pass over the PG coils on the fixed portion of the drum assembly. These pulses, coupled to the SY-2 board, permit normal tape and head movement. If the pulses are not going to the SY-2 board, the auto-stop function activates.

A scope check of the PG pulse path indicated that there was an intermittent signal being coupled to the SY-2 board. This discovery led to the PG pulse source. The "B" PG pulse was steady, but the "A" PG pulse was intermittent. The fixed coil from which this signal is developed was intermittently opening.

The Cause. Replacement of the complete drum cylinder assembly was necessary to replace the defective coil. This replacement solved the problem.

ZENITH KR9000

Trouble Symptom

On these machines, the main drive belt could come off as soon as the motor begins to run.

The Cause. The drum drive motor could be tilted. Use a washer under the rubber motor mounts to align the drive motor.

The motor belt drive pully might have to be changed. A newly designed pully is now available to solve this problem.

Trouble Symptom

The function buttons don't stay down. This problem might be intermittent or occur after the machine has warmed up. Look for this trouble on the SRP system control board.

The Causes. IC1 (905-103) could be defective. This chip is mounted on the SRP board, located on the front of the machine. Check the 200 kHz oscillators peak-to-peak voltage at TP-1 and TP-2.

If the machine shuts down during record, playback, or fast-forward, check TP-1 on the SRP board for 4.6 V p-p. Adjust or clean the VR-1 control if necessary. If it shuts down in the rewind mode, check TP-2 for 2.7 V p-p. Adjust or clean control VR-2 if necessary. These controls adjust the levels of the 200 kHz oscillator.

ZENITH VR9775 REMOTE CONTROL VCR

This machine uses microprocessors to operate all of the systems control circuits and modes.

Trouble Symptom

The VCR would turn on and the remote control functions were operating correctly. The video cassette would load into the machine properly. However, when any function button was pressed (play, record, etc.), the VCR would run for about 3 seconds and then shut down. If all systems are not "go" in this machine, the microprocessor shuts it down to avoid any(more) damage. Let's first take a brief look at this systems control operation.

Background Information (System Control Operation). The system control operations are mostly performed by three microcomputer chips. These are CPU-1, CPU-2, and CPU-3, as shown in the block diagram of Fig. 14-6. All are located on the SS-10 board. Inputs for system control are produced by the functon buttons on the front control panel, mechanical switches, malfunction detection signals from the servo block, the reel servo block, and input signals from the tuner timer block and camera. The CPUs make decisions based on the input signals and produce the correct outputs as per programs designed for them. The signals from the system

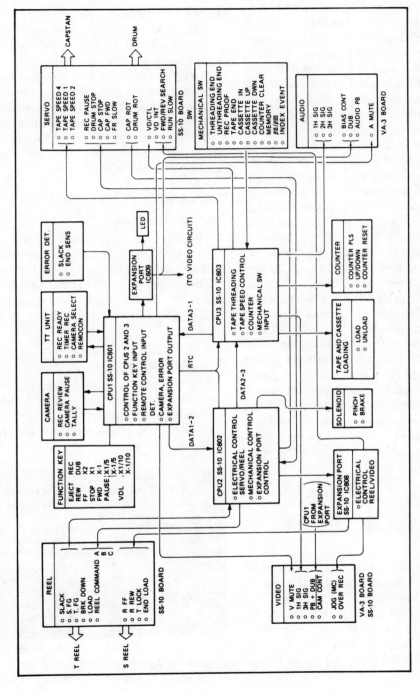

Fig. 14-6. *Zenith VR9775 VCR block diagram of microprocessor control system with Input/Output signals. Courtesy of Zenith Electronics Corporations.*

control are sent to the servo, video, reel servo, audio blocks, etc. Almost all operations of this unit are controlled by the signals from the CPUs. The block diagram outlines the system control and input and output signals.

The VCR has two separate motors that drive the supply and take-up reels. For this reason, the system control has almost no mechanical sections. Various modes are turned on by electrical signals that are controlled by the servo and reel blocks based on instructions issued by the CPUs. These electrical signals are sent to the motors. Mechanical sections controlled are the cassette loading and threading, the brake solenoid for reels, and the pinch roller solenoid. These solenoids are of the self-latching type; once a pulse voltage (200 ms) is given, the solenoid remains in the same mode. No sensor has been provided for detection of tape slack, as revolutions of the reel motors are being controlled by reel motor servo circuits on the SS-10 board.

The Cause. Therefore, when the microprocessor senses some problem, such as a jammed part or broken tape, it shuts the machine down. After a close inspection, the cylinder head was apparently jammed by a small sliver of metal. The microprocessor did not sense the PG pulses from the cylinder head (as it did not rotate) and thus shut all systems down and prevented any other damage. After the sliver of metal was removed from the cylinder head, the VCR operated properly.

PANASONIC PV1300 VCR

Symptom

The cylinder heads would search intermittently. The picture would shrink and then go very snowy and have ghosts.

The Cause. The problem was traced to an intermittent loss of the 30 Hz reference pulse from IC8004, the 3.58 oscillator/divider chip. Replacement of IC8004 returned the playback to normal operation.

RCA VBT200 VCR

Trouble Symptom

Bars of snow appeared across the screen, as shown in (Fig. 14-7). The trouble appeared to be a tape tension problem.

Service Technique. Look and see if both loading arms are going into the fully loaded position. If not, push them into place and note any difference. In this case, the problem cleared up.

The Cause. The right loading arm was not going to the fully loaded position. Replace the loading arm assembly.

PANASONIC PV3000 VCR

Trouble Symptom

The machine would loads but then immediately unloaded. It went into fast-forward and rewind modes, but it had no tape movement.

The Cause. A fuse (F1001) for the +9 V regulator circuit was open. You must completely disassemble the machine and shields to replace the fuse.

Trouble Symptom

The VCR would record and play back properly, except it would not playback video on the camera's viewfinder.

If the camera will record onto tape, the camera

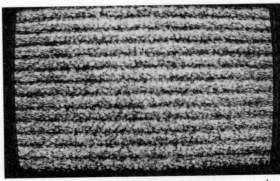

Fig. 14-7. *Bars of snow across the screen appeared to be caused by tape tension problem. Trouble was found to be a faulty loading arm assembly. Courtesy of RCA.*

is okay because the camera video (out) and viewfinder video (in) are on the same line.

The Cause. Testing indicated an open Q3020 video amplifier transistor. Q3020 amplifies video to the camera jack.

Trouble Symptom

A loss of head switching pulses caused horizontal lines to go across the monitor screen.

The Cause. Troubleshooting with the scope indicated that the PG pulses were mis-shaped. These mis-shaped PG pulses were caused by a defective (open) C2502 capacitor (located on the PG amplifier board).

RCA VET 650 VCR
Trouble Symptom (SLP Operation Only)

The customer description of the symptom related to a tape speed selection problem with the special effects VCR. The symptom was described as "locked in SLP only, both in playback and record." Other speeds could be selected in the record mode, and programs recorded in SP or LP were played back at the SLP speed. Also, the "stop" LED remained lighted in all modes.

The Cause. The most likely cause of the symptom described is a defective IC6007 (quad two-input AND gate) chip. Two of the AND gates in this chip are associated with circuitry that enables the 1-second SLP operation whenever search forward or search reverse modes are selected. A failure in either of these AND gates (part of IC 6007) can supply constant base bias to Q6061 resulting in continuous SLP operation.

RCA VET 250/450/650 VCRs
Trouble Symptom

One or two horizontal lines of "interference" appeared across the center of the picture of recorded material, whether played back on the same machine or a different one. The symptom was *not* evident when prerecorded programs (recorded on another machine) were played on the defective machine.

The Cause. The symptom could be the result of buffer oscillator frequency drift in the defective VCR machine. If the buffer oscillator is not locked to incoming vertical sync, incorrect control track pulses will be recorded. During playback, these incorrect control track pulses cause head switching pulses (interference), usually near the center of the picture.

Service Technique. Reset the buffer oscillator free run to a slightly slower frequency (35.8 ms rather than 35 ms). For models VET 250 and 450, adjust the buffer oscillator control R2015 (located on servo board) for a width of 35.8 ms (equal to 27.93 Hz). For model VET 650, adjust buffer oscillator control R6916 (located on the slow/still board) for a width of 35.8 ms (equal to 27.93 Hz).

RCA VDT150/VTE150 VCR
Symptom

Will not record tape.

Probable Cause. Defective buffer IC 907. Also check safety tab switch.

Symptom

Intermittent noise bands on playback (no cylinder lock in play).

Probable Cause. Check for a defective R501 tracking control.

Symptom

Poor recorded picture.

Probable Cause. Check record amplifier IC 205.

Symptom

No picture on playback.

Probable Cause. Switch transistors Q204, Q205, Q206, and Q207. Also, check video buffer transistor Q208.

Symptom

Grainy picture and no color.

Probable Cause. Video playback and no color.

Probable Cause. Video playback amplifier IC 202.

Symptoms

Trouble associated with upper D-D cylinder video heads. These problems might be snowy picture, poor-quality playback, no picture, or vertical jitter.

Probable Cause. Check Upper D-D cylinder —it might need to be cleaned. Cylinder head might be worn or defective.

Symptom

No picture on playback or E-E.

Probable Cause. Defective Q205 buffer transistor.

Symptom

No video in record or E-E.

Probable Cause. Switch IC603 defective.

Symptom

Snow on playback, no picture or no sound, no video information into TV set channel 3.

Probable Cause. Faulty RF converter.

Symptom Description

Will not record.

Probable Cause. Video record amplifier IC 201 defective.

Symptom

Noise in color picture.

Probable Cause. Phase comparator buffer IC 207 defective.

Symptom

No chroma.

Probable Cause. Signal amplifier IC 301.

Symptom

Noise in video

Probable Cause. Phase control (IC 501) or capstan speed control (IC 551) defective.

Symptom

In play, sound is from the tape, but video is from "off-the-air" channel.

Probable Cause. Zener diode D209.

Symptom

VCR problems associated with servo, such as capstan runs slow, audio wow, poor tracking, or no servo lock at any speed.

Probable Causes. Phase control (IC 501) or capstan speed control (IC 551) defective.

Symptom

No operation of any type.

Probable Cause. Fuse F901 blown or a defective IC 901 microprocessor chip.

Symptom

Repeated F901 fuse failure.

Probable Cause. Motor drive IC 902 or regulator transistor defective.

Symptom

No camera pause function.

Probable Cause. Defective microprocessor IC 902.

Symptom

Problems associated with reel motor, such as take-up reel stops or does not turn, eats tapes, or no fast-forward or reverse.

Probable Cause. Defective reel motor.

Symptom

Reel motor runs continuously.

Probable Cause. Motor drive IC 902 or motor drive IC 906 or motor drive IC 906 defective.

Symptom

No play, fast-forward, or rewind.
Probable Cause. Driver IC 911 defective.

Symptom

No pause function.
Probable Cause. Microprocessor IC 901 defective.

Symptom

Intermittent D-D cylinder speed.
Probable Cause. Defective D-D cylinder motor.

Symptom

No capstan operation in record mode. Playback mode operates properly.
Probable Cause. Phase match edit diode D668.

Symptom

The machine is in the play mode at all times.
Probable Cause. Defective microprocessor IC 903.

Symptom

Hum and/or garbled audio.
Probable Cause. Faulty record kill relay RL 401.

Symptom

Buzz in audio on E-E and record. Does not record audio.
Probable Cause. Switch IC 603 is defective.

Symptom

Distorted audio.
Probable Cause. Play/record amplifier IC 401 defective.

LATE MODEL FOUR-HEAD VCRs

Symptom

The trouble appears to be a tracking problem, and a beat appears in the picture. If you see a clear band of the picture at the top of the TV screen with machine in the SP mode, this symptom is probably caused by a faulty contact on the relay in the video head circuit that selects the SP or EP play heads. The relay function is to short out the unused heads for either mode.

The Cause. Bad relay contacts can inhibit shorting function, allowing both heads to pick up signals. Thus, the VCR transmits a beat to the screen between the signal picked up by both SP and EP heads. The EP heads are mounted on the drum to make contact with the tape after the SP heads so they don't pick up a signal until the SP heads have made part of their pass, which is the reason for the clear picture at the top of screen. To solve this problem, replace the relay.

VCR Glossary

ACC — *Automatic color control*, used to maintain an overall constant color signal level in the color circuits.

ACK — *Automatic color killer*.

adjacent track — This is the name of the video track to the immediate left or right of the track of concern.

AFC — *Automatic frequency control* used to phase-lock the color circuits to either the recording or playback color signal to achieve a stable color signal.

AFT — *Automatic fine tuning* is a special circuit found in most recent TVs and VCRs that makes the local oscillator of the tuner follow the channel of concern in order to produce a stable IF frequency. In other words, if for any reason the TV station being received changes frequency, the AFT circuit automatically compensates so that no interference is visible on the screen. No manual fining tuning is necessary.

AGC — *Automatic gain control* is used to maintain an overall constant picture level in the luminance circuits.

APC — *Automatic phase control*, used to help phase-lock the color circuits either to the recording or playback color signal to achieve a stable color signal.

azimuth — A term used to describe the left-to-right tilt of the gap of a recording head (viewed straight on).

balanced modulator — A circuit designed to output the frequency sum or frequency difference of its two signals. Any special characteristics of one of the input signals will be present in the output signal.

beats — The unwanted signals produced when two original signals are allowed to be mixed together.

bipolar PG — Pulse generator signals that have both positive and negative excursions.

burst — A short-time occurrence (8 cycles to 10 cycles) of the 3.58 MHz subcarrier signal that appears right after horizontal sync but is centered on the blanking portion of the video waveform. Burst is used to keep the color oscillator of a TV receiver locked to the TV broadcast station.

C signal — The color portion of a video signal.

capstan — A small rotating metal dowel that, in conjunction with a pressure roller, drives the recording tape to assure positive tape movement to the take-up reel.

chroma—The color portion of a video signal; the quality (hue, saturation) of a color.

chrominance—The color portion of a video signal; the difference between a reproduced color and a standard reference color.

clamp—The process of giving an ac signal a specific dc level.

control signal—A special signal recorded onto the videotape at the same time a video signal is being recorded; used during playback as a reference of the servo circuits.

converted subcarrier—This is the process of frequency shifting the color 3.58 MHz subcarrier and its sidebands down to 629 kHz.

cross talk—The name given to the unwanted signals obtained when a video head picks up information from an adjacent track.

cue—To scan the playback picture at a faster-than-normal speed in the forward direction.

DL—*Delay line.*

DDC—*Direct drive cylinder.* As used in VHS, this means that the video heads are driven by a self-contained brushless dc motor using no belts or gears. DD cylinders produce pictures with better stability.

dark clip—After emphasis, the negative-going spikes (undershoot) of a video signal might be too large in amplitude for safe FM modulation. A dark clip circuit is used to cut off these spikes at an adjustable level.

delta factor—A term used to indicate that a playback signal has some jitter or "wow and flutter." Delta factor, or a change in frequency, means that the color signal off the tape is not a stable frequency of 629 kHz, but rather a signal whose frequency at any instant is some small amount above or below 629 kHz.

deviation—A term used to describe how far the FM carrier swings when it is modulated. In VHS the upper limit is 4.4 MHz.

dew detector—A variable resistor whose resistance value depends upon ambient humidity.

dihedral—A term used to describe the relative position between the two video heads as they are mounted in the head cylinder. Perfect dihedral means that the tips of the heads are exactly 180 degrees apart.

dropout—A momentary absence of FM or color signal due to uneven oxide or a coating of dust on the tape or video heads.

duty cycle—In describing a rectangular waveform, the *duty* refers to the percentage of off-time and on-time for one complete cycle. A 50 percent duty cycle indicates that there are equal periods of off-time and on-time for one cycle; this is a square wave.

E-E—*E*lectronics-to-*e*lectronics. This is the picture viewed on the TV screen when a recording is being made. This picture goes through some but not all of the recorder's circuits and is used to test the operation of said circuits.

EQ—A shortened form of *equalization,* used in the audio record and playback circuits.

emphasis—The process of boosting the level of the high-frequency portions of the video signal.

FG—*F*requency *g*enerator, used in the servo circuits.

FL—An abbreviation for *filter.*

FM signal—The luminance portion of the video signal is used to control the frquency of an astable multivibrator. The output of this multivibrator is a frequency-modulated (FM) signal, shifting from 3.4 MHz to 4.4 MHz (pulse sidebands).

field—One-half of a television picture. A field consists of 262.5 horizontal scanning lines across the picture tube. Two fields are necessary to complete a fully scanned TV picture (frame). First, one field is scanned on the picture tube screen, starting at the top of the tube with line one, and ending at the bottom with line 262.5. Then, the next field begins at the top of the tube again with line 262.5 and ends at the bottom with line 525. The lines of the second field lie in between the lines of the first field. This property of falling in between lines is called *interlacing.* The two sweeps of the picture tube, or two fields, make up one complete TV picture or frame. Frame repetition is 30 Hz, therefore field repetition is 60 Hz.

flagwaving—This is the term used to describe a TV set's ability to accept unstable playback pictures from a videotape recorder. All home VTRs have some degree of playback instability before the active picture is scanned. This can cause a bending or flapping from side to side of the top inch or so of the screen. This movement is called *flagwaving.*

frame—One complete picture. For more details, refer to *field.*

gate—A circuit that delivers an output only when a

specific combination inputs is present. For use in analog or digital applications.

guard band — This is the space between video tracks on the video tape when in the SP mode. Guard bands contain no information.

hall effect IC — An external magnetic field causes current to flow in this type of device.

HD — *H*orizontal *d*rive signal.

head cylinder — A cylindrical piece of metal that houses the video heads. The tips of the heads protrude slightly from the surface of the cylinder so they can scan the tape as the cylinder spins.

head switching — The action of turning off the video head that is not in contact with the videotape. For example, a particular video head turns off 30 times per second.

head-switching pulse — The signal that is applied to the head amplifier to perform head switching. This is a square wave of 30 Hz with a 50 percent duty cycle.

helical — Describes a general type of VCR in which the tape wraps around the video head cylinder in the shape of a three-dimensional spiral, or helix. The video tracks are recorded as a series of slanted lines.

interchangeability — A term used to describe how well a particular VCR can play back a tape recorded on another VCR of the same type.

interlacing — The property of how scan lines of two TV fields lie between each other. Refer to *field*.

interleaving — A term used to indicate that the harmonics of the chrominance signal lie between the harmonics of the luminance portion of the video signal as it is viewed on a spectrum analyzer. This indicates that the color information of a video signal does not interfere with, although it is broadcast at the same time as, the luminance information. Signals that have this interleaving property are not readily seen on a TV screen because of their virtual cancellation characteristics.

jitter — The name given for the effect on the playback picture if a VCR has too much wow and flutter. The picture appears to have a rapid shaking movement.

luminance — The portion of video signal that contains the sync and black-and-white signal information.

MMV — Abbreviation for *m*onostable *m*ulti*v*ibrator. Usually, it's an IC device that gives a logic high or logic low output with a variable duration upon receipt of an input pulse or transition.

non-linear emphasis — This is similar to regular emphasis with the difference that smaller level, high-frequency portions of the signal are given more of a boost than higher level, high-frequency portions.

NTSC — Abbreviation for National Television Systems Committee. These four letters identify the United States color television standard.

PG — A *p*ulse *g*enerator used in VCR servo circuits.

Q — A term used to describe the graphic response of a filter or tuned amplifier.

review — To scan the playback picture at a faster-than-normal speed in the reverse direction.

rotary chroma — The name of the process used in VHS to change the phase of the chrominance signal at a rate of 15,734 Hz (same as the TV horizontal sync frequency).

rotary transformer — A device used to magnetically couple RF signals to and from the spinning video heads, thus eliminating the need for brushes.

sample and hold — A process used in a comparator circuits where the value of a particular signal is measured at a specific moment in time and is stored for later use.

search — To scan the playback picture at a faster-than-normal speed in either the forward or reverse direction.

servo — Short for servo mechanism. This is an electromechanical device whose mechanical operation (for instance, motor speed) is constantly being measured and regulated so that it closely matches or follows an external reference.

skew — Another term for *tension error*. Skew is actually the change of size or shape of the video tracks on the tape from the time of recording to the time of playback. This can occur as a result of poor tension regulation by the VCR or by ambient conditions that affect the tape.

subcarrier — The name of the 3.58 MHz continuous wave signal used to carry color information.

SS — *S*low and *s*till picture modes.

tension error — Refer to *skew*.

time base stability — A term describing how closely the playback video signal from a VCR matches an external reference video signal in regard to sync timing, rather than picture content.

tracking — This is the action of the spinning video heads during playback when they accurately track across the video RF information laid down during recording. Good tracking indicates that the heads are positioning themselves correctly and are picking up a strong RF signal. Poor tracking indicates that the heads are off track and picking up low-level RF signals and noise.

VCO — *Voltage-controlled oscillator*. An oscillator whose frequency of oscillation is governed by an external voltage.

video head — This is the electromagnet used to develop magnetic flux that will put RF information on the tape. In VHS systems, two video heads are mounted in a rotating cylinder around which the video tape is wrapped. As the cylinder spins, each video head is allowed to alternately scan the tape.

video track — One "strip" of RF information laid down during recorded onto a tape as the video head scans across the tape.

VHS — Abbreviation for the *video home system* type of VCR recording.

VTR or VCR — *Video tape recorder or video cassette recorder*.

V-V — *Video-to-video*, or the actual playback picture produced from a tape during playback.

VXO — *Voltage-controlled crystal (x) oscillator*. Similar to VCO except that a quartz crystal is used as a reference and can be varied.

white clip — After emphasis, the positive-going spikes (overshoot) of the video signal might be too large for safe FM modulation. A white-clip circuit is used to cut these spikes off, at an adjustable level.

XTAL — Abbreviation for *crystal*.

Y signal — The black-and-white portion of a video signal containing black-and-white information and sync.

Index

Index